Heterocyclic Chemistry I

Heterocyclic Chemistry

R. R. Gupta, M. Kumar, V. Gupta

Volume I: Principles, Three- and Four-Membered Heterocycles

Volume II: Five-Membered Heterocycles

Volume III: Six-Membered and Higher Heterocycles

Springer

Berlin
Heidelberg
New York
Barcelona
Budapest
Hong Kong
London
Milan
Paris
Singapur
Tokyo

R. R. Gupta, M. Kumar, V. Gupta

Heterocyclic Chemistry

Volume I:
Principles, Three- and Four-Membered Heterocycles

With 98 Figures and 34 Tables

 Springer

Prof. Radha Raman Gupta
Dr. Mahendra Kumar
Dr. Vandana Gupta

Department of Chemistry
University of Rajasthan
Jaipur-302004 / India

ISBN-13:978-3-642-72278-3   e-ISBN-13:978-3-642-72276-9
DOI: 10.1007/978-3-642-72276-9

Die Deutsche Bibliothek - CIP-Einheitsaufnahme
Gupta, Radha R.: Heterocyclic chemistry / R. R. Gupta ; M. Kumar ; V. Gupta. -
Berlin ; Heidelberg ; New York ; Barcelona ; Hong Kong ; London ; Milan ; Paris ; Singapur ;
Tokyo : Springer
Vol. 1. Principles, three- and four-membered heterocycles : with 34 tables. - 1998
    ISBN-13:978-3-642-72278-3

CIP data applied for

Typesetting: Camera ready by authors
Cover: E. Kirchner, Heidelberg
SPIN:10100048   66/3020 - 5 4 3 2 1 0 - Printed on acid-free paper

FOREWORD

Today, our world increasingly is conceived of as being molecular. An ever widening range of phenomena are described logically in terms of molecular properties and molecular interactions. The majority of known molecules are heterocyclic and heterocycles dominate the fields of biochemistry, medicinal chemistry, dyestuffs, photographic science and are of increasing importance in many others, including polymers, adhesives, and molecular engineering.

Thus, the importance of heterocyclic chemistry continues to increase and this three volume work by Drs. R. R. Gupta, Mahendra Kumar and Vandana Gupta is a welcome addition to the available guides on the subject. Its scope places it in a useful niche between the single-volume texts and monographs of heterocyclic chemistry and the multivolume treatises.

The authors have retained the well tried classical approach but have succeeded in placing their own individual spin on their arrangement. They have put together a well selected range from among the most important of the vast array of facts available. This factual material is ordered in a clear and logical fashion over the three volumes.

The present work should be of great value to students and practitioners of heterocyclic chemistry at all levels from the advanced undergraduate upwards. It will be of particular assistance in presenting a clear and modern view of the subject to those who use heterocycles in a variety of other fields and we wish it well.

Alan R. Katritzky
University of Florida

April 1998

PREFACE

Our aim is to bring out a text-cum-reference book on heterocyclic chemistry for undergraduate and graduate students, and research workers who wish to choose heterocyclic chemistry in their research careers. This will enable them to have a comprehensive knowledge of the syntheses, properties, reactions and their mechanisms and applications of heterocyclic compounds. Efforts have been made to include recent developments in heterocyclic chemistry and the subject is presented in a simple and lucid manner, so that those who have some background knowledge of organic chemistry can understand it well.

For the sake of convenience, subject matter covered in heterocyclic chemistry has been divided into three volumes:

Volume III : Six-Membered and Higher Heterocycles
Introduction
Six-Membered Heterocycles
Seven-Membered Heterocycles
Large Ring Heterocycles
Photochemistry of Heterocycles

Since there are detailed contents at the beginning of each chapter, the details for individual chapters are not given here.

We have freely consulted edited scientific works such as Comprehensive Heterocyclic Chemistry edited by A.R. Katritzky and C.W. Rees, Advances in Heterocyclic Chemistry edited by A.R. Katritzky and the Chemistry of Heterocyclic Compounds edited by A. Weissberger and E.C. Taylor. We are very grateful to editors, authors and publishers of these works.

We express our thanks to our colleagues and Ph.D. students who assisted us in proof reading and in several other ways. Our sincere thanks are also due to Dr. R. Stumpe and Dr. Joe P. Richmond editors at Springer Verlag for providing valuable suggestions for the preparation of the book.

Jaipur, India
April, 1998

R.R. Gupta
Mahendra Kumar
Vandana Gupta

Symbols and Abbreviations

α	Bond angle distortion or deviation / Coulomb integral (as stated)
A	Karplus constant
AIBN	1,2,2'-Azobisisobutyronitrile
β	Resonance integral
cm^{-1}	Wave length
conc.	Concentrated
DE	Delocalization energy
DRE	Dewar resonance energy
Λ	Diamagnetic susceptibility exaltation
$\Delta\upsilon(\delta_{ae})$	Chemical shift difference
DIBAL	Diisobutylaluminium hydride
Dil.	Diluted
DMF	N,N-Dimethylformamide
DMSO	Dimethyl sulfoxide
E	Bond angle strain / Torsional energy (as stated)
ERE	Empirical resonance energy
$Eu(fod)_3$	Tris-(6,6,7,7,8,8,8-heptafluoro-2,2-dimethyl-3,5-octanedionato)-europium
ε	Extinction coefficient
$\Delta G°$	Conformational free energy
$\Delta G\neq (G^{++})$	Free energy of activation
$\Delta H°$	Enthalpy difference between conformers
ΔH	Heat of formation (exp.)
HMPA/HMPT	Hexamethylphosphoric triamide
HOMO	Highest occupied molecular orbital
I_a and I_e	Band intensity in vibrational spectra for axial and equatorial conformers

J	Coupling cosntant
K_c	Rate of ring inversion
K_e	Equilibrium constant of interconvertible conformers
$K\theta$	Bond bending force constant
λ	Wave number
LDA	Lithium diisopropylamide
LTA	Lead tetraacetate
LUMO	Lowest unoccupied molecular orbital
MCPBA	m-Chloroperbenzoic acid
NBS	N-Bromosuccinimide
$h\nu$	Photochemical
nm	Nanometer (10^{-9})
N-PSP	N-Phenylselenophthalimide
pm	Picometer (10^{-12})
PPA	Polyphosphoric acid
R	Coupling constants ratio
REPE	Resonance energy per π-electron
ψ	Internal torsional angle
Tc	Coalescence temperature
TCNE	Tetracyanoethylene
tert	Tertiary
TFAA	Trifluoroacetic anhydride
THF	Tetrahydrofuran
THP	Tetrahydropyran
θ	Dihedral angle
TMEDA	N,N,N,N-Tetramethylenediamine
Tos / Tosyl	p-Toluenesulfonyl
VR	Vilsmeier reagent
ω	Dihedral angle deviation
χ_M	Diamagenetic susceptibility

CONTENTS

CHAPTER 1

INTRODUCTION

Heterocyclic Chemistry is an integral part of organic chemistry and constitutes a considerable part of the syllabus for undergraduate and graduate students through out the world. Heterocyclic chemistry deals with heterocyclic compounds which constitute about sixty-five percent of organic chemistry literature. Heterocyclic compounds are widely distributed in nature which are essential to life. Genetic material DNA is also composed of heterocyclic bases–pyrimidines and purines. A large number of heterocyclic compounds, both synthetic and natural, are pharmacologically active and are in clinical use. Several heterocyclic compounds have applications in agriculture as insecticides, fungicides, herbicides, pesticides etc. They also find applications as sensitizers, developers, antioxidants, copolymers etc. They are used as vehicles in the synthesis of other organic compounds. Chlorophyll–photosynthesizing and haemoglobin–oxygen transporting pigments are also heterocyclic compounds.

Huge amount of research work is generated on various phases of heterocyclic chemistry and for publication of such research work and review articles, there are three specialized journals devoted to heterocyclic chemistry:

1. Journal of Heterocyclic Chemistry
2. Heterocycles
3. Heterocyclic Communications

In addition to theses specialized journals, research papers on heterocyclic chemistry are being published in journals devoted to organic chemistry. There are three continuing series in heterocyclic chemistry:

1. Advances in Heterocyclic Chemistry (edited by A. R. Katritzky, Academic Press).
2. Chemistry of Heterocyclic Compounds (edited by E. C. Taylor, John Wiley).
3. General Heterocyclic Chemistry (edited by E. C. Taylor, Wiley-Interscience).

In addition to these series, annual reports on "Progress in Heterocyclic Chemistry" are being published under the sponsorship of International Society of Heterocyclic Chemistry (edited by H. Suschitzky and E. F. V. Scriven, Pergamon Press) to highlight new developments in the field of heterocyclic chemistry. A treatise "Comprehensive Heterocyclic Chemistry" (edited by A. R. Katritzky and C. W. Rees) has appeared in an eight-volume set in 1984 (Pergamon Press) and its second edition (edited by A. R. Katritzky and C. W. Rees) has appeared in 1996. A. R. Katritzky has introduced "Handbook of Heterocyclic Chemistry" (Pergamon Press, 1985) and a five-volume set on "Physical Methods in Heterocyclic Chemistry" (Academic Press). One of the authors (R. R. Gupta) as editor has also introduced a volume on "Physical Methods in Heterocyclic Chemistry" (Wiley-Interscience, 1984). From time to time several textbooks have appeared:

The Chemistry of Heterocyles (T. Eicher and S. Hauptmann, Thieme, 1995).

Heterocyclic Chemistry (T.L. Gilchrist, Pitman, 1st edn. 1985; 2nd edn. 1992).

Contemporary Heterocyclic Chemistry (George R. Newkome and William W. Paudler, Wiley-Interscience, 1982).

An Introduction to the Chemistry of Heterocyclic Compounds (R. M. Acheson, 3rd edn., John Wiley, 1976).

Heterocyclic Chemistry (D. W. Young, Longman, 1975).

Heterocyclic Compounds (K. Schofield, Ed., Butterworths, 1975).

Heterocyclic Chemistry (J. A. Joule and G. F. Smith, 2nd edn. Van Nostrand Reinhold, 1972; J. A. Joule, K. Mills and G. F. Smith, 3rd edn. Chapman & Hall, 1994).

Principles of Modern Heterocyclic Chemistry (L. A. Paquette, Benjamin, 1968).

Heterocyclic Chemistry (A. Albert, 2nd edn. Oxford University Press, 1968).

The Principles of Heterocyclic Chemistry (A. R. Katritzky and J. M. Lagowski, Academic Press, 1968).

The Structure and Reactions of Heterocyclic Compounds (M. H. Palmer, E. Arnold, 1967).

The Chemistry of Heterocyclic Compounds (A. A. Morton, McGraw-Hill, 1946).

We feel the need for a text-cum-reference book in heterocyclic chemistry which includes new developments in the field and can cater to the need of undergraduate and graduate students and research workers who have chosen heterocyclic chemistry, pharmaceutical chemistry and medicinal chemistry etc. in their research careers. In the present book efforts have been made to include each and every class of heterocyclic compounds with which students are concerned in their courses devoted to heterocyclic chemistry. The subject matter has been divided into sixteen chapters spreading over in three volumes (as mentioned in the preface). Each chapter except the first (introduction) contains a detailed table of contents. At the end of each chapter references have been cited, so that readers who are interested in having more information regarding synthetic pathways, reactions and their mechanisms, properties, application etc. of heterocyclic compounds may be able to locate the information they require.

CHAPTER **2**

NOMENCLATURE OF HETEROCYCLES

CONTENTS

1 GENERAL

The systematic names of the chemical structures require essentially certain nomenclature systems and for each nomenclature system the name and numbering require appropriate nomenclature rules. Different nomenclature systems[1–12] are available in the literature, however the most relevant systems recommended by IUPAC for naming heterocycles are discussed.

2 NOMENCLATURE SYSTEMS

The most systematic nomenclature of heterocycles is based on the names of carbocyclic analogs and, therefore, the nomenclature systems are modifications of those used for carbocycles. However, the nomenclature system which is more commonly used for heterocycles involves the combination of trivial and systematic names.

2.1 Systematic Nomenclature System (Hantzsch–Widman System)

This is the most widely used systematic method and is used for naming three- to ten-membered monocyclic heterocycles of various degree of unsaturation containing one or more heteroatoms. This nomenclature system specifies the ring size and the nature, type and position(s) of the heteroatom(s). The heteromonocycles are named by using following revised rules[13] of Hantzsch–Widman nomenclature system.

(1) Combination of prefix(es) with stem

The heteromonocyclic system is named by combining one or more 'a' prefixes for the heteroatom(s) with a stem indicating the size of the ring. If the stem begins with vowel, the terminal letter 'a' of the 'a' prefix is dropped.

Prefixes : The prefixes indicate the heteroatoms present in the heterocyclic systems. The prefixes for different heteroatoms are presented in Table 1 in the preferential order.

Table 1. Prefixes for heteroatoms ('a' prefixes in decreasing order of priority)

Heteroatom	Symbol (Valence)	Prefix
Oxygen	O (II)	Oxa
Sulfur	S (II)	Thia
Selenium	Se (II)	Selena
Tellurium	Te (II)	Tellura
Nitrogen	N (III)	Aza
Phosphorus	P (III)	Phospha
Arsenic	As (III)	Arsa
Antimony	Sb (III)	Stiba
Bismuth	Bi (III)	Bisma
Silicon	Si (IV)	Sila
Germanium	Ge (IV)	Germa
Tin	Sn (IV)	Stanna
Lead	Pb(IV)	Plumba
Boron	B (III)	Bora
Mercury	Hg (II)	Mercura

Stems : The stems are used to indicate the size of the ring and the saturation or unsaturation in the heteromonocyclic systems and are summarized in Table 2.

Table 2. Stems for three- to ten-membered heterocycles

Ring size		Unsaturation	Saturation
Three-membered		-irene[i]	-irane[ii]
Four-membered		-ete	-etane[ii]
Five-membered		-ole	-olane[ii]
Six-membered	A*	-ine	-ane
	B*	-ine	-inane
	C*	-inine	-inane
Seven-membered		-epine	-epane
Eight-membered		-ocine	-ocane
Nine-membered		-onine	-onane
Ten-membered		-ecine	-ecane

However, the following points should be considered while giving names to the heteromonocycles :

(i)   the stem 'irine' is used for three-membered nitrogen-containing unsaturated heteromonocycles.

(ii) the stems 'iridine', 'etidine' and 'olidine' are used for nitrogen-containing saturated three-, four- and five-membered heteromonocycles respectively.

(iii) the terminal vowel of a numerical prefix is not dropped even when the prefix begins with the same vowel, e.g., cyclotetraazoxane.

(iv) the ending of the name depends on the presence or absence of nitrogen.

(v)   unsaturated stems are used for the rings with maximum numbers of non-cumulative double bonds possilbe, when the heteroatoms have the normal valences given in Table 1.

(vi) saturated stems are used for the rings without double bond(s)

(vii) if the stems are not specified for partialy or completely saturated heteromonocycles, the prefixes 'dihydro-', 'tetrahydro-', *etc.* should be used.

(viii) the terminal 'e' used in all the stems is optinal (stems without terminal 'e' for unsaturated non-nitrogenous rings with six or more ring members are used in CAS index nomenclature, for example, dioxin, dithiin and oxathiin.

(ix) the stems 'etine' and 'oline', which were formerly used for nitrogen containing four- and five-membered rings respectively with one double bond have no longer been recommended by IUPAC.

(x)   the stems for six-membered rings depend on the least preferred heteroatom in the ring, i.e., the heteroatom immediately preceding the stem. To determine the proper stem for a six-membered ring, the following set containing least preferred heteroatom is selected :

      6A* = O, S, Se, Te, Bi, Hg

      6B* = N, Si, Ge, Sn, Pb

      6C* = B, P, As, Sb

(xi) the stems (syllables) indicating ring sizes (3, 4, 7, 8, 9 and 10) are considered to be derived from numerical prefixes (Table 3).

(xii) trivial names e.g., pyrrole, pyrazole, imidiazole, pyridine, pyridazine, pyrimidine, *etc.*, permitted for some heteromonocyclic systems by IUPAC should be prefered over the systematic names.

(xiii) oxine must not be used for pyran because it has been used as a trivial name for quinolin-8-ol.

(xiv) azine must not be used for pyridine because of its use as a class name of the compounds containing $=N-N=$ group.

Table 3. Stems (syllables) for different heterocyclic rings

Ring size	Syllable	Derived from
Three-membered	-ir	tri
Four-membered	-et	tetra
Seven-membered	-ep	hepta
Eight-membered	-oc	octa
Nine-membered	-on	nona
Ten-membered	-ec	deca

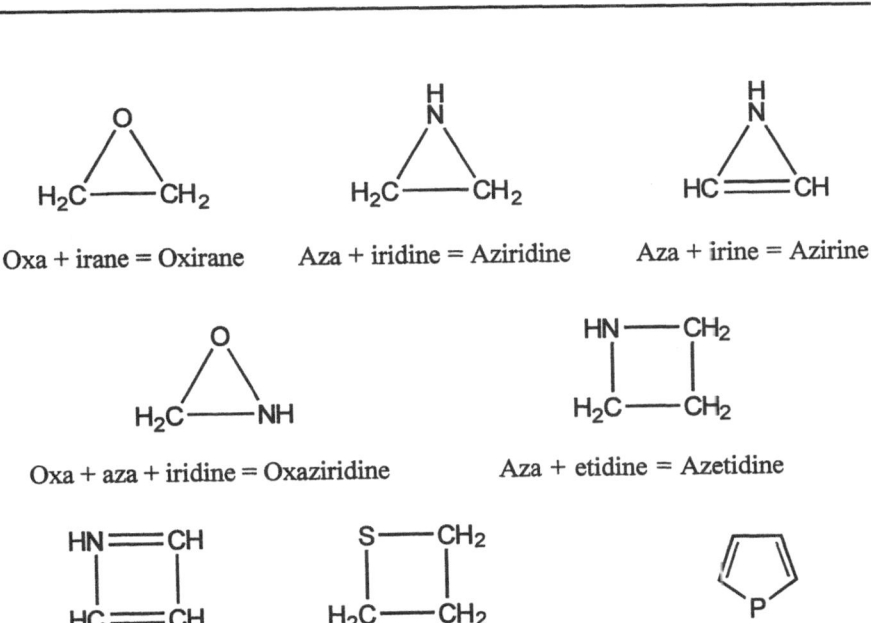

Oxa + irane = Oxirane   Aza + iridine = Aziridine   Aza + irine = Azirine

Oxa + aza + iridine = Oxaziridine   Aza + etidine = Azetidine

Aza + ete = Azete   Thia + etane = Thietane   Phospha + ole = Phosphole

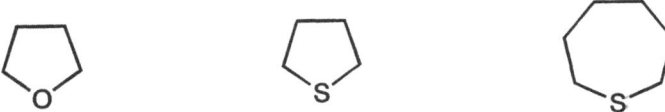

Oxa + olane = Oxolane   Thia + olane = Thiolane   Thia + epane = Thiepane

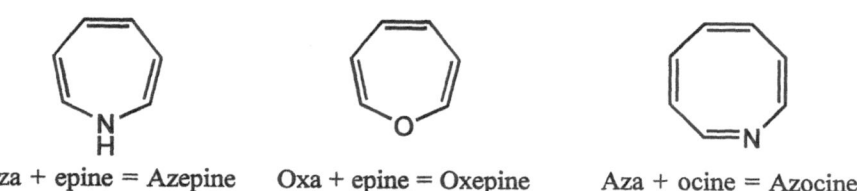

Aza + epine = Azepine   Oxa + epine = Oxepine   Aza + ocine = Azocine

(2) Presence of identical heteroatoms (Multiplicity of heteroatoms)

When two or more heteroatoms of the same type are present in a ring, the prefixes, di-, tri-, tetra, *etc.* are used and placed before the prefix used for the heteroatom.

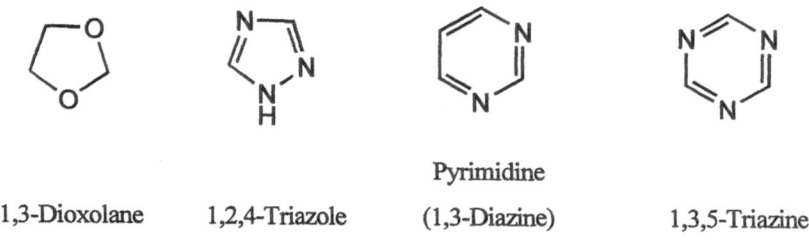

		Pyrimidine	
1,3-Dioxolane	1,2,4-Triazole	(1,3-Diazine)	1,3,5-Triazine

(3) Presence of different heteroatoms (Order of preference)

When two or more different heteroatoms are present in the same ring, the prefixes of heteroatoms are combined in the order of their appearance in Table 1.

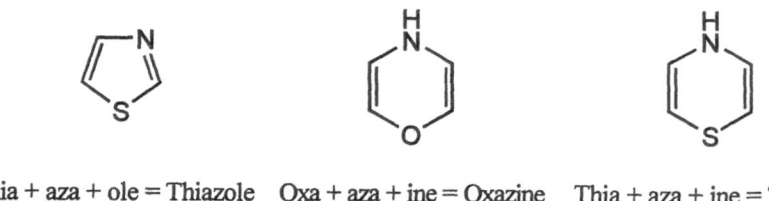

Thia + aza + ole = Thiazole   Oxa + aza + ine = Oxazine   Thia + aza + ine = Thiazine

(1,3-Thiazole)   (1,4-Oxazine)   (1,4-Thiazine)

(4) Numbering

(i) **With one heteroatom :** The numbering in the ring starts from the heteroatom giving the position-1 and proceeds in such a way as to give the lowest possible locant to the substituent if present.

| Pyridine | 2,5-Dimethylpyridine | 3-Methyloxepin(e) | Azocine |

(ii) With two or more identical heteroatoms : When two or more identical heteroatoms are present in a ring, the ring is numbered in such a way that the heteroatoms are assigned the lowest possible set of number locants.

| [1,3] | [1,4] | [1,2,4] | [1,3,4] |

correct numbering incorrect numbering correct numbering incorrect numbering

(iii) With two or more different heteroatoms : When heterocyclic ring contains two or more different heteroatoms, the numbering starts from the heteroatom with the highest preference as in Table 1 : O, S, N (oxygen takes precedence over sulfur and sulfur over nitrogen). The remaining heteroatoms are given the lowest number locants.

| 1,3-Thiazole | 1,3-Oxazole | 1,2-Oxathiolane |

1,2,4-Thiadiazole (correct)   1,3,5-Thiadiazole (incorrect)

(5) Presence of saturated atom (Indicated hydrogen)

(i)   When a heterocyclic ring with maximum number of noncumulative double bonds contains a saturated atom, its position is given the lowest possible locant and is numerically indicated by an italic capital *H* before the name of heterocyclic ring system.

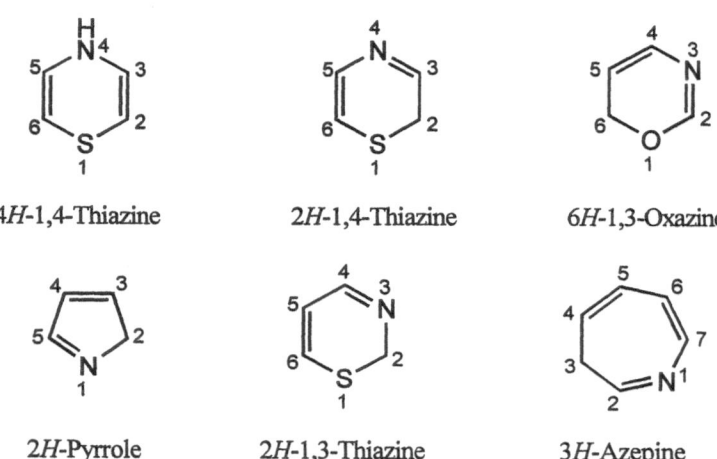

4*H*-1,4-Thiazine   2*H*-1,4-Thiazine   6*H*-1,3-Oxazine

2*H*-Pyrrole   2*H*-1,3-Thiazine   3*H*-Azepine

(ii) However, the heterocyclic system in which a carbon atom of the ring is involved in the carbonyl group, the indicated hydrogen is normally cited as an etalic capital *H* in parentheses after the locant of the additional structural feature.

Pyridin-2(1*H*)-one   Pyrazin-2(3*H*)-one

2.2 Trivial System

This system of nomenclature is based on the trivial and semitrivial names of the heterocycles which were given before their structural identifications. The trivial and semitrivial names of the heterocycles were given on the basis of their characteristic properties or on the sources from which they are obtained. Therefore, the trivial and semitrivial names provide hardly any structural information. However, trivial and semitrivial names of some heterocycles are recognized by IUPAC and their names are retained in the fusion nomenclature system. The following Table 4 lists trivial and semitrivial names of the heterocycles which are recognized by IUPAC and retained in IUPAC recommendations[13].

Table 4. Heterocycles with their recognized trivial and semitrivial names

Structure	Trivial name
	Pyrrole
	Furan
	Thiophene
	Selenophene
	Tellurophene
	Pyrazole

(1*H*-isomer)

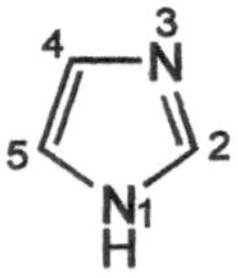

Imidazole

(1*H*-isomer)

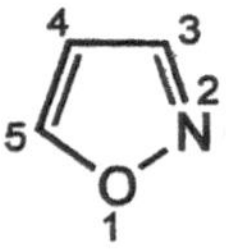

Isoxazole

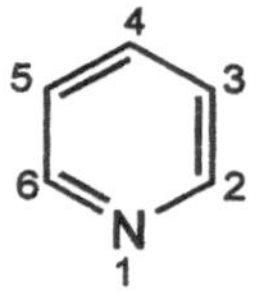

Pyridine

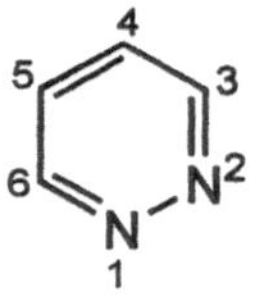

Pyridazine

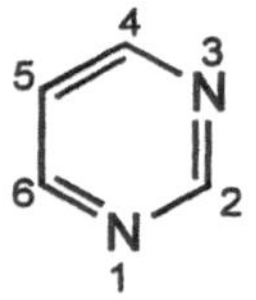

Pyrimidine

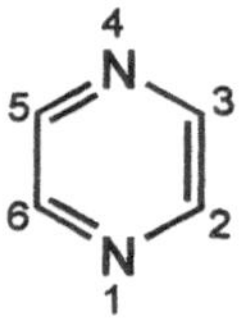

Pyrazine

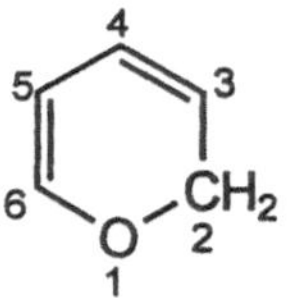

Pyran

(2*H*-isomer)

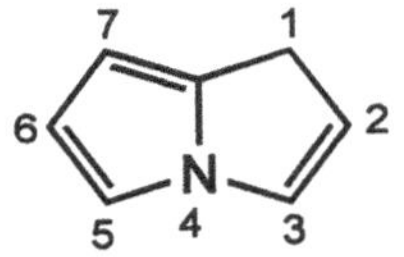

Pyrrolizine

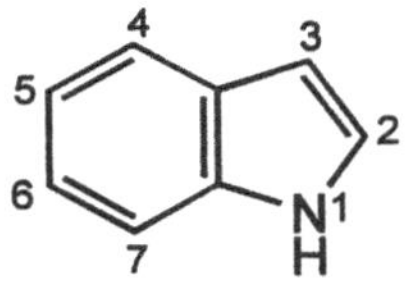

(1*H*-isomer)

Indole

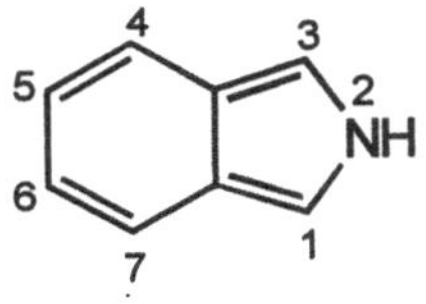

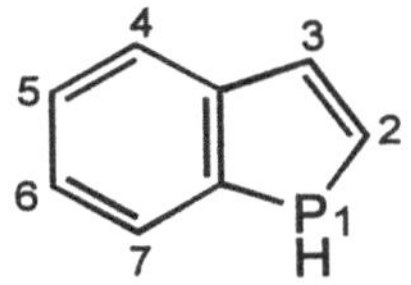

(1*H*-isomer)

Isoindole

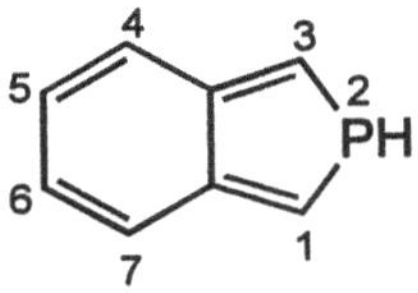

(2*H*-isomer)

Phosphindole

(1*H*-isomer)

Isophosphindole

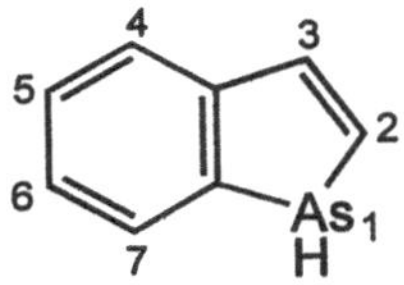

(2*H*-isomer)

Arsindole

(1*H*-isomer)

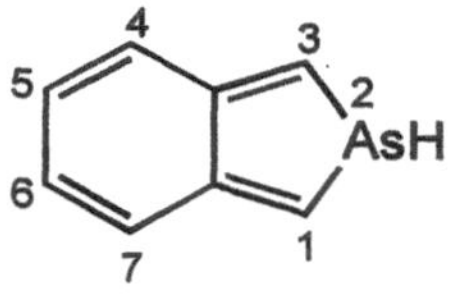

Isoarsindole

(2*H*-isomer)

Indazole

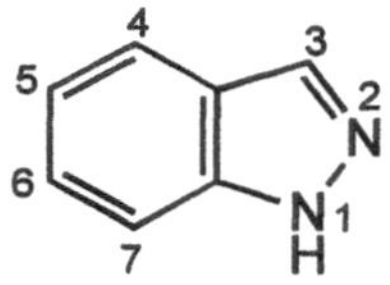

(1*H*-isomer)

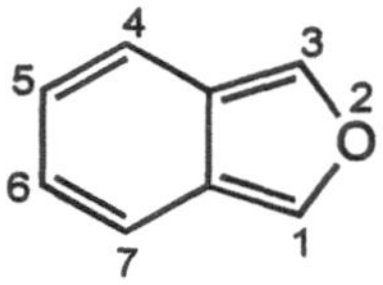

Isobenzofuran

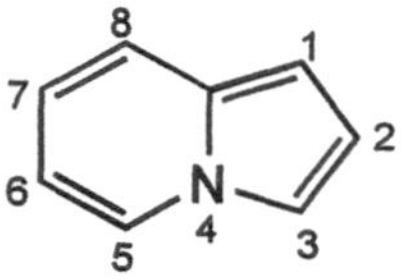

Indolizine

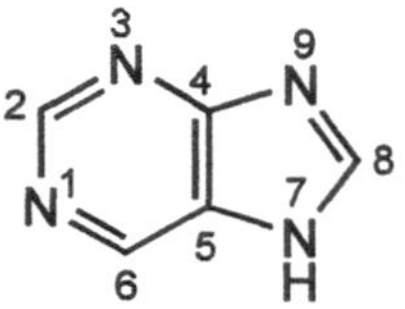

Purine

(exception to systematic
numbering)

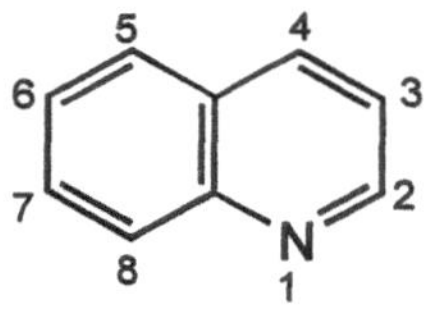

Quinoline

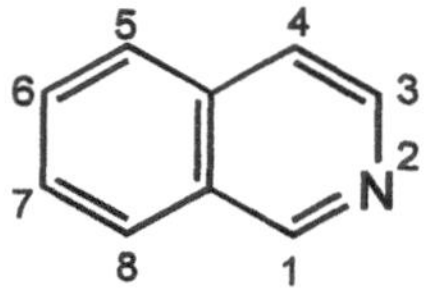

Isoquinoline

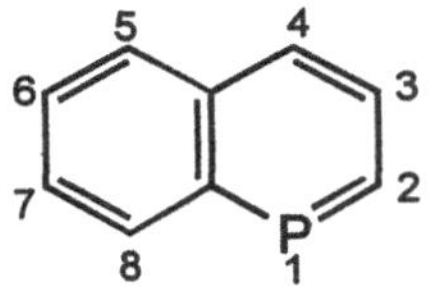

Phosphinoline

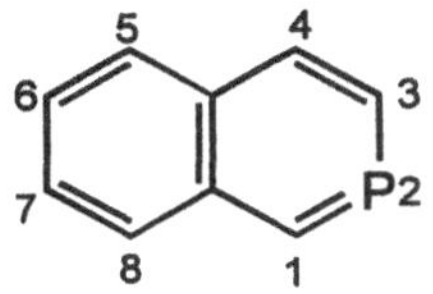

Isophosphinoline

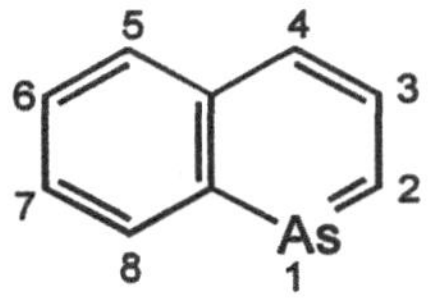

Arsinoline

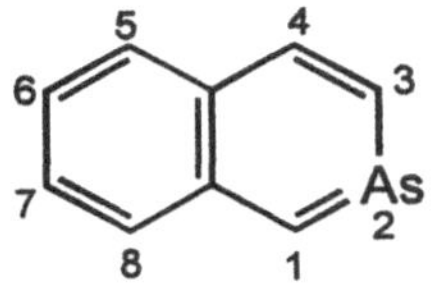

Isoarsinoline

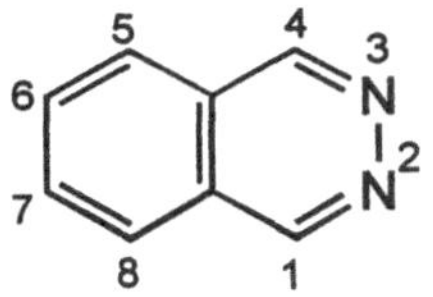

Phthalazine

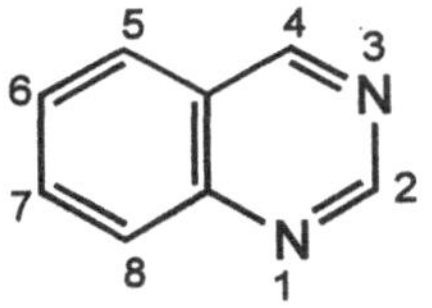

Quinazoline

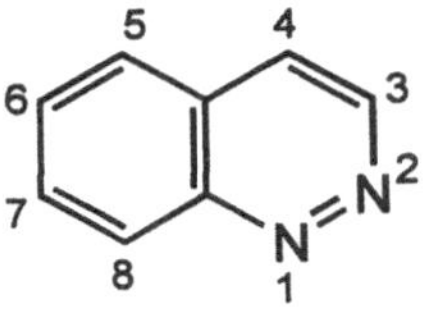

Cinnoline

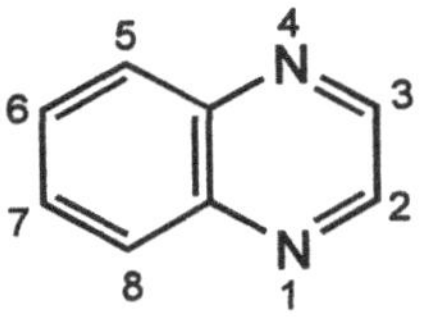

Quinoxaline

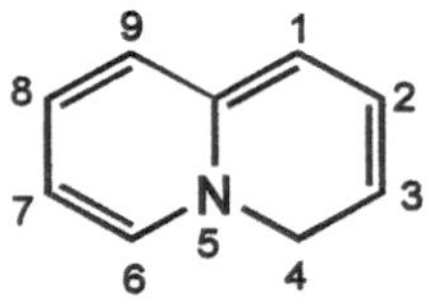

Quinolizine

(4*H*-isomer)

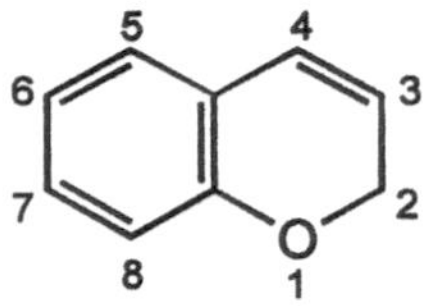

Chromene

(2*H*-isomer)

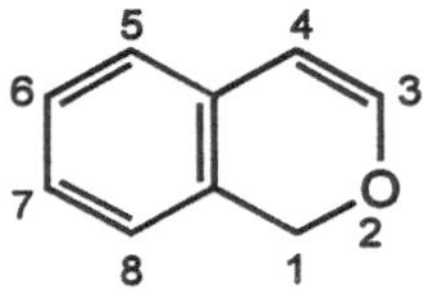

Isochromene

(1*H*-isomer)

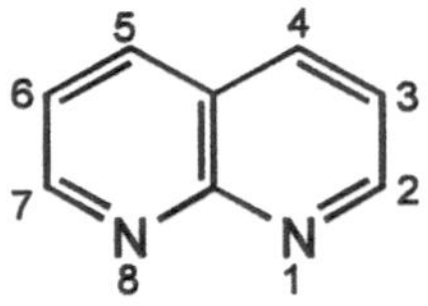

1,8-Naphthyridine

(1,8-isomer)

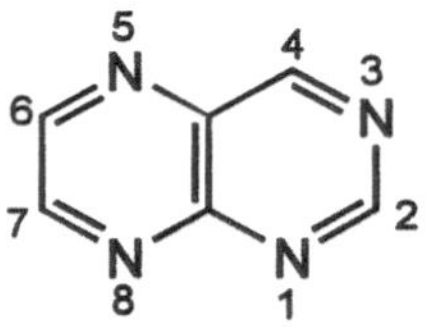

Pteridine

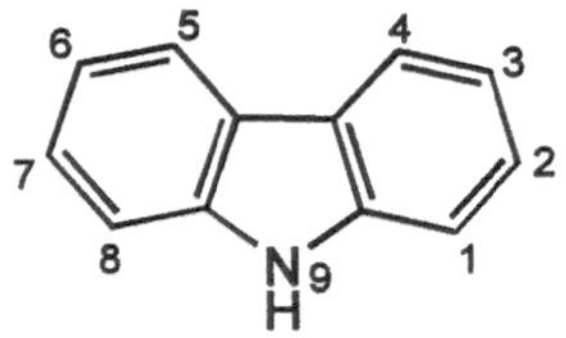

Carbazole

(9*H*-isomer)
(exception to systematic numbering)

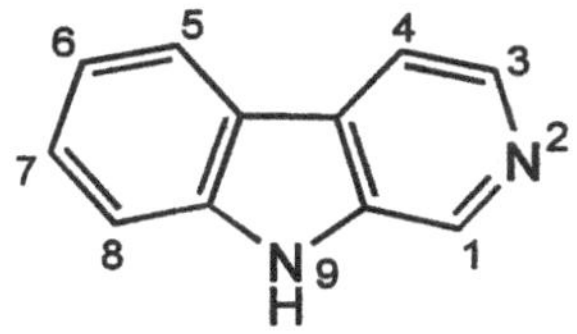

β-Carboline

(9*H*-isomer)

Acridine

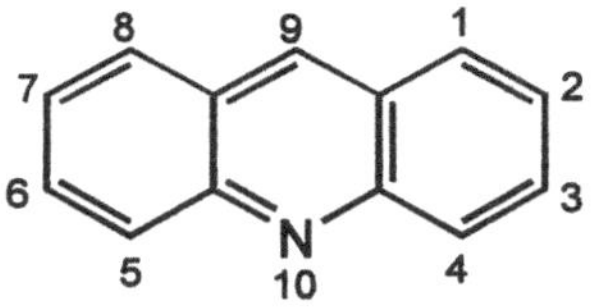

(exception to systematic numbering)

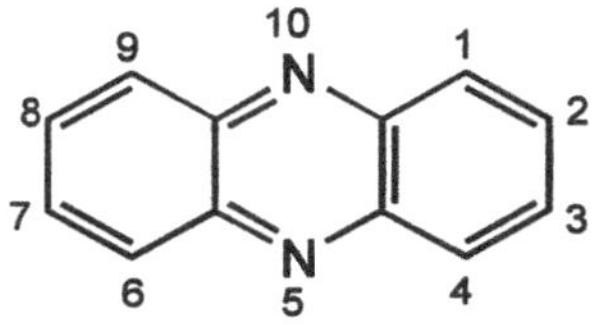

Phenazine

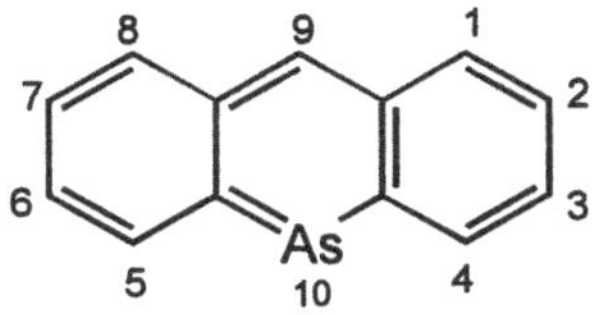

Acridarsine

(exception to systematic numbering)

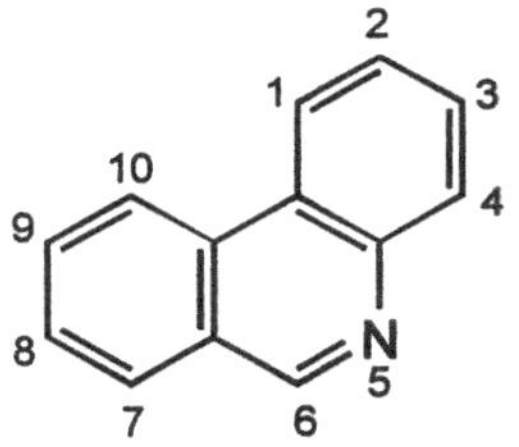

Phenanthridine

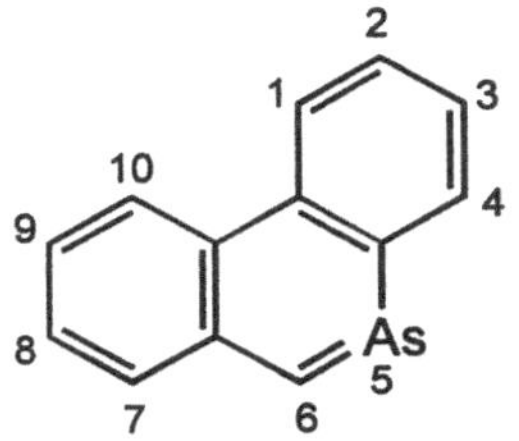

Arsanthridine

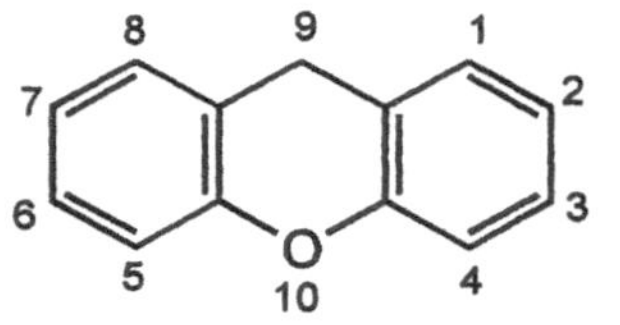

Xanthene

(9*H*-isomer)
(exception to systematic numbering)

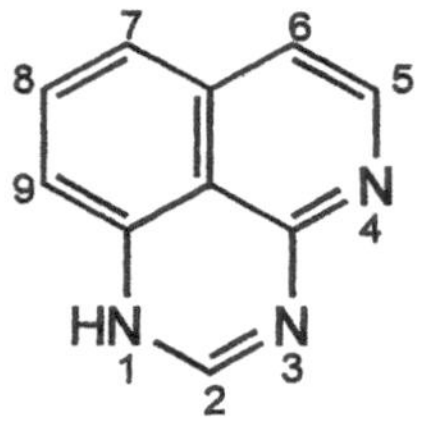

Perimidine

(1*H*-isomer)

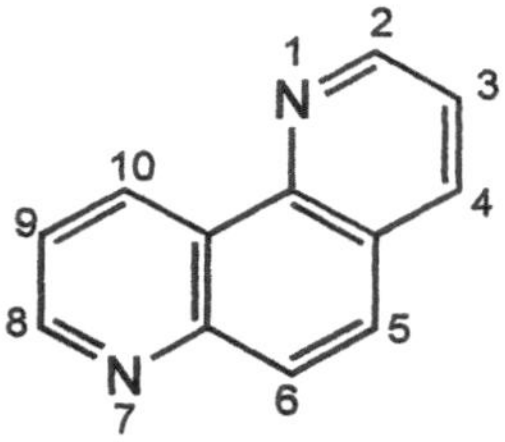

Phenanthroline

(1,7-isomer)

2.3 Fusion Nomenclature System

The systematic names of the fused heterocycles are given by this nomenclature system. The fused heterocyclic system is considered to be constructed by the combination of two or more cyclic structural units (components). The cyclic structural units contain maximum number of non-cumulative double bonds and are fused in such a way that each structural unit (component) has one bond common with other. The names of the structural units can be trivial or systematic.

The fused heterocyclic system are named according to the following rules :

1.   The fused heterocyclic system is dissected into its components in which one is base component and other(s) is attached component(s).

Benzothiazole   Benzene   Thiazole

Benzimidazole   Benzene   Imidazole

2. The components are given their recognized trivial names, if possible, which are selected from the Table of trivial names (Table 4). If monocyclic component has no recognized trivial name, the systematic name is used.

3. Base component should be a heterocyclic system. If there is choice, the base component is determined by the order of preference.

Selection of Base Component :

(i) Nitrogen containing component : a nitrogen containing component is selected as the base component.

Base component : Pyridine   Base component : Pyrrole

(ii) Component(s) with heteroatom(s) other than nitrogen : if nitrogen is absent in the heterocyclic ring(s), the ring with other heteroatom(s) is selected as the base component.

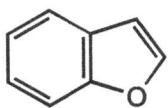

Base component : Furan   Base component : Thiophene

If both the components contain heteroatoms other than nitrogen, the component which contains heteroatom appearing highest in Table 1 is selected base component (component with heteroatom of highest group in periodic table and with heteroatom of higher atomic number if heteroatoms are of the same group).

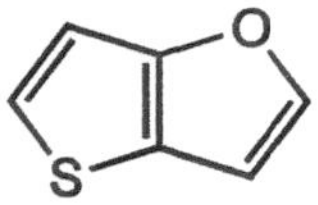

Base component : Furan

(iii) Component with greatest number of rings : a component with as many rings as possible is selected as the base component (bicyclic condensed systems or polycyclic systems with recognized trivial names).

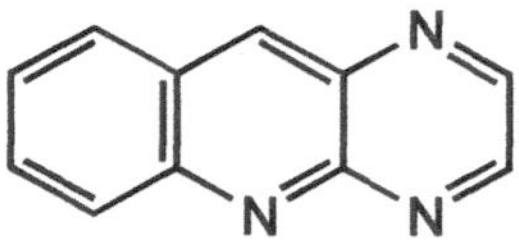

Base component : Quinoline

(iv) Rings of unequal size : if fused heterocyclic system contains rings of unequal size, the component with the largest size of the ring is selected as a base component.

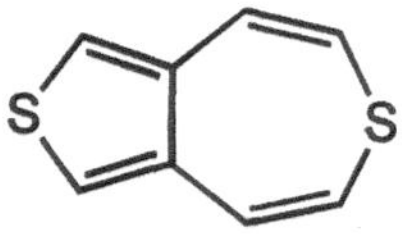

Base component : Thiepine 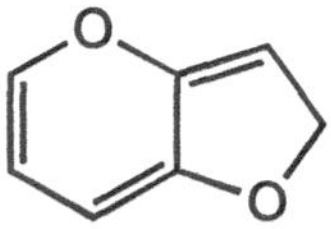   Base component : Pyran

(v) Rings of equal size with different number of heteroatoms : if the heterocyclic system containing rings of equal size with different number of heteroatoms, the ring with greater number of heteroatoms of any kind is considered as a base component.

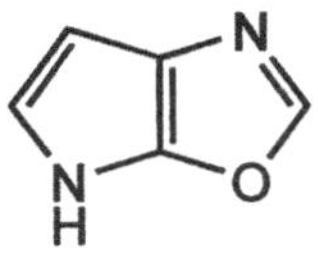

Base component : Oxazole

(vi) Rings of equal size with equal number of different heteroatoms : if both the components contain rings of same size with equal number of different heteroatoms, the component containing ring with greatest variety of heteroatoms is selected as a base component.

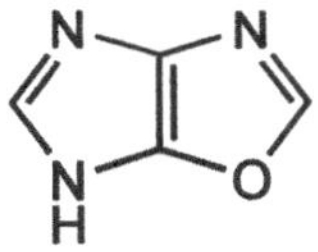

Base component : Oxazole

If two heteroatoms of the same group are present, the component containing the ring with heteroatoms appearing first in Table 1 is preferred as the base component.

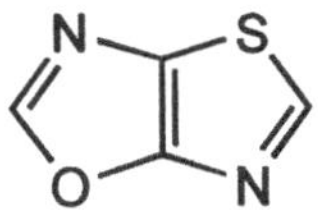

Base component : Oxazole 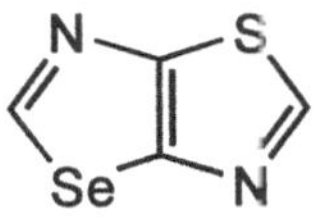   Base component : Thiazole

(vii) Rings of same size with same numbers and same kinds of heteroatoms : if the components contain rings of the same size with same numbers and same kinds of heteroatoms, the component containing ring with heteroatoms which have the lowest locant numbers is preferred as a base component.

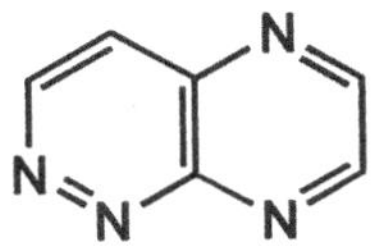

Base component : Pyridazine 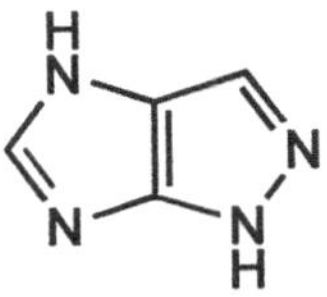   Base component : Pyrazole

4. The attached component (second component) is added as a prefix to the name of the base component. The prefix designating an attached component is formed by changing terminal 'e' of a trivial or Hantzsch-Widman name of a component into 'o'. This 'o' is not dropped when followed by a vowel.

Pyrazine	Pyrazino-
Pyrazole	Pyrazolo-
Thiazole	Thiazolo-

However, there are some exceptions to this rule. The prefixes for some common heterocycles used in the fused nomenclature are presented in Table 6.

Table 6. Prefixes for some common heterocycles

Heterocyclic compound	Prefix
Pyridine	Pyrido-
Quinoline	Quino-
Isoquinoline	Isoquino-
Furan	Furo-
Thiophene	Thieno-
Imidazole	Imidazo-

5. The bonds of the base component are alphabetized with consecutive italic letters starting with 'a' for 1,2- bond, 'b' for 2,3-bond, 'c' for 3,4-bond 'd' for 4,5-bond and so on.

6. The atoms of ring system of second component (attached component) are numbered in the normal way; 1, 2, 3, 4, 5, *etc.*, observing the principle of the lowest possible numbering.

7. The atoms common to both rings (side of fusion) are indicated by the appropriate letters and numbers and are enclosed in a square bracket and placed immediately after the prefix of the attached component. The numbers (positions of attachment) of the second component are placed in the sequence in which they are attached to the base component.

Thieno[2,3-*b*]furan = Thiophene (attached component) + Furan (base component)

Benzopyrano[3,4-*b*]benzothiazine   [1,4]Benzothiazine   Benzopyran
                               (base component)   (attached component)

Pyrazino[2,3-*c*]pyridazine   Pyridazine   Pyrazine
                          (base component)   (attached component)

8.   **Common heteroatom** : If a position of fusion is occupied by a heteroatom, both the components (ring systems) are considered to possess that heteroatom.

Imidazo[2,1-*b*]oxazole   Imidazole   Oxazole
                         (attached component)   (base component)

9.   **Numbering of fused heterocyclic system :**

(i)   Fused heterocyclic system is numbered independently of the combining components (base and attatched components). The numbering is started from the atom adjacent to the bridgehead position with the lowest possible locant(s) to the heteroatom(s). If there is choice, the heteroatom appearing highest in Table 1 is preferred.

Benzo[*b*]furan   3,1-Benzoxazepine

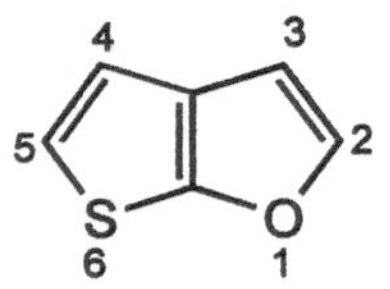

Thieno[2,3-*b*]furan 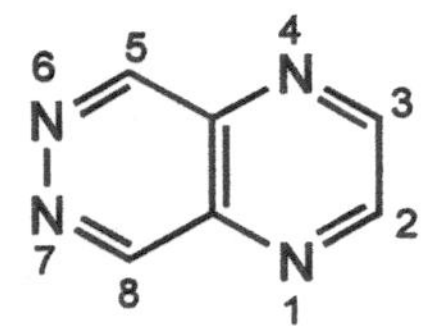   Pyrazino[2,3-*d*]pyridazine 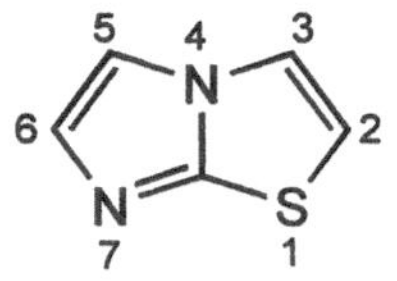   Imidazo[2,1-*b*]thiazole

(ii) Carbon atom common to two rings is given the lowest possible position, but not numbered. However, the heteroatom at a position of fusion of two rings (common heteroatom) is numbered.

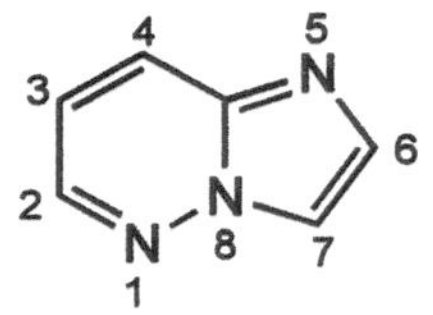

Imidazo[1,2-*b*]pyridazine 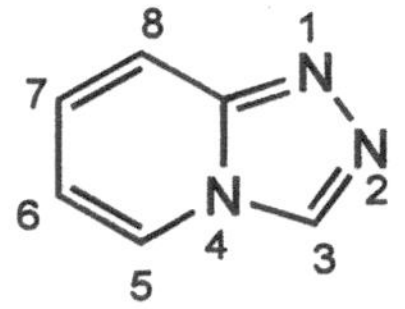   1,2,4-Triazolo[4,3-*a*]pyridine

(iii) The position of a saturated atom is indicated by an italic hydrogen and is given the lowest possible number locant.

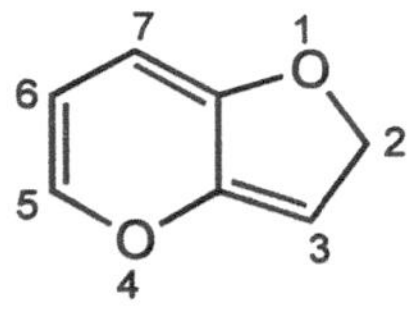

2*H*-Furo[3,2-*b*]pyran

9.   Benzofused heterocycles :

(i) If a benzene ring is fused to the heterocyclic ring, the compound is named by placing number(s) indicating position(s) of the heteroatom(s) before the prefix benzo- (from benzene) followed by the name of the heterocyclic component.

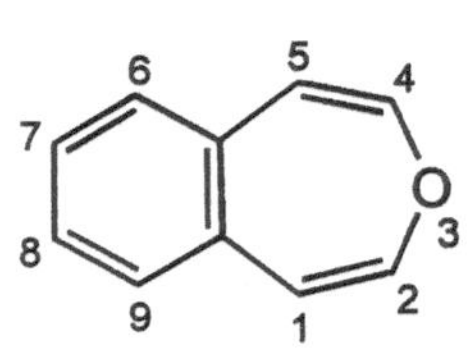

3-Benzoxepin 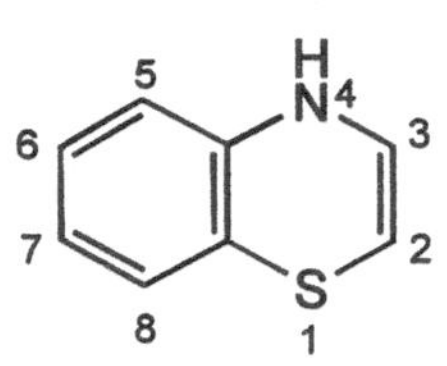   4*H*-1,4-Benzothiazine 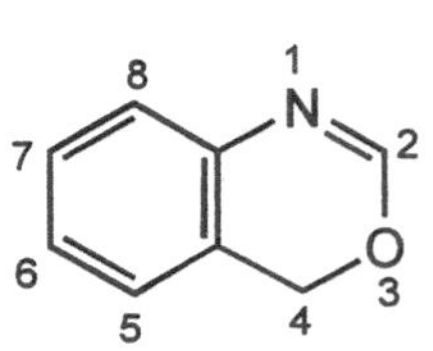   4*H*-3,1-Benzoxazine

(ii) If two benzene rings are ortho-fused to a six-membered 1,4-dihetero-monocyclic ring containing different heteroatoms, the heterocyclic system is named by adding the prefix 'pheno-' to the Hantzsch–Widman name of the heteromonocycle.

10*H*-Phenoxazine 10*H*-Phenothiazine Phenoxathiine

(iii) However, the heterocyclic system in which two benzene rings are ortho-fused to a six-membered 1,4-diheteromonocycle containing the same heteroatoms are named by adding the replacement prefix for the heteroatom (Table 1) to the term '-anthrene' with elision of an 'a'.

Thianthrene Phenazine
(exception to this rule)

2.4 Replacement Nomenclature System

Heterocyclic compounds are considered to be derived from the carbocyclic compounds by the replacement of one or more carbon atoms by the heteroatom(s). This forms the basis of replacement nomenclature of heterocycles and, therefore, the rules of the replacement nomenclature are similar to those applied for the carbocyclic compounds. This is the most systematic nomenclature and is used for the heterocycles containing unusual heteroatom(s) and also for spiro- and bridged heterocyclic systems.

2.4.1 Monocyclic Heterocycles

1. The corresponding carbocyclic ring (the ring obtained from the heterocyclic compound by replacing heteroatom(s) by CH_2, CH, or C according to the valence of heteroatom(s)) is named by IUPAC rules.

2. The type of heteroatom is indicated by a prefix according to Table 1. Since all the prefixes end with the letter 'a' the replacement nomenclature is also known as 'a' nomenclature. The position and prefix for each heteroatom are placed before the name of the corresponding carbocyclic ring.

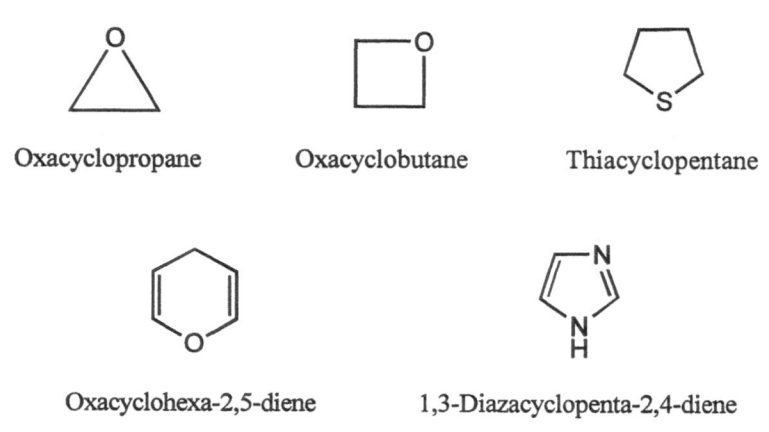

Oxacyclopropane   Oxacyclobutane   Thiacyclopentane

Oxacyclohexa-2,5-diene   1,3-Diazacyclopenta-2,4-diene

3. The replacement names derived from benzene are retained only if three double bonds are present, otherwise the names with -ene, -diene, *etc.* as necessary are used.

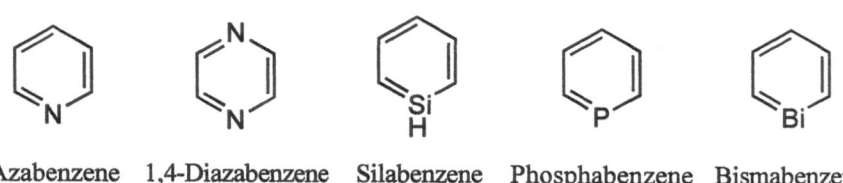

Azabenzene 1,4-Diazabenzene Silabenzene Phosphabenzene Bismabenzene

Oxacyclopenta-2,4-diene   Silacyclopenta-2,4-diene

4. The sequence and numbering of the heteroatoms follow the rules as in systematic nomenclature. However, the following preferential order is followed :

    (i)   first preference   : heteroatom appearing highest in Table 1
    (ii)   second preference   : other heteroatoms according to Table 1
    (iii) third preference   : multiple bonds
    (iv) fourth preference   : substituents, if there is choice alphabetical order is considered

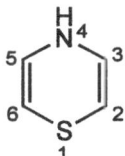

1-Oxa-3-azacyclopenta-2,4-diene   1-Thia-4-azacyclohexa-2,5-diene

1-Thia-4-aza-2-silacyclohexane   1,4-Dithiacyclohexa-2,5-diene

2.4.2 Fused Heterocycles

In replacement nomenclature system following rules are applied for naming fused heterocycles :

1. The position(s) and prefix(es) for the heteroatom(s) are written in front of the name of the corresponding carbocyclic ring.

2,5-Diazaanthracene   3,9-Diazaphenanthrene

2. The numbering of the corresponding carbocyclic ring is retained irrespective of the position(s) of the heteroatom(s). However, the lowest possible number locants are assigned to the heteroatoms in the complete fused heterocyclic system and, if choice, Table 1 is followed.

3. The numbering of fused heterocyclic system in the replacement system of nomenclature is frequently different from that given in fusion nomenclature system. The heteroatom at the ring junction is assigned individual number in the fusion nomenclature system, whereas in the replacement nomenclature system the heteroatom at the ring junction is given the same locant as an adjacent nonconjunction carbon atom but with a suffix 'a' or 'b', *etc.*

Imidazo[2,1-b]thiazole
(Fusion nomenclature)

1-Thia-3a,6-diazapentalene
(Replacement nomenclature)

4. The corresponding carbocyclic ring without maximum number of non-cumulative double bonds is named without using prefix for hydrogen atom. The ring is named in its state of hydrogenation.

2-Oxa-3-thia-1,5-diazaindane

5. If the corresponding carbocyclic ring is without maximum number of non-cumulative double bonds, the heterocyclic system is considered to be with maximum number of conjugated or isolated double bonds. The corresponding carbocyclic ring is named as it contains maximum number of non-cumulative double bonds.

1,4-Dithianaphthalene   1-Oxa-4-thianaphthalene

The additional hydrogen atom, if present, is indicated by an italic capital '*H*' with its position and placed before the position(s) and the prefix(es) of the heteroatom(s).

1*H*-2-Oxachrysene

2.4.3 Spiro Heterocycles

The compounds in which two rings are fused at a common point are known as spiro compounds and the common atom which is quaternary in nature is designated as spiro atom. The spiro compounds may be classified according to the number of spiro atoms; (i) monospiro- (ii) dispiro- and (iii) trispiro ring systems.

Spiro heterocycles are named according to the rules adopted for naming spiro hydrocarbons :

1. Spiro heterocycles with one spiro atom consisting of one or both heterocyclic rings are named by prefixing spiro to the name of normal alkane with same number of carbon atoms. The number of atoms in each ring are indicated by arabic numbers separated by a full stop and enclosed in a square bracket in ascending order and are placed between spiro prefix and the name of hydrocarbon. The heteroatoms are indicated by the prefixes and are prefixed with their positions to the name of spiro hydrocarbon.

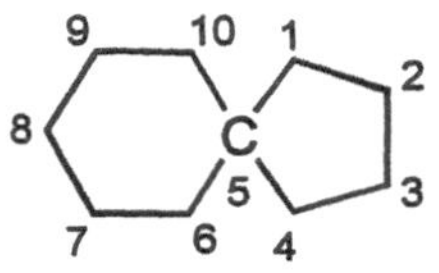

Spiro hydrocarbon

Spiro[x.y]alkane
x = number of atoms other than
    spiro atom in smaller ring.
y = number of atoms other than
    spiro atom in larger ring
                ↓
Spiro[4.5]decane

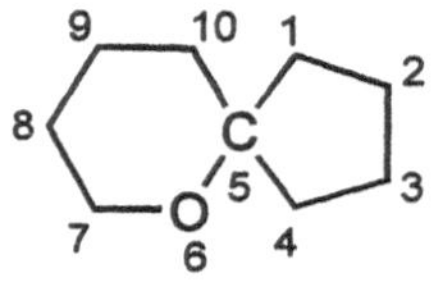

Spiro heterocycle

Spiro[x.y]alkane
x = four atoms
    (spiro atom is not included)
y = five atoms
    (spiro atom is not included)
alkane : total number of atoms
    (including heteroatom)
    are = 10 : decane
Prefix for heteroatom : oxa
                ↓
6-Oxaspiro[4.5]decane

2. The numbering starts from the ring atom of the smaller ring (if rings are of unequal size) attached to the spiro atom and proceeds first around the smaller ring and then around the larger ring through the spiro atom. The heteroatoms are assigned the lowest possible number locants.

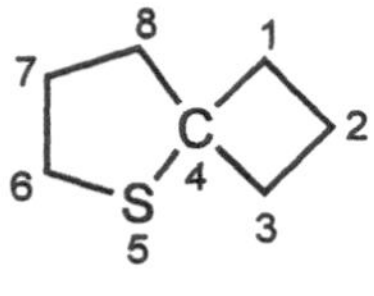

5-Thiaspiro[3.4]octane

If there is choice between two different heteroatoms, the preferential numbering is decided according to the appearance of the heteroatoms in Table 1.

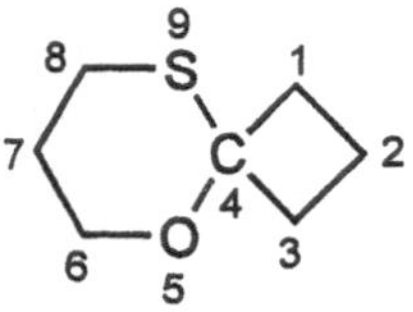

5-Oxa-9-thiaspiro[3.5]nonane

3. The heterocyclic ring is preferred over the carbocyclic ring of the same size. If both the rings are heterocyclic, the preference is given to the heterocyclic ring with heteroatom appearing first in Table 1.

2-Oxaspiro[5.5]undecane   1-Oxa-6-thiaspiro[4.4]nonane

4. If the unsaturation is present in a ring, the pattern of numbering remains the same but the direction around the ring remains in such a way that the multiple bond is given number as low as possible. However, the heteroatom is preferred over the multiple bond.

1-Oxaspiro[4.5]dec-6-ene   6-Oxaspiro[4.5]dec-9-ene

5. When one or both the components of spiro heterocycle are fused polycyclic system, the names of both the components are cited after prefix 'spiro' in square bracket in alphabetical order and are separated by the numbers of spiro atom. The components in such spiro heterocyclic system retain their numbering, but the second component is numbered by primed numbers.

Spiro[cyclopenta-2,4-diene-1,3'-3*H*-indole]   Spiro[piperidine-4,9'-xanthene]

6. If both the heterocyclic components are the same in spiro heterocyclic system, 'spirobi-' is prefixed to the name of heterocyclic component.

3,3'-Spirobi(3*H*-indole)

2.4.4 Bridged Heterocycles

Bridged heterocyclic systems are named according to the rules for bridged hydrocarbons. The heteroatoms with their locants are prefixed to the name of bridged hydrocabron.

2.4.4.1 Bicyclic Systems

1. Bridged heterocyclic system consisting of two rings with two or more common atoms is given the name of acyclic hydrocarbon with the same total number of carbon atoms, which is preceded by the prefix 'bicyclo-' with the descending order of numbers separated by a full stop in square bracket indicating number of atoms separating bridgehead atoms. The prefixes indicating heteroatoms with their locants are prefixed to the name of bridged hydrocarbon.

2. The numbering starts from one of the bridgehead atoms and proceeds through the longest possible route to the second bridgehead atom and then by the second longest route to the first bridgehead atom and finally by shortest route from one bridgehead atom to the second bridgehead atom.

3. The heteroatom is given number as low as possible.

4. When there is choice between heteroatom and multiple bond, the heteroatom is preferred.

5. If there is comparison between heteroatoms, the preference of numbering is given according to their appearance in Table 1 and their prefixes with locants are arranged alphabetically.

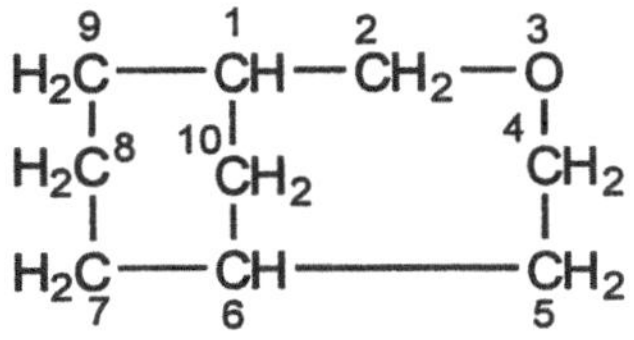

3-Oxabicyclo[4.3.1]decane

Total number of atoms = 10 : acyclic hydrocarbon – decane derivative
Number of rings = 2 : prefix 'bicyclo-' is used
Number of atoms separating bridgehead atoms
in three routes and arranged in descending order : [4.3.1]
Prefix for heteroatom with its locant : 3-oxa

Thus, the name of heterocyclic compound is : 3-Oxabicyclo[4.3.1]decane.

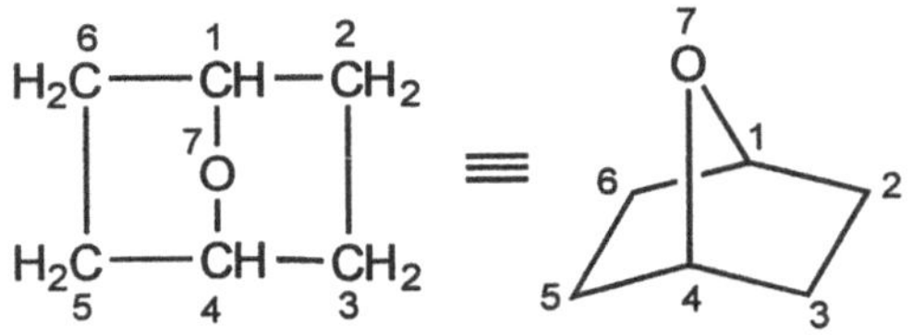

7-Oxabicyclo[2.2.1]heptane

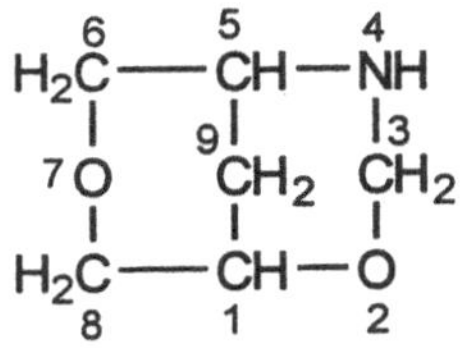

4-Aza-2,7-dioxabicyclo[3.3.1]nonane

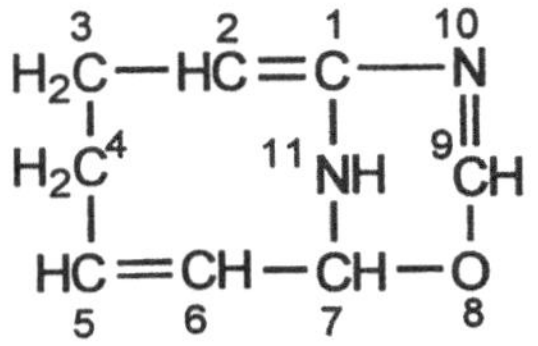

10,11-Diaza-8-Oxabicyclo[5.3.1]undeca-1,5,9-triene

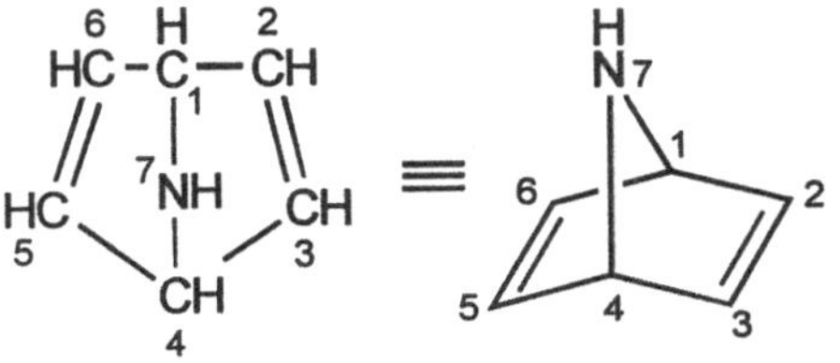

7-Azabicyclo[2.2.1]hepta-2,5-diene

2.4.4.2 Polycyclic Systems

1. Polycyclic bridged heterocycles are also named according to the rules adopted for the bicyclic bridged heterocycles. However, the prefix tricyclo-, tetracyclo-, *etc.*, depending on the number of rings is used.

2. The prefix (cyclo-) is followed by the numbers separated by full stops in square bracket in decreasing order indicating :

(i) the number of atoms of two branches of the main ring containing maximum number of atoms,

(ii) the number of atoms in main bridge excluding bridgehead atoms and

(iii) the number of atoms in secondary bridges.

3. The location of each secondary bridge (the position of bridgehead atoms involved in the secondary bridges) is indicated by superscripts to the numbers indicating its length and are seprated by a comma.

4. The prefix (tri- or tetracyclo- depending on the number of rings) with numbers separatd by full stop in square bracket is followed by the name of acyclic hydrocarbon of the carbon atoms equivalent to the total number of atoms in the bridged heterocyclic system.

5. The prefixes for the heteroatoms and the substituents with their positions are indicated and placed as usual.

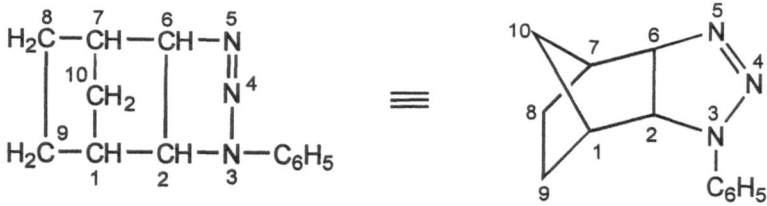

3-Phenyl-3,4,5-triazatricyclo[5.2.1.0^{2,6}]decane

Total number of atoms = 10 : acyclic hydrocarbon – decane
Number of rings = 3 : prefix tricyclo- is used
Number of heteroatoms = 3 (nitrogen at 3,4 and 5 positions) : prefix- 3,4,5-triaza-
Number of atoms in two branches of main ring : 5 and 2
Number of atoms in main bridge : 1
Number of atoms in secondary bridge : 0
Superscripts for the location of secondary bridge : 2,6 (positions of attachment
                                                   of secondary bridge)
Thus, the name of
heterocyclic compound is as   : 3-Phenyl-3,4,5-triazatricyclo[5.2.1.0^{2,6}]decane.

10-Methoxy-6-azatricyclo[4.3.1.0^{3,8}]decane

3-Thia-8,10-diazatricyclo[5.3.2.0^{2,6}]dodecane

2.4.5 Heterocyclic Ring Assemblies

Two or more heterocyclic ring systems, single or fused, joined to each other by single or double bonds are called heterocyclic ring assemblies.

Following rules are adopted for naming heterocyclic ring assemblies :

1. The heterocyclic ring assembly of two identical heterocyclic systems is named by prefixing 'bi-' before the name of heterocyclic system or its corresponding radical.

2. The numbering of heterocyclic ring assembly is similar to that of the heterocyclic system or its corresponding radical. One heterocyclic system is assigned with unprimed numbers and other with primed numbers. The points of attachment are indicated and placed before the name.

2,2'-Bipyridine or 2,2'-Bipyridyl   2,2'-Bifuran or 2,2'-Bifuryl

3. If there is choice, the heterocyclic ring with lower numbered point of attachment is assigned with unprimed numbers.

2,3'-Bipyridine or 2,3'-Bipyridyl   2,3'-Bifuran or 2,3'-Bifuryl

2,3'-Biquinoline or 2,3'-Biquinolyl

4. The locants of the substituents are placed in the ascending order with the preference of unprimed numbers over primed numbers (unprimed numbers are considered as lower numbers).

2,2',5,5'-Tetramethyl-1,1'-bipyrrolyl

5. Non-identical heterocyclic ring assemblies are named by selecting one heterocyclic system as a base component and other as a substituent. The base component is numbered with unprimed numbers and the substituent ring is with primed numbers. The heterocyclic ring assembly with non-identical heterocyclic rings is named as :

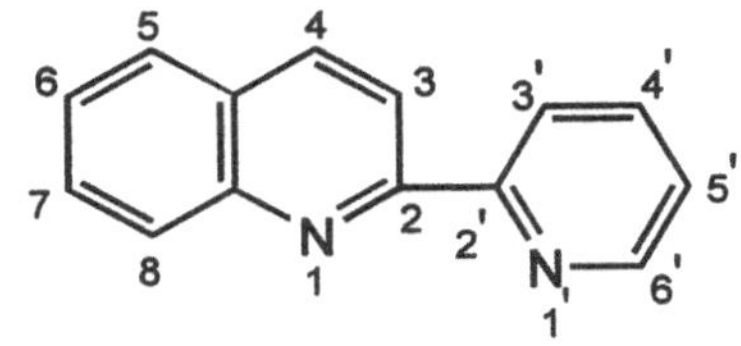

2-(2'-Pyridyl)quinoline

6. The heterocyclic ring assemblies constructing of three or more identical heterocyclic systems are named as follows :

(i) the appropriate numerical prefix, ter-, quater-, *etc.*, is placed before the name of the corresponding heterocyclic system.

(ii) unprimed numbers are assigned to the one of the terminal heterocyclic systems and others are assigned single, double, triple, *etc.* primed numbers.

(iii) the points of attachment are assigned the lowest possible numbers.

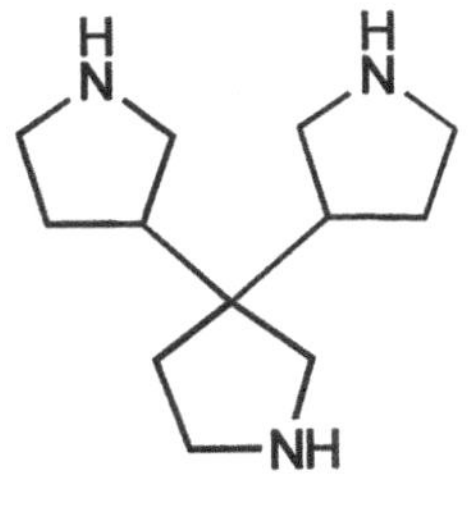

3,3',3"-Terpyrrolidine

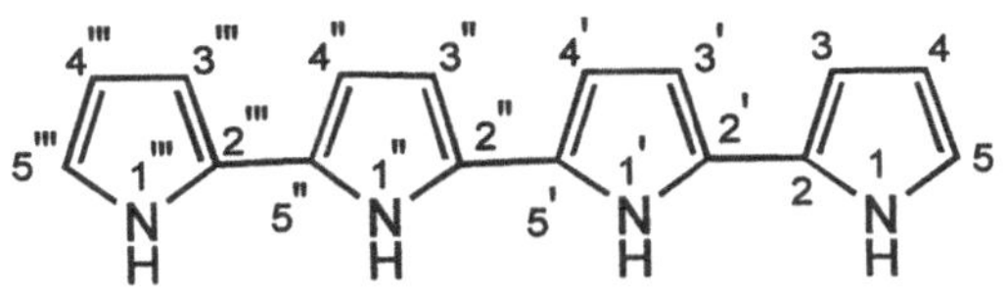

2,2':5',2":5",2'''-Quaterpyrrole

REFERENCES

1. A. Hantzsch and J.H.Weber, *Ber.* **20**, 3119 (1887).

2. O. Widman, *J. Prakt. Chem.* **38(2)**, 185 (1888).

3. A. M. Patterson and L. T. Capell, *The Ring Index (2nd Edn.)*, Am. Chem. Soc., Washington, D. C., 1960 and Supplements I, 1963; II, 1964; III, 1965.

4. *IUPAC Nomenclature of Organic Chemistry*, Definitive Rules for sections A-C (3rd Edn.), Butterworths, London, 1971.

5. J. H. Fletcher, O. C. Dermer and R. B. Fox (Eds.), *Nomenclature of Organic Compounds*, *Adv. in Chem. Ser.* **126,** Am. Chem. Soc., Washington, D. C., 1974.

6. R. S. Chan, *Introduction to Chemical Nomenclature*, Butterworths, London, 1974.

7. A. D. McNaught in A. R. Katritzky and A. J. Boulton (Eds.), *Adv. Heterocycl. Chem.* Vol. **20**, Academic Press, New York, 1976, pp. 175.

8. J. Rigaudy and S. P. Klesney, *IUPAC Nomenclature of Organic Chemistry*, Definitive Rules, Sections A to H, Pergamon Press, Oxford, 1979.

9. IUPAC Commission on Nomenclature of Organic Chemistry, *Pure Appl. Chem.* **55**, 409 (1983).

10. A. R. Katritzky, *Handbook of Heterocyclic Chemistry*, Pergamon Press, Oxford, 1985.

11. *CRC Handbook of Chemistry and Physics* (78th Edn.), C.R.C. Press, Boca Raton, 1997.

12. A. D. McNaught and P. A. S. Smith, in A. R. Kartritzky and C. W. Rees (Eds.), *Comprehensive Heterocyclic Chemistry* Vol. **1**, Pergamon Press, Oxford, 1984, pp. 7.

13. R. Panico, W. H. Powell and J. –C. Richer, *A Guide to IUPAC Nomenclature of Organic Compounds, Recommendations 1993*, Blackwell Scientific Publications, Oxford, 1993.

AROMATIC HETEROCYCLES

CONTENTS

1 GENERAL

Aromatic heterocycles are planar or nearly planar cyclic conjugated heterocycles and are associated with (4n + 2) delocalized π-electrons. These are characterized by their (i) enhanced stability, (ii) substitution reactions rather than addition reactions with the retention of aromaticity, and (iii) special spectroscopic characteristics.

The aromatic heterocycles are considered to be derived from two aromatic ring systems; benzene **1** and cyclopentadienyl anion **2**.

1. **Six-membered aromatic heterocycles:** These heterocycles are considered to be derived from benzene by replacing CH group of the benzene ring by isoelectronic N, O^+ or S^+ (scheme-1).

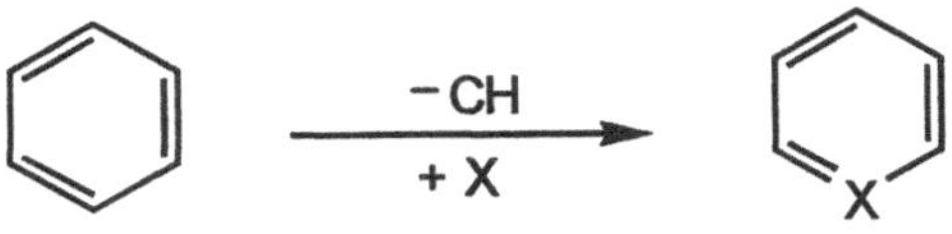

Scheme-1

2. **Five-membered aromatic heterocycles :** These heterocycles are considered to be derived from cyclopentadienyl anion by the replacement of CH group by NH, O and S (scheme-2).

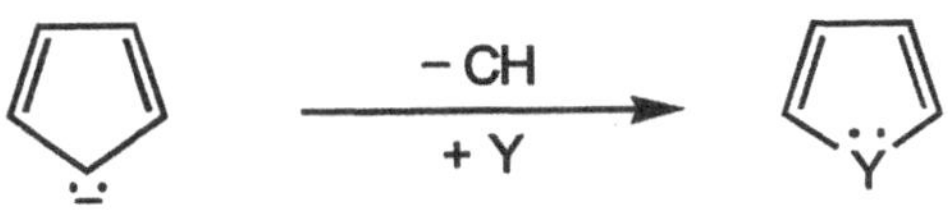

Scheme-2

The multiple replacement of CH groups in the six- and five-membered ring systems is also possible with the retention of aromatic character, since the trivalent and divalent heteroatoms contribute one or two electrons to the aromatic system.

1.1 Chemical Behaviour of Aromatic Heterocycles (Characteristics of Heteroatom in Ring)

1.1.1 Six-Membered Aromatic Heterocycles

Since the six-membered aromatic heterocycles are considered to be derived from the benzene ring by the replacement of CH group by the electronegative heteroatom, the chemical behaviour of these heterocycles can be considered to be similar to that of benzene. The symmetry of benzene ring is, however, perturbed by the introduction of a heteroatom. The electronegative heteroatom causes withdrawal of electrons from the ring carbon atoms and thus affects localization of electrons with the result of electron deficiency on the ring carbon atoms. The six-membered heterocycles are, therefore, known as π-deficient aromatic heterocycles. The introduction of further heteroatom(s) into six-membered ring reinforces the electron-withdrawing effect and results in the further electron deficiency on the ring carbon atoms. Thus, the chemistry of these heterocycles **5, 6** and **7** is affected by the electron deficiency on the ring carbon atoms caused by the electron-withdrawing effect of the heteroatom(s) (Fig. 1).

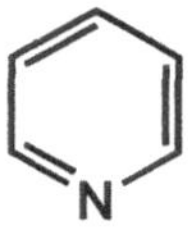

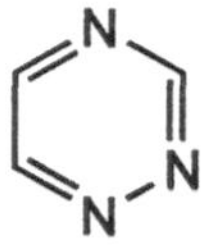

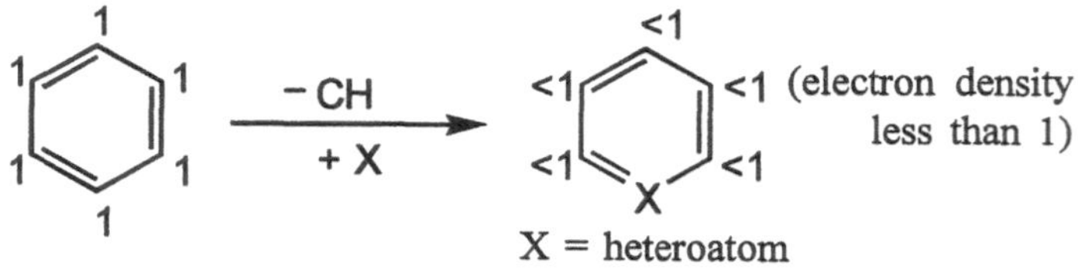

Fig. 1. Representation of electron density in six-membered heterocycles

The electron density on each carbon atom in the benzene ring is considered to be one (1), the electron density in six-membered heterocycles, therefore, will be less than one (1) due to the electron-withdrawal effect of the electronegative heteroatom. The molecular orbital calculations also confirm this prediction as in the following structures **5** and **6** (Fig. 2) :

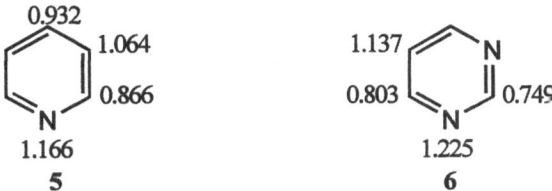

Fig. 2. Distribution of π-electron density in six-membered heterocycles

Resonance theory also supports the theoretical predictions and leads to the same conclusion (Fig. 3).

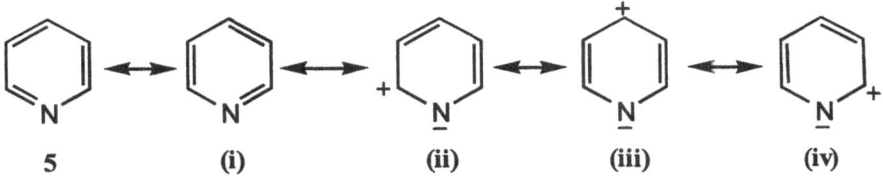

Fig. 3. Resonating structures of pyridine

The π-deficiency in six-membered aromatic heterocycles is also reflected in their characteristic spectroscopic and chemical properties.

1.1.2 Five-Membered Aromatic Heterocycles

Five-membered aromatic heterocycles are considered to be derived from cyclopentadienyl anion **2** and the lone pair on the heteroatom is involved in the cyclic delocalization of π-electrons. Six π-electrons are, therefore, delocalized over five atoms. The electrons density on each carbon atom in the ring is approximated to be greater than one (1) (6/5 = 1.2) in comparison to the benzene ring, a π-neutral system in which one electron is on each carbon atom. The five-membered aromatic heterocycles are, therefore, referred to as π-excessive aromatic heterocycles. The electron-donor characteristic of heteroatom and the π-excessiveness in five-membered aromatic heterocycles can be evidenced by their resonating structures (Fig. 4) :

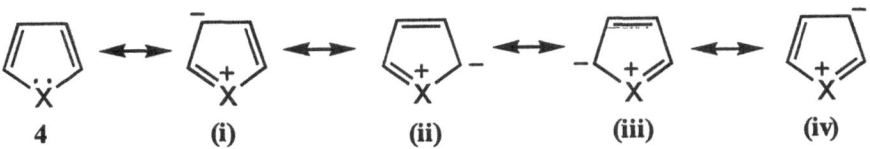

Fig. 4. Resonating structures in five-membered aromatic heterocycles

The molecular orbital calculations also support the theoretical predictions (Fig. 5).

1.064   1.036

1.030   0.912

N   S
H

8   **9**

Fig. 5. Distribution of π-electron density

1.1.3 Mixed Aromatic Heterocycles
(π-Deficient + π-Excessive Aromatic Heterocycles)

Five-membered aromatic heterocycles **10** containing both types of heteroatoms (heteroatom with electron-withdrawing effect as in six-membered heterocycles and heteroatom with electron-releasing effect as in five-membered heterocycles) are considered to be mixed systems with the characteristics of both π-deficient and π-excessive aromatic heterocycles. The chemistry of these heterocycles shows the similarities with six-membered as well as with five-membered aromatic heterocycles with one heteroatom.

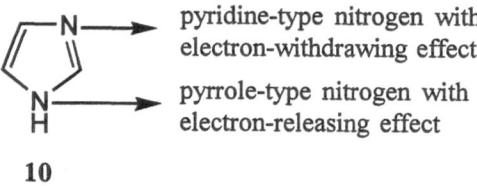

pyridine-type nitrogen with electron-withdrawing effect

pyrrole-type nitrogen with electron-releasing effect

10

2 STRUCTURAL TYPES

2.1 Six-Membered Aromatic Heterocycles

These are heterocyclic analogs of benzene in which CH group(s) of the benzene ring is replaced by the heteroatom(s) with the distortion in symmetry due to carbon–heteroatom bond(s). These heterocycles are planar with a complete and uninterrupted cycle of p-orbitals and are associated with (4n+2) π or 6π-cyclically delocalized electrons **5–7, 11–15.** Other fifth group elements (P, As, Sb and Bi) can also replace CH group, but the aromatic character decreases with increasing size of the heteroatom **16–18.** Oxygen and sulfur can also replace CH group of the benzene ring forming positively charged ionic species, pyrylium **19** and thiopyrylium **20** cations, respectively (Fig. 6).

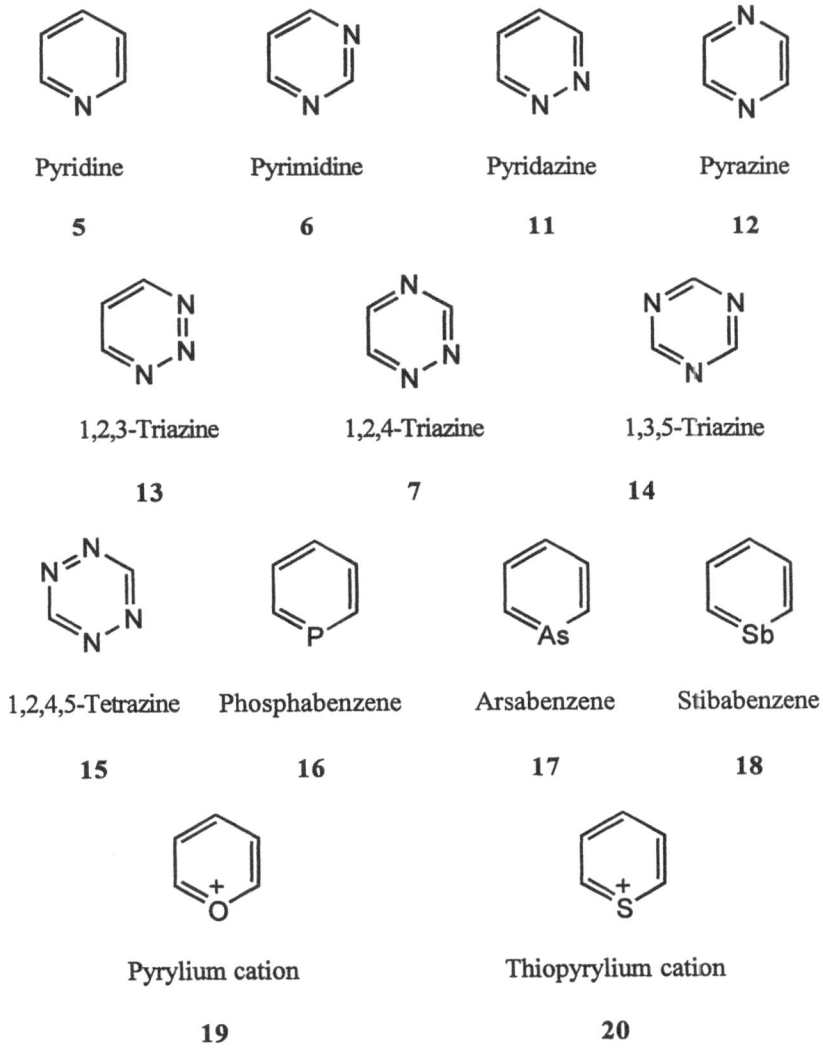

Pyridine   Pyrimidine   Pyridazine   Pyrazine

5   **6**   **11**   **12**

1,2,3-Triazine   1,2,4-Triazine   1,3,5-Triazine

13   **7**   **14**

1,2,4,5-Tetrazine   Phosphabenzene   Arsabenzene   Stibabenzene

15   **16**   **17**   **18**

Pyrylium cation   Thiopyrylium cation

19   **20**

Fig. 6. Six-membered aromatic heterocycles

2.2 Five-Membered Aromatic Heterocycles

These are heterocyclic analogs of cyclopentadienyl anion and are considered to be derived by replacing CH group by divalent heteroatom (**8, 9** and **21**). These heterocycles are planar with sp²-hybridized atoms and involve a lone pair of electrons on the heteroatom in cyclic delocalization. The aromaticity of these heterocycles depends on the extent of involvement of lone pair on heteroatom in cyclic delocalization which, in turn, depends on the electronegativity of the heteroatom.

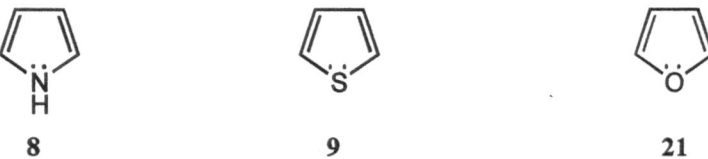

8   **9**   **21**

Nitrogen is trivalent and it may be substituted for a CH group in the five-membered heterocycles with one heteroatom. Thus, the analogs of pyrrole, furan and thiophene (**10, 22–26**) are possible (Fig. 7).

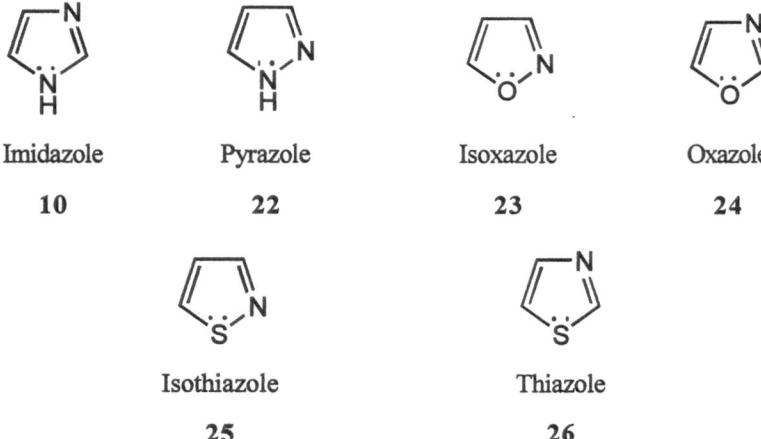

Imidazole   Pyrazole   Isoxazole   Oxazole

10   **22**   **23**   **24**

Isothiazole   Thiazole

25   **26**

Fig. 7. Five-membered heterocycles with two heteroatoms

The pyridine-type nitrogen –N= is not involved in maintaining aromaticity and induces electron-withdrawing effect similar to the meta directing effect in benzene. The singly bonded heteroatom donates electrons to the ring carbon atoms and the lone pair on the heteroatom is involved in the aromaticity and thus exerts α-directing effect.

2.3 Benzo-fused Aromatic Heterocycles

The fusion of a benzene ring to an aromatic heterocyclic ring retains aromaticity in the modified form. These aromatic heterocycles exhibit bond alternation involving partial localization of the bonds.

2.3.1 Benzo-fused Six-Membered Aromatic Heterocycles

Benzo-fused six-membered heterocycles (27–34) (Fig. 8) are related to naphthalene in the same way as pyridine with benzene. The fusion of a benzene ring, however, causes decrease in the aromaticity due to the bond alternation.

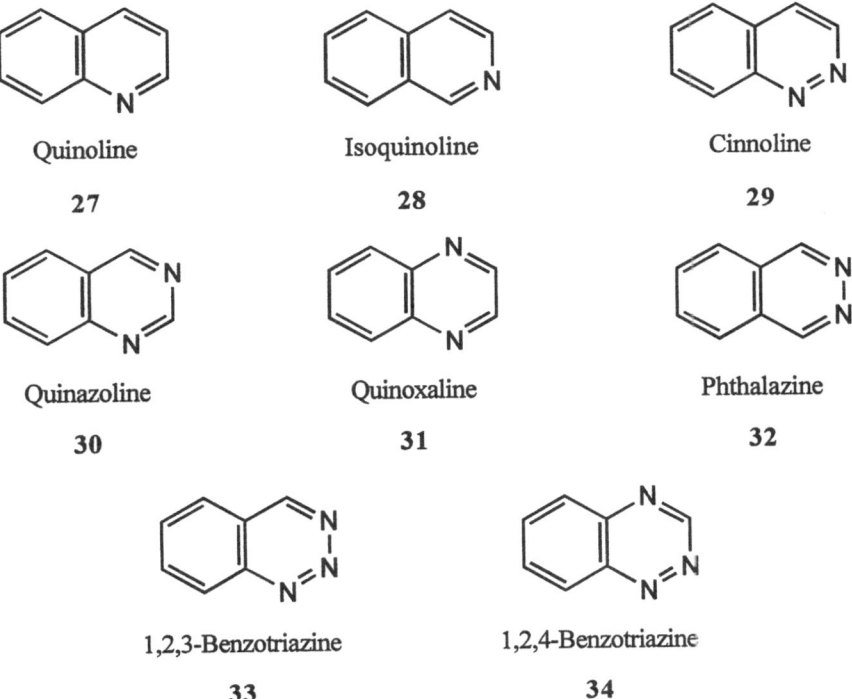

Quinoline	Isoquinoline	Cinnoline
27	**28**	**29**

Quinazoline	Quinoxaline	Phthalazine
30	**31**	**32**

1,2,3-Benzotriazine	1,2,4-Benzotriazine
33	**34**

Fig. 8. Benzo-fused six-membered aromatic heterocycles

2.3.2 Benzo-fused Five-Membered Aromatic Heterocycles

Benzo-fused five-membered aromatic heterocycles 35–38 (Fig. 9) also show some bond alternation. However, the heterocyclic compounds 39–41 in which a benzene ring is fused to a five-membered aromatic heterocyclic ring by a carbon–carbon single bond are less aromatic due to considerable bond alternation.

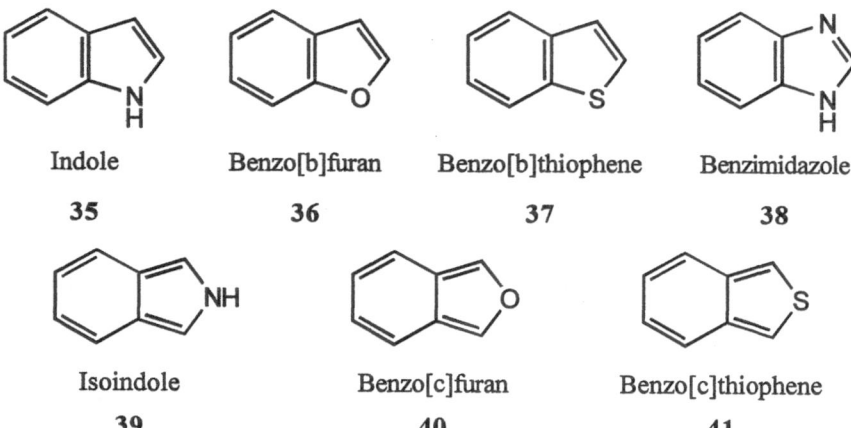

Indole	Benzo[b]furan	Benzo[b]thiophene	Benzimidazole
35	**36**	**37**	**38**

Isoindole	Benzo[c]furan	Benzo[c]thiophene
39	**40**	**41**

Fig. 9. Benzo-fused five-membered aromatic heterocycles

2.4 Other-fused Aromatic Heterocycles

The fusion of two six-membered aromatic heterocyclic rings through a carbon-carbon single bond results in the formation of heterocyclic analogs of naphthalene, **42** and **43**. Similarly, the fusion of a six-membered aromatic heterocyclic ring with a five-membered aromatic heterocyclic ring forms a bicyclic heterocycle **44** which is also considered to be included in the class of aromatic heterocycles.

1,5-Naphthyridine	Pteridine	Purine
42	**43**	**44**

The fusion of two rings across the C–N bond in such a way that the nitrogen atom is common to both the rings also leads to the formation of bicyclic aromatic heterocycles **45** and **46**. These heterocycles are planar and possess 10π-electrons required for the aromaticity according to the $(4n + 2)$ π-rule.

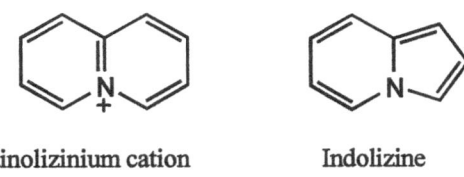

Quinolizinium cation   Indolizine

45   46

3 AROMATICITY IN HETEROCYCLES

3.1 Relationship with Carbocyclic Aromatic Compounds

Carbocyclic aromatic compounds are planar cyclic conjugated systems with $(4n + 2)$ π-electrons. These compounds have specially stabilized π-electron system in which all the bonding molecular orbitals are completely filled and the antibonding orbitals remain vacant. Thus, the concept of aromaticity is not particularly related to the nature and the number of the ring atoms, but is associated with the molecular orbital energies.

The main characteristic features of the carbocyclic aromatic system are :

(i) unusual stability,

(ii) tendency to undergo substitution reactions with the retention of aromatic character,

(iii) uniformity in bond lengths in the rings, and

(iv) special spectroscopic characteristics, particularly NMR-spectra.

All these characteristic features are also exhibited to a greater or lesser extent by the aromatic heterocycles. Six-membered aromatic heterocycles are related to benzene as these are derived by the replacement of one or more CH group(s) of the benzene ring by the trivalent heteroatom(s), but the structural geometry of the benzene ring is perturbed with the introduction of heteroatom. Similarly, five-membered aromatic heterocycles are related to the cyclopentadienyl anion. Benzo-fused aromatic heterocycles are considered to be analogs of naphthalene and are related to naphthalene similarly as the six-membered aromatic heterocycles with benzene. To correlate the aromaticity of aromatic heterocycles with carbocyclic analogs, it is desirable to discuss the structures of benzene, naphthalene and cyclopentadienyl anion.

3.1.1 Structure of Benzene

In benzene trigonally hybridized (sp²) carbon atoms form a strainless planar ring. Thus, there are six σ C–H bonds, six σ C–C bonds and six $2p_z$ orbitals (one on each carbon atom). Six $2p_z$ orbitals are perpendicular to the plane of the ring. Six $2p_z$ orbitals overlap in two ways and thus two equivalent structures **1a** and **1b** are obtained. Each $2p_z$ orbital, however, overlaps with its neighbouring $2p_z$ orbital equally to form a completely delocalized molecular orbital embracing all the six carbon atoms (Fig. 10).

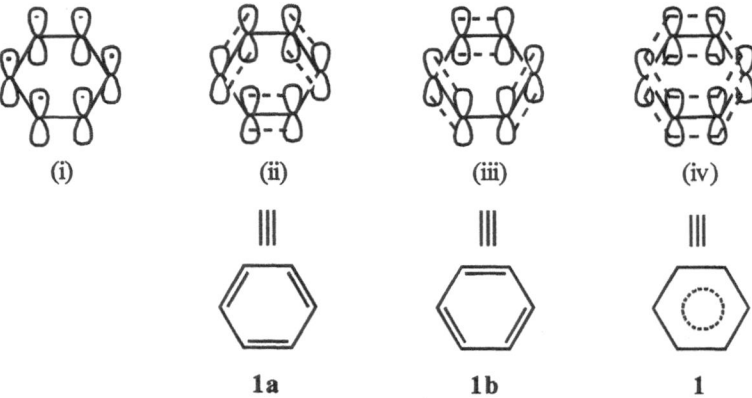

Fig. 10. Structure of benzene

Since six $2p_z$ orbitals are involved, six-molecular orbitals are formed; three bonding and three antibonding orbitals. Six $2p_z$ electrons, in pairs, occupy three bonding molecular orbitals which are of the lowest energy and the antibonding orbitals remain vacant. The energy of three pairs of the delocalized π-electrons is less than that of three pairs of the localized π-electrons and hence benzene molecule is stabilized by resonance and has high delocalization energy (resonance energy or stabilization energy). Thus, benzene has a regular hexagonal structure with equal carbon–carbon bond distances (1.39Å) which are intermediate between carbon–carbon single bond (Csp²–Csp² = 1.46Å) and carbon–carbon double bond (Csp²=Csp² = 1.34Å).

1

3.1.2 Structure of Naphthalene

Ten carbon atoms of naphthalene are at the corners of two fused hexagons. All the carbon atoms are sp²-hybridized and are attached to three other atoms by σ-bonds with all carbon atoms and hydrogen atoms in a single plane. Unhybridized $2p_z$ orbitals with one electron on each carbon atom are perpendicular to the plane of the ring. These $2p_z$ orbitals on each carbon atom are combined in three ways resulting in three structures **47a**, **47b** and **47c**. However, $2p_z$ orbitals are overlapping equally with their adjacent $2p_z$ orbitals and a cloud of π-electrons is formed which is considered as two partial overlapping sextets with a common pair of π-electrons (Fig. 11). Because of containing delocalized 10π-electrons (according to the Huckel's rule i.e. (4n+2) π-electrons in a planar cyclic structure), naphthalene is an aromatic molecule.

Naphthalene can be considered to be a resonance hybrid of following three resonating structures **47a**, **47b** and **47c** (Fig. 12) which are not usually equivalent as central double bond in **47a** is different from that in other two resonating

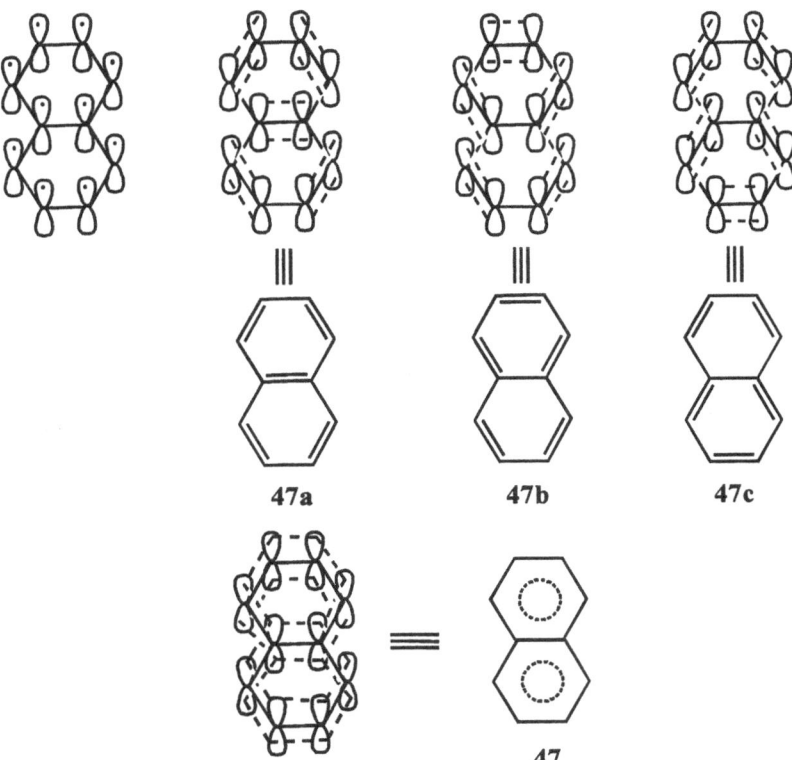

| | 47a | 47b | 47c |

47

Fig. 11 Structure of Naphthalene

structures **47b** and **47c**. However, if three resonating structures are assumed to contribute equally to the naphthalene structure; C_1-C_2 bond has more double bond character than C_2-C_3 bond. Molecular approximations also show bond orders of 1.724 and 1.603, respectively as compared to in benzene (1.667). C_1-C_2 Bond is expected to have more double bond character than the single bond and C_2-C_3 bond to have more single bond character than the double bond. In agreement with these predictions, C_1-C_2 and C_2-C_3 bond distances are 1.36Å and 1.415Å, respectively. The non-equivalency of the bonds is called partial bond fixation and is usually observed in all the fused aromatic systems.

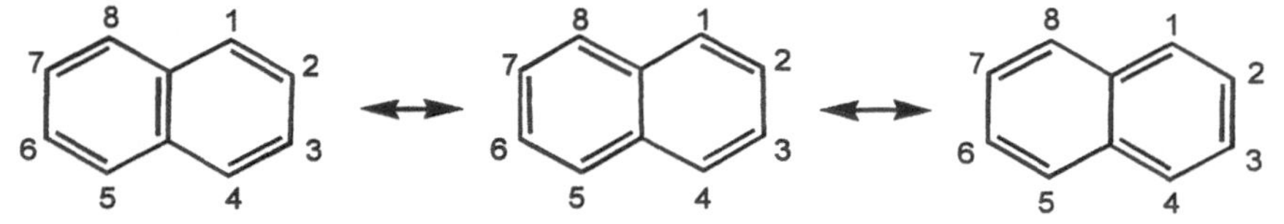

Fig. 12. Resonating structures of naphthalene

3.1.3 Cyclopentadienyl Anion System

Cyclopentadienyl anion **2** has planar cyclic structure with delocalized six π-electrons in three bonding molecular orbitals. The relative stability of this system is reflected by the unexpected acidity of cyclopentadiene as the resulting cyclopentadienyl anion is greatly stabilized by resonance (Fig. 13).

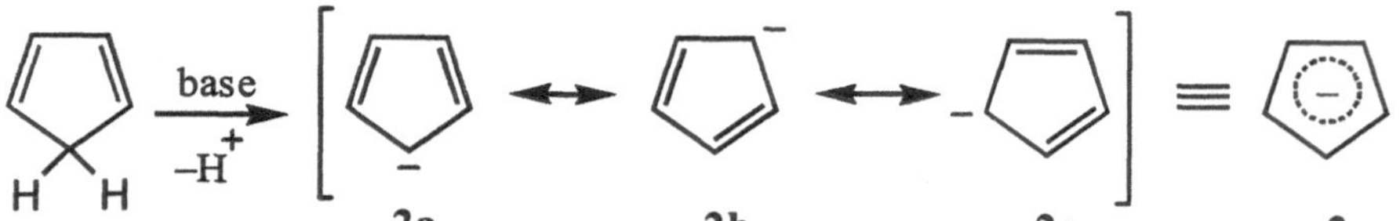

Fig. 13. Resonating structures of cyclopentadienyl anion

Cyclopentadienyl anion is an isoelectronic with five-membered aromatic heterocycles; pyrrole, thiophene and furan. These heterocyclic systems are aromatic by virtue of an aromatic sextet. The planar cyclic conjugated systems with (4n+2) π-electrons are stabilized by delocalization of π-electrons and are said to be aromatic, while the systems with $4n\pi$-electrons are considered antiaromatic.

3.2 Criteria of Aromaticity in Heterocycles[1,2]

Aromaticity can be defined qualitatively as a cyclic planar system associated with (4n+2) π-electrons is said to be an aromatic. The degree of aromaticity depends on the involvement of lone pair electrons on the heteroatom in cyclic delocalization of π-electrons. The quantitative estimation of aromaticity is although difficult, however it is based on the delocalization of π-electrons and on the physico-chemical properties relating to the aromaticity particularly, molecular geometry, enhanced stability and diamagnetic susceptibility exaltations.

3.2.1 Structural Criteria

3.2.1.1 Bond Lengths

Delocalization of π-electrons in an aromatic system causes carbon–carbon bond to have typical bond length between that of a carbon–carbon single bond [Csp^2–Csp^2] and carbon-carbon double bond [Csp^2=Csp^2]. Delocalization of π-electrons in benzene causes the bond lengths to be equal (1.39Å), but in the acyclic conjugated system the cyclic delocalization is not possible and the bond lengths between sp^2-hybridized carbon atoms are not equal and thus exhibits bond alternation. It can be generalized that the compounds with non-alternating bond lengths are considered to be aromatic, whereas the compounds with bond alternations are non-aromatic. In heteroaromatic systems, carbon–heteroatom bond is generally shorter than that in its saturated analogs.

Some single and double bond lengths in acyclic polyenes[3] with sp^2-hybridized atoms are given for the comparison with the bond lengths in aromatic heterocycles:

single bonds		double bonds	
C–C	= 1.48 Å	C=C	= 1.34 Å
C–N	= 1.45 Å	C=N	= 1.27 Å
C–O	= 1.36 Å	C=O	= 1.22 Å
C–S	= 1.75 Å	C=S	= 1.64 Å
N–N	= 1.41 Å	N=N	= 1.23 Å

3.2.1.1.1 Five-Membered Aromatic Heterocycles

Five-membered aromatic heterocycles exhibit cyclic delocalization of π-electrons by which carbon–carbon single bonds acquire double bond character and carbon–carbon double bonds acquire the character of single bond [Csp^2–Csp^2]. But as the different heteroatoms of differing electronegativity are involved in the five-

membered heterocycles, bond localization is also observed which is consistent-
with the localized structures. The degree of bond localization depends on the
electronegativity of the heteroatom. Bond lengths in five-membered heterocycles
are summarized in Tables 1 and 2.

Table 1. Bond lengths in five-membered heterocycles with one heteroatom

X	X–C$_2$	C$_2$–C$_3$	C$_3$–C$_4$
O	1.362 Å	1.361 Å	1.431 Å
S	1.714 Å	1.370 Å	1.420 Å
NH	1.370 Å	1.380 Å	1.417 Å

3.2.1.1.1.1 Aromaticity Order

The aromaticity in five-membered heterocycles depends on the availability of the
lone pair on heteroatom for the involvement in cyclic delocalization and hence
depends on the electronegativity of heteroatom. The oxygen-heterocycles exhibit
considerable degree of bond localization because of higher electronegativity of
oxygen than nitrogen and sulfur. Thus, oxygen atom has greater hold on the lone
pair of electrons and causes localization of the electrons which results in bond
alternation. The order of aromaticity in five-membered heterocycles is as :

sulfur-heterocycles > nitrogen-heterocycles > oxygen-heterocycles.

3.2.1.1.1.2 Bond Alternation

The aromaticity in five-membered heterocycles can be assessed by the bond
alternation which is the ratio of bond length of C$_2$–C$_3$ to C$_3$–C$_4$. The closeness of
the ratio to unity is observed in the aromatic compounds. Thus, from the bond
alternation ratio it can be concluded that sulfur-heterocycles are normally more
aromatic than the nitrogen- and oxygen-heterocycles.

Table 2. Bond lengths in five-membered heterocycles with two heteroatoms

Compound	a	b	c	d	e
	1.349 Å	1.331 Å	1.416 Å	1.373 Å	1.359 Å
	1.349 Å	1.326 Å	1.378 Å	1.358 Å	1.369 Å
	-	1.309 Å	1.425 Å	1.356 Å	-
	1.357 Å	1.293 Å	1.395 Å	1.352 Å	1.370 Å
	1.724 Å	1.304 Å	1.372 Å	1.367 Å	1.713 Å

$$\text{Bond alternation ratio } R = \frac{C_2 - C_3}{C_3 - C_4}$$

If $R \approx 1$ the compound will be aromatic

thiophene > pyrrole > furan
(0.964)   (0.959)   (0.950)

The fusion of a benzene ring to the five-membered aromatic heterocycles reduces the aromatic character. Indole **35** is nearly planar and shows bond alternation. However, in the heterocycles in which benzene ring is fused across the carbon–carbon single bond, the aromaticity is further reduced due to the considerable degree of bond alternation. Thus, isoindole **39** and benzo[c]furan **40** show much lower degree of aromatic character and exhibit diene-type character.

Indole   Benzo[b]furan   Benzo[b]thiophene

35   **36**   **37**

Isoindole   Benzo[c]furan   Benzo[c]thiophene

39   **40**   **41**

3.2.1.1.2 Six-Membered Aromatic Heterocycles

The bond lenghts in six-membered heterocycles are intermediate between single and double bonds due to the cyclic delocalization. The further introduction of heteroatom causes reduction in the aromatic character because more π-electrons are partially localized on the heteroatoms and thus cause bond localization. The bond lengths in six-membered nitrogen-heterocycles are summarized in Table 3.

The fusion of a benzene ring to the six-membered heterocyclic ring reduces aromatic character causing partial bond localization as in naphthalene.

Table 3. Bond lengths in six-membered azaheterocycles

Compound	Bond lengths					
	1–2	2–3	3–4	4–5	5–6	6–1
(pyridazine)	1.340 Å	1.395 Å	1.394 Å	-	-	-
(pyrimidine)	1.317 Å	-	1.344 Å	1.358 Å	-	-
(pyrazine)	1.314 Å	1.358 Å	-	-	-	-
(1,2,4-triazine)	1.335 Å	1.314 Å	1.339 Å	1.317 Å	1.401 Å	1.317 Å
(1,3,5-triazine)	1.319 Å	-	-	-	-	-

3.2.1.1.3 Relationship of Bond Lengths with Bond Orders

Bond order is related to the bond length and provides a measure of the ring aromaticity. The intermediate bond lengths caused by the cyclic delocalization of π-electrons are associated with the fractional bond orders. If C–C bond is assumed to have bond order of unity and C=C and C≡C bonds of two and three respectively, a graph can be plotted between bond orders and bond lengths and can be used to predict fractional bond order for the intermediate bond length between C–C and C=C bonds. The cyclic conjugated compound is said to be aromatic, if it shows neither strong first order nor second order double bond fixation (Fig. 14).

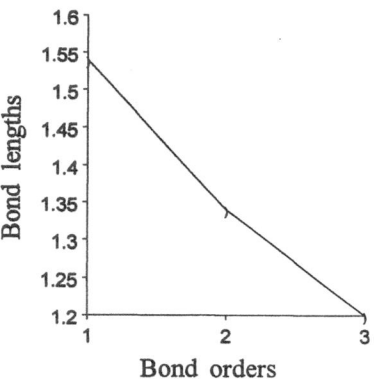

Fig. 14. Representation of relationship between bond orders and bond lengths

3.2.2 Electronic Criteria

3.2.2.1 Ultraviolet Spectroscopy[4]

The ultraviolet spectra of aromatic heterocycles are compared with their carbocyclic analogs for the similarity in bonding pattern in order to assess aromaticity qualitatively.

3.2.2.1.1 Six-Membered Aromatic Heterocycles

3.2.2.1.1.1 Monocyclic Azaheterocycles

Six-membered aromatic nitrogen-heterocycles possess basic π-electron system of the benzene ring with the addition of non-bonding lone pair electrons on the nitrogen atom and, therefore, two types of electronic transitions are involved :

(i) n→π* transitions and (ii) π→π* transitions. The lone pairs on the heteroatom are responsible for the weak transitions, known as n→π* transitions, at the longer wavelength (270–340nm). These absorptions are very weak as compared to π→π* transitions of π-electrons, but the bands due to n→π* transitions are prominent in the heterocycles with two or three nitrogen atoms. π→π* Transitions which are similar to as in aromatic six-membered carbocyclic analogs range from 240–260nm. π→π* And n→π* transitions in six-membered azaheterocycles are summarized in Table 4 alongwith π→π* transitions of benzene for comparison.

Table 4. UV-Absorption bands in six-membered azaheterocycles

Compound	π→π* bands		n→π* bands	
	λ_{max} (nm)	log ε	λ_{max} (nm)	log ε
Benzene	254	2.04	-	-
Pyridine	251	3.30	270	2.65
Pyridazine	246	3.11	340	2.50
Pyrimidine	243	3.31	298	2.51
Pyrazine	260	3.75	328	3.02

The aromatic characters of phosphabenzene, arsabenzene, and stibabenzene can be evidenced by their similar UV-spectra as of nitrogen analogs. The UV-spectra of arsabenzene, stibabenzene and phosphabenzene exhibit bands at 219 nm and 268 nm; 230nm and 312nm; and 213nm and 246 nm, respectively. These absorptions are attributed to the π→π* transitions similarly as in nitrogen analogs. The shifting of bands to the longer wavelength due to π→π* transitions is attributed to the weaker bonding in heteroaromatics with heavier atoms.

3.2.2.1.1.2 Bicyclic Azaheterocycles

Bicyclic azaheterocycles have much more complex electronic spectra with the overlapping of π→π* and n→π* transitions. But π→π* transitions are much more intense and cover n→π* transitions. The transition energies related to the π→π* transitions correspond fairly well with the aromatic carbocyclic analogs, however, the band intensities are different. The absorption spectra of naphthalene, quinoline, isoquinoline and quinolizinium ion are very similar except in the intensities of the long-wavelength bands (Table 5).

Table 5. UV-Spectral bands in bicyclic azaheterocycles

Compound	$\pi \rightarrow \pi^*$ λ_{max} (nm)		
	(log ε)		
Naphthalene	219	275	311
	(5.10)	(3.75)	(2.39)
Quinoline	225	270	313
	(4.48)	(359)	(3.37)
Isoquinoline	217	266	317
	(4.57)	(3.61)	(3.49)
Quinolizinium ion	220	267	311
	(4.61)	(3.45)	(3.32)

3.2.2.1.2 Five-Membered Aromatic Heterocycles

Five-membered aromatic heterocycles with one heteroatom exhibit a band of moderate intensity followed by a band of high intensity at shorter wavelength. A significant feature, the absence of bands due to promotion of an electron from the lone pair orbital (non-bonding) to a π-orbital of the ring (n$\rightarrow\pi^*$), is attributed to the large s-character due to smaller ring angle in the five-membered rings. The shifting of absorption to the longer wavelengths follows the sequence :

thiophene > pyrrole > furan.

The positions and the energy splitting of $\pi\rightarrow\pi^*$ transitions in the five-memberd heteroaromatic rings are similar to that in benzene and are very different to those for conjugated dienes (Table 6).

Table 6. UV-Spectral bands in five-membered aromatic heterocycles

Compound	λ_{max} (log ε)
Pyrrole	210 (4.20)
Furan	207 (3.90)
Thiophene	215 (3.80), 231 (3.87)

3.2.2.1.3 Benzo-fused Aromatic Heterocycles

The annelation increases complexity of the spectra as in carbocyclic analogs. UV spectra of benzo[b] heterocycles follow the same trend in the shifting to larger wavelengths as observed in the heteromonocycles.

The absorption of benzo[c] heterocycles at longer wavelength than their benzo[b] counterparts reflects lower aromaticity in benzo[c] heterocycles[6,7].

3.2.2.2 Infrared Spectroscopy[8-10]

Infrared spectra also provide information on the aromaticity in heterocycles. The vibrational frequencies in infrared spectra are related to the bond strengths which in turn depend on the bond lengths. The vibrational frequency is expected to increase, if the single bond acquires partial double bond character by conjugation and similarly decreases, if the double bond acquires partial single bond character. Infrared spectra, therefore, reflect conjugation and cyclic delocalization in aromatic compounds.

In the six-membered aromatic heterocycles, vibrational modes of the ring skeleton are related and approximate in the positions to those found in the benzene derivatives. The bending modes of the ring hydrogen atoms are similar to those of the corresponding arrangement of the adjacent hydrogen atoms on a benzene ring and these generalizations are applicable to all the azines. The bands relating to four ring stretching modes for pyridine and pyrimidine are summarized in Table 7 alongwith the corresponding bands of monosubstituted benzene.

Table 7. Approximate positions of ring stretching modes for pyridine, pyrimidine and benzene.

Compound	A	B	C	D
Monosubstituted benzene	1610–1600	1590–1580	1520–1417	1460–1440
Pyridine	1610–1595	1570–1550	1520–1480	1420–1410
Pyrimidine	1600–1545	1575–1540	1510–1410	1470–1330

Aromatic stretching frequency can be distinguished from the saturated C–H absorption as it gives absorption above 3000 cm^{-1}. Characteristic C–H bending vibrations in the six-membered heteroaromatics appear in the region 1300–1000 cm^{-1} (in-plane C–H bending) and in the region 700–1000 cm^{-1} (out-of-plane).

3.2.2.3 Photoelectron Spectroscopy[11,12]

Photoelectron spectroscopy involves the ejection of electrons from the occupied molecular orbitals with the interaction of high energy ultraviolet radiations with the molecule. The energy of an ejected electron can be measured and the difference between the energy of radiation used and that of an ejected electron is the ionization potential of an electron. A molecule containing electrons of differing energy can lose any one electron if its ionization potential is less than the energy of radiation used. The photoelectron spectra consists of a series of bands and each of which corresponds to the orbital of differing energy. The photoelectron spectra provide identification of the orbitals of differing energies from which the electrons are ejected. Thus, photoelectron spectra can predict the changes in the π-bonding orbitals energy levels in a series of aromatic heterocycles.

The π-bonding electronic structure in benzene is very similar to that in pyridine because of the similarity in π-bonding orbitals. As the interaction of six p-orbitals in benzene results in the formation of six molecular orbitals; three bonding and three antibonding, similarly six molecular orbitals are formed in pyridine. But the difference is that the energies of the π-orbitals in pyridine are lowered as compared to in benzene because of the electronegativity of nitrogen atom. π_2 And π_3 which are degenerate in benzene become non-degenerate in pyridine in which π_2 being of lower energy (Fig. 15). The energy of π_2-orbital depends on the electronegativity of the heteroatom and increases as the electronegativity decreases.

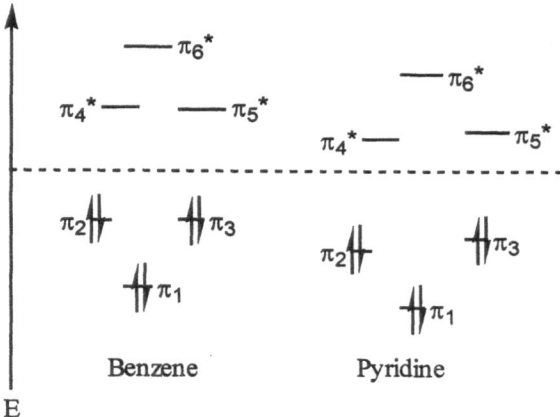

Fig. 15. Representation of orbital energies in benzene and pyridine

In the heterocyclic analogs of pyridine, in which nitrogen is replaced by the elements of fifth group (P, As, Sb, Bi), the energy of π-orbital decreases with increasing size of the heteroatom.

Similarly, π-bonding orbitals of the cyclopentadienyl anion and pyrrole show similarity, but the introduction of the heteroatom causes splitting of π_2 and π_3 molecular orbitals (non-degenerate with lower energy) (Fig. 16).

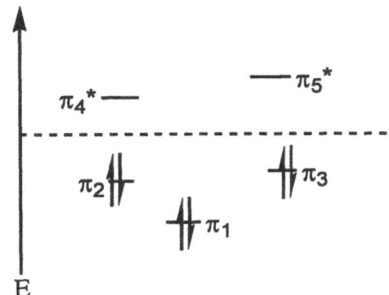

Fig. 16. Representation of orbital energies in pyrrole

3.2.2.4 Dipole Moments[13-15]

The quantitative estimation of aromaticity by dipole moments is difficult because the dipole moment is the sum of the σ-bond, π-bond and lone pair moments. However, the dipole moments in aromatic heterocycles provide significant information on the involvement of the heteroatom in the electronic distribution and, in turn, indicate the presence of cyclic delocalization in the aromatic heterocycles.

3.2.2.4.1 Six-Membered Aromatic Heterocycles

Six-membered aromatic heterocycles exhibit greater dipole moments than those in their corresponding saturated analogs. The larger dipole moments in six-membered aromatic heterocycles is attributed to the cyclic delocalization.

The larger dipole moment of pyridine **5** (2.20D) than that of piperidine **48** (1.57D) supports the structure with charge-separation and confirms dipolar structures as the contributing resonating structures in pyridine due to cyclic delocalization of π-electrons (Fig. 3).

2.20 D   1.57 D

Pyridine   Piperidine

5   **48**

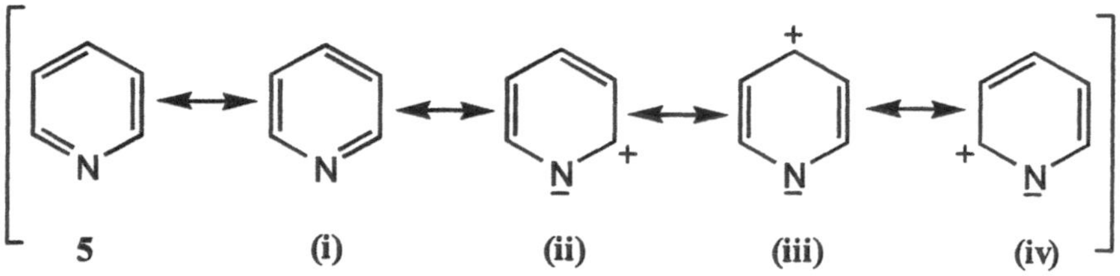

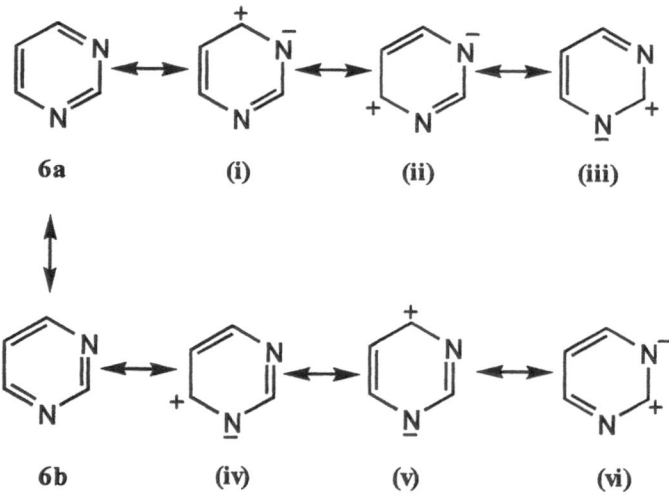

Fig. 3. Representation of charge separation in pyridine

The large dipole moment in pyrimidine (2.40D) also predicts the delocalization of π-electrons and thus confirms the contribution of the following resonance structures to the ground state of pyrimidine **6** (Fig. 17).

Fig. 17. Resonating structures in pyrimidine representing charge separation

Pyridazine **11** has appreciably high dipole moment (4.0D). The high value of the dipole moment in pyridazine is attributed to the fact that both the nitrogen atoms are present on the same side of the ring and, therefore, there is a greater pull of the electrons towards the side of nitrogen atoms.

3.2.2.4.2 Five-Membered Aromatic Heterocycles

The dipole moments of the five-membered aromatic heterocycles are lower than those of the corresponding tetrahydro derivatives. The lower dipole moments in these heterocycles are attributed to the counteraction of two effects; inductive effect and mesomeric effect. Thus, the existence of two opposing structural effects supports the contributing resonating structures involving cyclic delocalization of π-electrons in the five-membered aromatic heterocycles.

The tetrahydro derivatives have negative end at the heteroatom because of the electron pull towards electronegative heteroatom due to the inductive effect. In five-membered aromatic heterocycles; pyrrole, thiophene and furan, the inductive effect is still operating, but this effect is superimposed by the mesomeric effect operating in the opposite direction (Fig. 18).

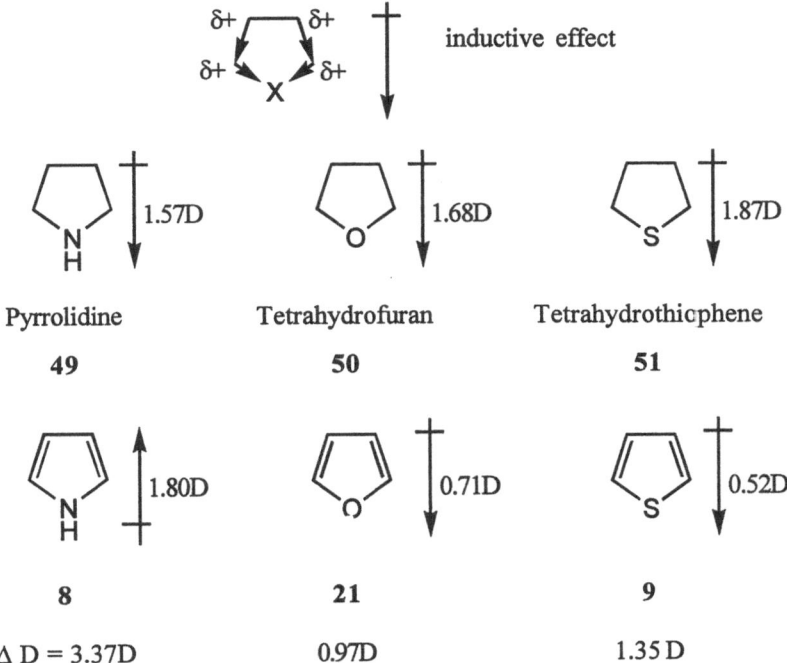

Fig. 18. Dipole moments in five-membered heterocycles

The larger difference in thiophene reflects the involvement of d-orbitals of sulfur. The direction of the dipole in pyrrole is reverse of that in pyrrolidine and thus very large difference (3.37D) is due to the large contribution of the mesomeric effect.

The dipole moments in five-membered aromatic heterocycles are assumed to be the evidence of charged resonating structures involving cyclic delocalization of the π-electrons (Fig. 4).

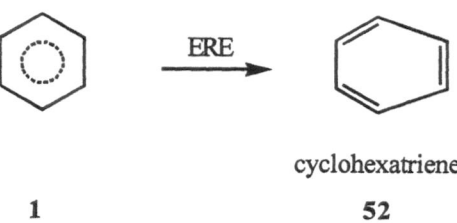

Fig. 4

3.2.3 Energetic Criteria

The energetic criteria are widely used for the quantitative estimation of aromaticity in the aromatic compounds. The enhanced stability of delocalized structure over the same localized structure is attributed to the resonance energy. The different energy terms used for the stabilization in the aromatic compounds are; stabilization energy, empirical and vertical energy, delocalization energy, conjugation energy and binding energy. These all energy terms are nearly the same, the modified treatment of the experimental data, however, leads to the different terminology of the resonance energy.

Delocalization energy is the energy term calculated by the molecular orbital theory, while the energy derived by the experimental methods is known as empirical resonance energy.

Empirical resonance energy (ERE) : The difference between the energy of the actual structure and the energy of the hypothetical localized structure is known as empirical resonance energy and is represented as (Fig. 19) :

ERE = Energy of the real structure − Energy of the hypothetical structure
      (delocalized structure)   (localized structure)

cyclohexatriene

1   52

Fig. 19. Empirical resonance energy

Vertical resonance energy : The difference between the energy of the resonance hybrid and the energy of one of the resonating structures contributing maximum to the resonance hybrid is known as vertical resonance energy (Fig. 20).

Vertical resonance energy = Energy of resonance hybrid – Energy of one of the
resonating structures

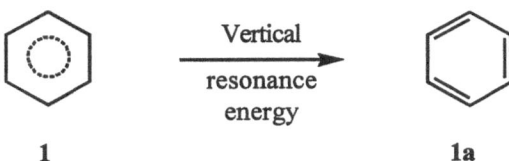

Fig. 20. Vertical resonance energy

Distortion energy : The difference between empirical resonance energy and vertical resonance energy is known as distortion energy and this energy is required to compress the single bond and to stretch the double bond. The distortion energy is represented as (Fig. 21) :

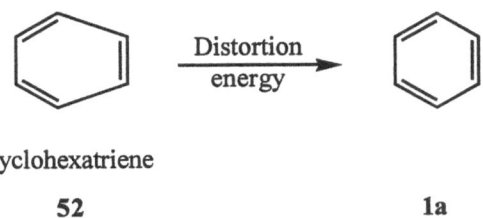

cyclohexatriene

52   1a

Fig. 21. Distortion energy

These energy terms are inter-related and their correlationship is represented in the schematic form (Fig. 22).

3.2.3.1 Empirical Resonance Energy[16,17]

Empirical resonance energies are determined by the thermochemical quantities including the hybridization energy term because the energy term for the carbon–carbon single bond between Csp^2–Csp^2 is of different energy from that between Csp^3–Csp^3.

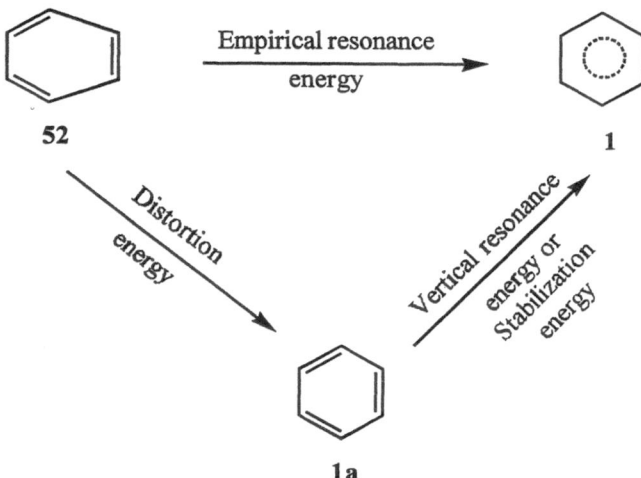

Fig. 22. Energy cycle involving different energy terms

Empirical resonance energies for the heteroaromatic systems are determined from:

(i) heat of combustion and (ii) heat of hydrogenation

3.2.3.1.1 Heat of Combustion

Heat of combustion is the heat evolved when one mole of the substance is completely burnt in oxygen. Thus the heat change associated with the combustion of pyridine is :

$$C_5H_5N \text{ (g)} + \frac{25}{4} O_2 \longrightarrow 5CO_2 \text{ (g)} + \frac{5}{2} H_2O + \frac{1}{2} N_2 \text{ (g)} \qquad .....(1)$$

The heats of combustion of heterocycles can be determined experimentally from which heats of formation are determined. The heat of formation (ΔH) of a compound is equal to the sum of the heats of formation of the products obtained by the combustion of the compound minus the heat of combustion of the compound as :

ΔH (obsd.) =   Sum of the heats of formation of $-$ Heat of combustion
                      the products (obtained by   of the compound
                      combustion of the compound)   ....(2)

The heat of formation is calculated theoretically by adding bond energy terms (ΣE) for the bonds in localized structure. The empirical resonance energy (ERE) of a compound can be calculated by the heat of formation.

ERE = Observed heat of formation $-$ Calculated heat of formation   .....(3)

Thus, the difference in the heat of formation determined from the experimental value of the heat of combustion and the theoretically calculated value is a measure of the stabilization in the delocalized system and is known as empirical resonance energy and represented as :

$$\Delta H = \Sigma E + ERE \qquad .....(4)$$

$$ERE = \Delta H - \Sigma E \qquad .....(5)$$

ΔH = Heat of formation determined from the experimental heat of combustion

ΣE = Heat of formation calculated by bond energy terms

The value of the empirical resonance energy depends on the bond energy terms and thus suffers from some shortcomings :

(i) Same bond energy term is used for the carbon–carbon single bond for both the Csp^3–Csp^3 and Csp^2–Csp^2 single bonds.

(ii) Hybridization energy for carbon–heteroatom bond : the same bond energy is used for the carbon–heteroatom bond (C–X) without giving due consideration to the hybridization.

(iii) There are energy difference between primary, secondary and tertiary C–H bonds.

(iv) Strong 'next-nearest-neighbour' interactions in the oxygen ring compounds.

However, these differences are accounted for by the energy term known as conjugation energy. Empirical resonance energies of some aromatic heterocycles calculated from the heats of combustion are summarized in Table 8.

Table 8. Empirical resonance energies of aromatic heterocycles (from heats of combustion)

Compound	ERE	
	kcal/mol	kJ/mol
Benzene (for comparison)	35.9	150
Pyridine	27.9	117
Quinoline	48.4	200
Pyrrole	21.6	90
Indole	46.8	196
Thiophene	29.1	122
Furan	16.2	68

3.2.3.1.2 Heat of Hydrogenation

Heat of hydrogenation can also be used for the determination of empirical resonance energy in the aromatic compounds. The empirical resonance energy is the difference between the heat of hydrogenation of a compound and that of the hypothetical compound of similar structure with localized bonds. For example in benzene :

$$ERE = \text{Heat of hydrogenation} - \text{Heat of hydrogenation}$$
$$\text{of benzene} \qquad \qquad \text{of cyclohexatriene} \qquad \qquad .....(6)$$

Heat of hydrogenation for cyclohexene :

$$\bigcirc + H_2 \longrightarrow \bigcirc + 28.6 \, kcal/mol$$

Heat of hydrogenation for cyclohexadiene :

$$\bigcirc + 2H_2 \longrightarrow \bigcirc + 28.6 \times 2 = 57.2 \, kcal/mol$$

Heat of hydrogenation for cyclohexatriene :

$$\bigcirc + 3H_2 \longrightarrow \bigcirc + 28.6 \times 3 = 85.8 \, kcal/mol$$

Difference in heats of hydrogenation =

Heat of hydrogenation – Heat of hydrogenation
of benzene   of cyclohexatriene

=   (–49.8)   –   (–85.8)
=   36.0 kcal/mol (150 kJ/mol)

This difference is the empirical resonance energy of benzene.

However, certain objections are raised against this method;

(i) Strain energy of localized structure : Since cyclohexene is taken as the reference compound in the determination of empirical resonance energy of benzene, the strain energy of the cyclohexene ring which makes heat of hydrogenation more favourable appears quite irrelevant as a component of the resonance energy of benzene.

(ii) Hybridization change : The hydrogenation involves conversion of Csp^2–H to Csp^3–H bonds. The energy difference associated with this change has not been included in deriving empirical resonance energies.

Both the classical procedures do not evaluate the energy that is the property of a molecule, but an energy that is the enthalpy change for the particular reaction.

Thus, the modified method[16] has been developed for the calculation of empirical resonance energies of the aromatic heterocycles including the contributions of structural factors contributing empirical resonance energies. In this method the empirical resonance energies are calculated by using the contributions of the composite bond energy terms. The composite bond energy terms are evaluated from the heats of atomization determined from the experimental heats of combustion and are used in the estimation of empirical resonance energies of the aromatic heterocycles.

However, inspite of the certain objections against the determination of empirical resonance energies (ERE) by thermochemical quantities, the empirical resonance energies (ERE) have provided widest ranging energy data for the quantitative assessment of the aromaticity in heteroaromatic systems.

3.2.3.2 Delocalization Energy[16,17]

Delocalization energy is the additional bonding energy which results from the delocalization of π-electrons originally constrained in the isolated double bonds and corresponds to the vertical resonance energy. For the determination of delocalization energy, the π-electrons of the planar heterocycles are considered independently as they have no interaction with σ-electrons because of the different planes of σ- and π-orbitals. The interaction between two types of electrons is therefore neglegible in a planar molecule.

The molecular orbitals for π-electrons in the conjugated system may be treated independently of σ-electrons. The energies of π-electrons in a π-molecular orbital are expressed in terms of two quantities; α and β. The first term α is the coulomb integral and represents approximately the energy of an electron in an isolated p-orbital before overlapping. The second term β is called resonance integral and expresses the degree of stabilization resulting from the π-orbital overlapping.

The energies of six π-electrons in six π-orbitals in benzene can be calculated by Hückel molecular orbital treatment in terms of α and β constants. The energies can be represented as a series of energy levels above and below an energy zero (α). The energies of six molecular orbitals (three bonding orbitals and three antibonding orbitals) are in the increasing order; $\alpha + 2\beta$, $\alpha + \beta$, $\alpha + \beta$; $\alpha - \beta$, $\alpha - \beta$, $\alpha - 2\beta$. Thus, the total energy of six π-electrons in the occupied bonding molecular orbitals is $6\alpha + 8\beta$ (as there are two electrons in each occupied orbital).

The energy of π-electrons in a localized carbon–carbon double bond is : $2(\alpha + \beta) = 2\alpha + 2\beta$. The energy of six π-electrons in three isolated double bonds is : $6(\alpha + \beta) = 6\alpha + 6\beta$. The delocalization energy of benzene is : $(6\alpha + 8\beta) - (6\alpha + 6\beta) = 2\beta$. The arrangement of π-electrons in benzene is, therefore, more stable by 2β and it is associated with the delocalization of six π-electrons (Fig. 23).

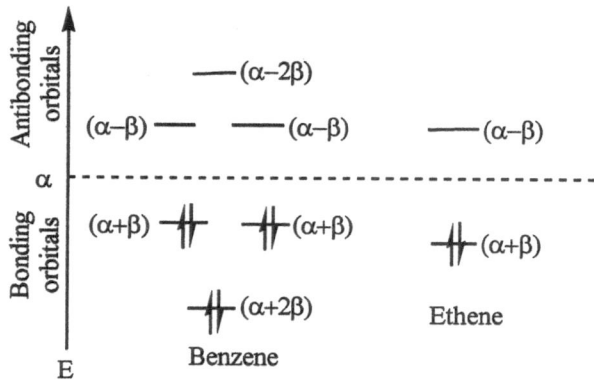

Fig. 23. Energy level diagram showing delocalization energy

Since pyridine is considered to be derived from benzene by replacing –CH= by a nitrogen atom, pyridine is assumed to have the same delocalization energy as in benzene (2β). However, the introduction of a nitrogen atom, because of its more electronegativity than carbon, causes uneven distribution of the electrons in pyridine. In the calculation of delocalization resonance energies of the aromatic heterocycles, appropriate parameters are added to coulomb integral for the heteroatom and to resonance integral for carbon–heteroatom bond to include the structural changes. The resonance energies in units of β are summarized in Table 9.

Table 9. Resonance energies in units of β for heterocycles

Compound	Total π-energy
Benzene (for comparison)	8.000
Pyridine	7.249
Thiophene	5.186
Furan	4.598
α-Pyrone	7.081
γ-Pyrone	6.999

3.2.3.3 Dewar Resonance Energy[17,19,20]

The resonance energy determinations by the conventional methods involve pure single and double bonds without giving due consideration to the contributions associated with the change in hybridization. Dewar redefined the resonance energy in order to differentiate clearly aromatic, nonaromatic and antiaromatic conjugated molecules. Dewar based his theory on the reference energy relative to which aromatic stabilization is calculated. The reference energy of a reference structure uses 'double' and 'single' bond energies appropriate to the nonaromatic systems (rather than those for non-conjugated systems).

Acyclic polyenes exhibit both effective energies; $E_{C=C}$ of the carbon-carbon double bond and E_{C-C} of the carbon-carbon single bond. The π-energy of a linear polyene is directly proportional to the chain length and each addition of a carbon–carbon single bond and a carbon–carbon double bond contributes similarly as in butadiene or octatetraene. Conjugation has the same energetic consequences (per bond) in all the nonaromatic systems.

The energy of any acyclic polyene is an additive function of the individual bond energy terms. With the same bond energies, the reference energy for a cyclic conjugated system is estimated. The reference energy of a compound represents the energy if the compound is absolutely olefinic in nature. The difference in the π-energies is termed as Dewar resonance energy (DRE) and can be represented as :

$$DRE = E - E^\circ \qquad\qquad .....(7)$$

E = Energy of the cyclic conjugated system determined by the heat of combustion

E° = Energy for reference structure

The Dewar resonance energies can also be evaluated for the aromatic heterocycles. On the basis of Dewar resonance energies, the following assumptions are made for the clear distinction of aromatic, nonaromatic and antiaromatic heterocycles :

(i)   the cyclic systems with appreciable positive resonance energy (additional π-energy) are termed as aromatic and the additional stabilization energy has been called Dewar resonance energy[18].

(ii) cyclic systems with negative value of resonance energy or with less π-energy than that of reference structure are termed as antiaromatic.

(iii) cyclic systems having resonance energy close to zero (within 1 or 2 kcal/ mol) are classified as nonaromatic.

Dewar resonance energies (DRE) for the heterocyclic compounds are summarized in Tables 10 and 11.

Table 10. Dewar resonance energy (DRE) data for six-membered heterocycles

Compound	DRE	
	(kcal/mol)	(kJ/mol)
Benzene (for comparison)	22.60	94.56
Pyridine	23.10	96.65
Pyrazine	17.10	71.55
Pyrimidine	20.20	84.52
Naphthalene	33.60	140.58
Quinoline	34.10	142.67
Isoquinoline	34.10	142.67
1,5-Naphthyridine	33.20	138.91
1,8-Naphthyridine	36.40	152.30
Quinoxaline	28.10	117.57
Quinazoline	30.39	127.15
Acridine	41.30	172.80

Table 11. Dewar resonance energy (DRE) data for five-membered heterocycles

Compound	DRE	
	(kcal/mol)	(kJ/mol)
Pyrrole	5.30	22.18
Indole	23.80	99.58
Isoindole	11.60	48.53
Carbazole	40.90	171.13
Imidazole	15.43	64.65
Benzimidazole	30.90	129.29
Thiophene	6.50	27.20
Benzo[b]thiophene	24.80	103.76
Benzo[c]thiophene	9.30	38.91
Dibenzothiophene	44.60	186.61
Furan	4.30	17.99
Benzo[b]furan	20.30	84.94
Benzo[c]furan	2.40	10.04
Dibenzofuran	39.90	166.94

In order to compare aromaticity in different heterocycles, the resonance energies are calculated in the framework of HOMO and the total π-energy is divided by the number of π-electrons to calculate energy contribution of per π-electron. The resonance energies per π-electron are summarized in Table 12.

Table 12. Resonance energies per π-electron (REPE) of some heterocycles

Compound	REPE (β)
Pyridine	0.058
Pyrimidine	0.049
Pyrazine	0.049
Pyrrole	0.039
Thiophene	0.032
Furan	0.007
Pyrazole	0.055
Imidazole	0.042
Quinoline	0.052
Isoquinoline	0.051
Indole	0.047
Benzo[b]thiophene	0.044
Benzo[b]furan	0.036
Isoindole	0.029
Benzo[c]thiophene	0.025
Benzo[c]furan	0.002

3.2.4.6 Magnetic Criteria

3.2.4.6.1 PMR-Spectra[21-24]

The ring current in an aromatic compound due to cyclic delocalization of π-electrons causes deshielding of the proton and results in the shifting of its signal to the low field. This forms a basis for diagnostic criterion of aromaticity.

Shielding and Deshielding :

The electrons around the nucleus create secondary magnetic field that opposes the applied magnetic field. The nuclei in the region of high electron density experience a field proportionately weaker than that in a region of low electron density and higher field (low frequency) has to be applied to bring them into resonance. Such nuclei are said to be shielded by the electrons. The high electron density shields nucleus and causes resonance to occur relatively at high field (with low δ-value). The low electron density causes resonance to occur relatively at low field (higher δ-value) and nucleus is said to be deshielded.

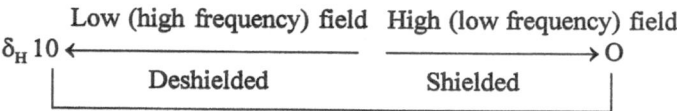

Ring Current and Chemical Shift in Aromatic System :

The π-electrons of an aromatic molecule are delocalized in the extended circular orbitals above and below the plane of the ring. A magnetic field applied at the right angle to the plane of the molecule causes the highly mobile delocalized electrons to circulate around their orbitals much in the same way as do electrons in a loop of wire and a ring current of much greater magnitude than the small circuit current associated with σ-electrons is induced. The induced ring current, in turn, produces a secondary magnetic field that opposes the applied field inside the ring, but outside the ring the applied field is reinforced. Thus, the proton lying on the periphery of the ring are deshielded and their signals are shifted to the lower field (high frequency with high δ-value) (Fig. 24).

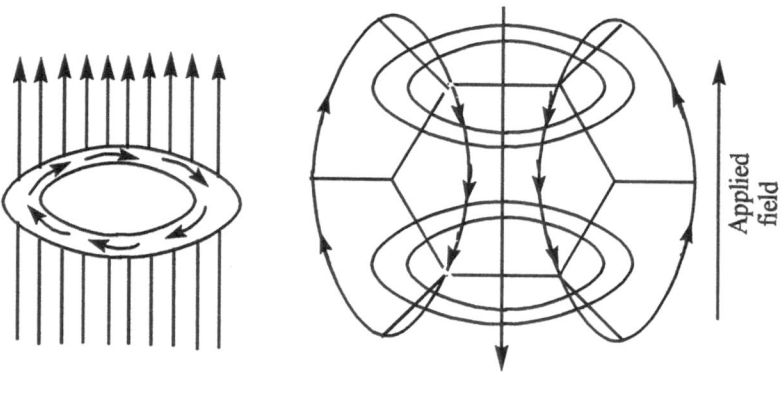

Aplied field

Induced field

(Applied field causes π-electrons to circulate around the orbital producing ring current)

(The induced magnetic field due to ring current strengthens applied field around the protons out-side the ring giving higher δ-value)

Fig. 24. Ring current in aromatic system

The ring current effects are increased with the size of the ring and are higher for the six-membered ring compared to the five-membered ring. The effect of the ring current can qualitatively be observed by comparing aromatic with non-aromatic compounds. The proton chemical shift depends on the density of the electrons at a ring carbon to which hydrogen is attached. Ring current follows the order :

benzene > pyridine > thiophene > pyrrole > furan

3.2.4.1.1 Six-Membered Aromatic Heterocycles

The deshielding effect due to aromatic ring current in the benzene ring causes the chemical shift of protons to δ 7.24 ppm. The same ring current persists in the azabenzenes. But the introduction of electronegative atom(s) in the benzene ring providing six-membered heterocycle(s), nitrogen in pyridine, exerts a strong deshielding effect on the α-hydrogens and a similar, but smaller, on the γ-hydrogens. The protons in the β-position in pyridine are slightly shifted to the higher field. Further introduction of the nitrogen atom produces similar effects, but strict additivity is not observed. Two adjacent nitrogen atoms, as in pyridazine, exert a much larger deshielding effect on the α-protons than the sum of α- and β-effects of a single nitrogen atom. Proton chemical shifts in the six-membered azaheterocycles are summarized in Table 13.

3.2.4.1.2 Five-Membered Aromatic Heterocycles

PMR-spectra of the parent five-membered aromatic heterocycles exhibit two multiplets of which one at the lower field (high δ-value) is assigned to the α-protons. The chemical shift for β-protons, except for pyrrole, increases with decreasing electronegativity of the heteroatom. However, the chemical shifts due to α-protons do not show any trend because of the paramagnetic shielding contributions which increase with the involvement of d-orbitals. The proton chemical shifts for parent five-membered aromatic heterocycles are summarized alongwith their saturated counterparts for the comparison in Table 14.

Proton chemical shifts for imidazole and pyrazole are comparable and indicates the existence of ring current in imidazole and pyrazole rings. The ring protons of thiazole are deshielded relative to those of imidazole. Isothiazole ring has been observed to have the same degree of aromatic character as in the benzene ring. ^{1}H NMR spectral data for the five-membered heterocycles with two heteroatoms are summarized in Table 15.

Table 13. ^{1}H NMR spectral data in six-membered azaheterocycles

Compound	δ ppm*				
	H-2	H-3	H-4	H-5	H-6
(for comparison)	7.24	7.24	7.24	7.24	7.24
	8.52	7.16	7.55	7.16	8.52
	N	9.17	7.52	7.52	9.17
	9.26	N	8.78	7.36	8.78
	8.60	8.60	N	8.60	8.60
	N	9.63	N	8.53	9.24
	9.18	N	9.18	N	9.18

*In CDCl$_3$ solution using TMS as internal reference.

Table 14. ¹H NMR spectral data for five-membered heterocycles with one heteroatom

	Pyrrole (Pyrrolidine)	Furan (Tetrahydro-furan)	Thiophene (Tetrahydro-thiophene)	Selenophene (Tetrahydro-selenophene)	Tellurophene (Tetrahydro-tellurophene)
H-2	6.68 (2.77)	7.29 (3.62)	7.18 (2.75)	7.88 (2.79)	8.87 (3.10)
H-3	6.22 (1.63)	6.24 (1.79)	6.99 (1.92)	7.22 (1.96)	7.78 (2.03)

Table 15. ¹H NMR spectral data for five-membered heterocycles with two heteroatoms

Compound	δ ppm			
	H-2	H-3	H-4	H-5
Pyrazole	N	7.61	7.31	7.61
Imidazole	7.86	N	7.25	7.25
Isothiazole	N	8.54	7.26	8.72
Thiazole	8.88	N	7.98	7.41

3.2.4.2 Proton-Proton Coupling Constants and Bond Order

The variation of vicinal proton-proton coupling constants with bond order has been used to assess the aromaticity in five- and six-membered aromatic heterocycles[25,26]. The vicinal coupling constants provide evidence for the bond fixation and delocalization[27]. The ratio of the ortho coupling constants $J_{(ratio)} = J_{ab} : J_{bc} = 1$ for the heteroaromatic rings fused to the benzene ring may be considered as the reference value for the complete delocalization, whereas the value 0.55 is considered as the reference value for the complete localization. However, the vicinal coupling constants are considerably influenced by the ring angles and the inductive effects alongwith the effects of heteroatoms (Fig. 25).

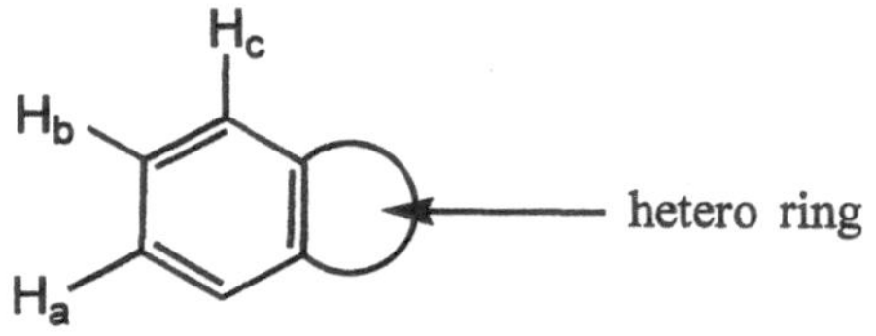

Fig. 25. Coupling constant in fused aromatic heterocycles

The ratio of the vicinal coupling constants can also be correlated qualitatively with the resonance energy per π-electron (REPE) to know about the cyclic delocalization in the heteroaromatic systems[28]. The ratio of the coupling constants and REPE for the azaheterocycles are summarized in Table 16.

It is obvious from the table that REPE increases with increasing $J_{(ratio)}$ and the linear relationship between two quantities has been observed. This relationship can, therefore, give an indication of delocalization of the electrons.

3.2.4.3 ^{13}C NMR Spectroscopy[29,30]

^{13}C NMR spectra can also be used to assess qualitatively the aromaticity in aromatic heterocycles. In ^{13}C NMR spectra the correlation exists between π-electron density and the ^{13}C-chemical shifts as the ^{13}C-shielding of the ring carbon atoms are approximately proportional to the π-electron density.

In the five-membered aromatic heterocycles (with one heteroatom), the carbon atom signals appear at more shielded positions (δ 100–150 ppm), while in the six-membered aromatic heterocycles the signals appear in the range δ 150–170 ppm.

3.2.4.3.1 Six-Membered Aromatic Heterocycles

The ring carbon atoms α- to the heteroatom are highly deshielded and the carbon atoms γ- to the heteroatoms are also deshielded relative to those in benzene ring. The introduction of second nitrogen atom causes further deshielding at the α- and γ- carbons of approximately 10 ppm and 3 ppm, respectively and the shielding

Table 16. $J_{(ratio)}$ and REPE of nitrogen heterocycles

Conjugated system	J_{ab}/J_{bc}	REPE (β)
	0.61	0.007
	0.71	0.027
	0.74	0.029
	0.71	0.029
	0.91	0.047

effect at the β-carbon is of approximately 3 ppm. The ^{13}C-chemical shifts in six-membered aromatic heterocycles are summarized in Table 17.

Table 17. ^{13}C-chemical shifts in six-membered azaheterocycles*

Compound	δ ppm				
	C-2	C-3	C-4	C-5	C-6
	128.5	128.5	128.5	128.5	128.5
	150.6	124.5	136.4	124.5	150.6
	N	152.8	127.6	127.6	152.8
	159.5	N	157.5	122.1	157.5
	145.6	145.6	N	145.6	145.6
	N	163.6	N	142.0	150.2
	167.5	N	167.5	N	167.5

*In CDCl$_3$ solution using TMS as internal reference.

3.2.4.3.2 Five-Membered Aromatic Heterocycles

In ^{13}C NMR spectra of five-membered π-excessive heterocycles, the carbon atom resonances appear at much shielded positions (100–150 ppm). The signal for pyrrole α-carbon atom is broadened because of the coupling with an adjacent nitrogen at the position-1. The frequencies for β-carbon atoms show systematic shifting to the higher field with increasing electronegativity of the heteroatom. All the shifts are comparable with benzene ($\delta = 128.5$ ppm). ^{13}C-chemical shifts for the five-membered aromatic heterocycles are summarized in Table 18.

Table 18. ^{13}C-chemical shifts in π-excessive heterocycles*

Compound	δ ppm			
	C-2	C-3	C-4	C-5
benzene	128.5	128.5	128.5	128.5
cyclopentadienyl anion	103	103	103	103
pyrrole (N–H)	118.5	108.2	108.2	118.5
furan (O)	142.6	109.6	109.6	142.6
thiophene (S)	125.4	127.2	127.2	125.4
selenophene (Se)	127.3	129.8	129.8	127.3
tellurophene (Te)	127.3	138.0	138.0	127.3

*In CDCl$_3$ solution using TMS as internal reference.

3.2.4.4 Diamagnetic Susceptibility Exaltations[31,32]

Diamagnetic susceptibility is an additive property and can be calculated by adding susceptibility contributions of the atoms and bonds. The aromatic compounds possess larger diamagnetic susceptibilities than the expected from the comparison with the values of alkenes. The diamagnetic susceptibility of an aromatic compound can be expressed as :

$$\chi_M = \chi_{additive} + \Lambda \qquad\qquad .....(8)$$

where Λ is diamagnetic susceptibility exaltation and is considered to be associated with the specific aromatic properties of the aromatic compounds.

The diamagnetic susceptibility exaltation can be defined as the difference between the observed diamagnetic susceptibility and that calculated from the atomic and bond susceptibility contributions. Thus, diamagnetic susceptibility exaltation can be expressed as :

$$\Lambda = \chi_M - \chi_{M'} \qquad\qquad .....(9)$$

where χ_M = observed diamagnetic susceptibility and
$\chi_{M'}$ = estimated diamagnetic susceptibility.

$\chi_{M'}$ can be estimated by using several empirical methods.

The compounds with diamagnetic susceptibility exaltation ($\chi_M - \chi_{M'} = \Lambda > 0$) are classified as aromatic. The aromatic compounds are found to exhibit large diamagnetic susceptibility exaltations and the non-aromatic compounds exhibit zero exaltation ($\chi_M - \chi_{M'} = \Lambda = 0$) and this forms a basis for using diamagnetic susceptibility exaltation as a criterion of aromaticity. The diamagnetic susceptibility exaltations of aromatic heterocycles are summarized in Table 19.

From the diamagnetic susceptibility exaltations, it can be inferred that there exists a correlation between the diamagnetic susceptibility exaltations and the aromaticity. These results clearly justify to conclude that the diamagnetic susceptibility exaltation is the valid criterion of aromaticity and provides clear demarcation between aromatic and non-aromatic heterocycles.

3.2.4.5 Faraday Effect (Magneto-optical Rotation)[33]

The rotation of the plane of the polarization of light by transparent substance placed in a magnetic field is called magneto-optical rotation and this effect is known as Faraday effect. The magneto-optical rotation is an additive property and can be calculated by the contributions of the rotation of the bonds. The compounds with delocalized π-electrons exhibit magneto-optical exaltation over the calculated rotation. The aromatic compounds are characterized by the large magneto-optical rotation which is influenced by the nature of the substituents. The substitution of hydrogen by the electron-withdrawing groups decreases magneto-optical rotation exaltation, while the substitution by the electron-releasing groups increases exaltation.

Table 19. Diamagnetic susceptibility exaltations in aromatic heterocycles

Compound	χ_M (exp.)	$\chi_{M'}$ (calcd.)	Λ
		$(10^{-6} \text{ cm}^3/\text{mol})$	
Benzene (for comparison)	54.8	41.1	13.7
Pyridine	49.2	35.8	13.4
Pyrazine	37.6	30.5	7.1
Naphthalene	91.9	61.4	30.5
Quinoline	86.0	56.1	29.9
Isoquinoline	83.9	56.1	27.8
Pyrrole	47.6	37.4	10.2
Thiophene	57.4	44.4	13.0
Furan	43.1	34.2	8.9
Pyrazole	42.6	36.0	6.6
1,3-Thiazole	50.6	38.3	12.3
1,3,4-Thiadiazole	37.3	32.2	5.1
3,5-Dimethyloxazole	59.7	51.5	8.2

The aromaticity order in the six-membered aromatic heterocycles on the basis of magneto-optical rotation exaltation is as follows :

benzene > pyridine > pyridine N-oxide

The aromaticity order in the five-membered aromatic heterocycles has been observed to be as :

thiophene > pyrrole > furan

3.3 Heteroaromatic Ring Systems[34-36]

3.3.1 Five-Membered Heteroaromatic Ring Systems

3.3.1.1 Five-Membered Heteroaromatics with One Heteroatom

Five-membered heteroaromatics are heterocyclic analogs of the cyclopentadienyl anion, which is planar symmetrical pentagon with five sp^2-hybridized carbon atoms containing six π-electrons to form an aromatic sextet. Similarly, in the five-membered heteroaromatics with one heteroatom, five sp^2-hybridized atoms sustain six π-electron system. Each carbon atom contributes one electron, while heteroatom provides two electrons to the aromatic sextet (Fig. 26). Thiophene, with sulfur heteroatom, expends its valence shell by using vacant d-orbitals in hybridization. Thiophene, instead of six electrons in five-orbitals, has six electrons in six-orbitals using pd^2 hybrid orbitals.

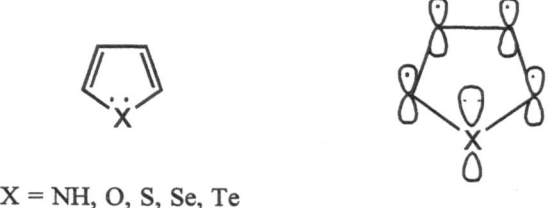

X = NH, O, S, Se, Te

Fig. 26. Structural representation of five-membered heterocycles

Five-membered heteroaromatics with one heteroatom are π-excessive as six-electrons are distributed over five atoms. These heterocycles possess considerable aromatic character as these are characterized by high degree of reactivity towards electrophilic substitution reactions rather than addition reactions.

The aromatic nature of these heterocycles is evidenced by their bond lengths, dipole moments and resonance energies. The shorter carbon–heteroatom bond lengths in these heterocycles indicate partial double bond character.

The aromatic character in these heterocycles depends on the availability of the lone pair on the heteroatom for delocalization and in turn depends on the electronegativity of the heteroatom. The order of electronegativity in the heteroatoms is : oxygen > nitrogen > sulfur. The oxygen atom because of its high electronegativity has strong attraction on the lone pair and restricts the lone pair upto some extent to take part in the delocalization of π-electrons causing partial localization of the π-electrons. The order of aromaticity on the basis of electronegativity of heteroatoms is : thiophene > pyrrole > furan.

The resonance energy in five-membered aromatic heterocycles follows the order : thiophene > pyrrole > furan. The higher degree of stabilization energy and thus aromaticity of thiophene is attributed to the following reasons :

(i) release of angle strain due to larger bonding radius of sulfur than nitrogen and oxygen,

(ii) sulfur being less electronegative than the nitrogen and oxygen atoms and

(iii) use of d-orbitals of sulfur for bonding.

The following resonating structures also contribute to the resonance hybrid (Fig. 27).

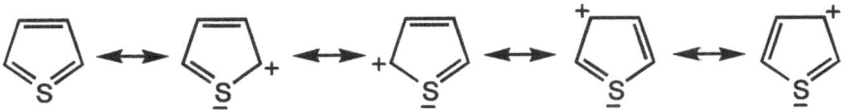

Fig. 27. Additional resonating structures in thiophene using d-oribitals

Thus, in the five-membered heteroaromatics the greatest aromaticity is associated with thiophene and the aromaticity falls along the series : thiophene, pyrrole and furan.

3.3.1.2 Five-Membered Heteroaromatics with Two Heteroatoms

Five-membered heteroaromatics with two heteroatoms (azoles) are of two types depending on the relative position of pyridine-type nitrogen.

(i) 1,2-Azoles : containing additional nitrogen atom (pyridine-type) at the position-2.

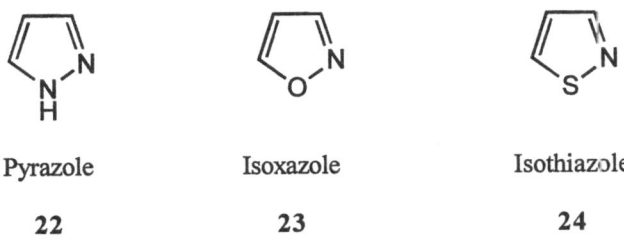

Pyrazole	Isoxazole	Isothiazole
22	**23**	**24**

(ii) 1,3-Azoles : with additional nitrogen atom (pyridine-type) at the position-3.

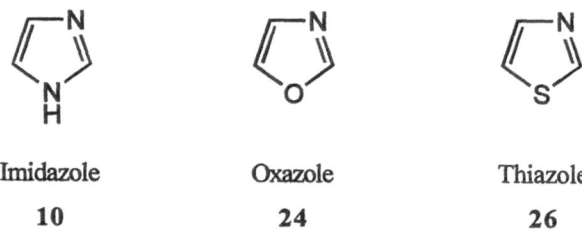

Imidazole	Oxazole	Thiazole
10	**24**	**26**

These heterocycles are aromatic as they have similar electronic structures as in pyrrole, thiophene and furan and possess an aromatic sextet of six π-electrons. Each carbon atom and the pyridine-type nitrogen contribute one electron, while the other heteroatom, similar to that in pyrrole, thiophene and furan, contributes a lone pair of electrons to the aromatic sextet. The lone pair on nitrogen (pyridine-type) is situated orthogonally to the π-electron cloud and does not participate in the cyclic delocalization, but provides basic character to the azoles. These heterocycles are considered as mixed system with the features of π-excessive and π-deficient heterocycles, and are contributed by the following resonating structures (Figs. 28 and 29) :

1,2-Azoles :

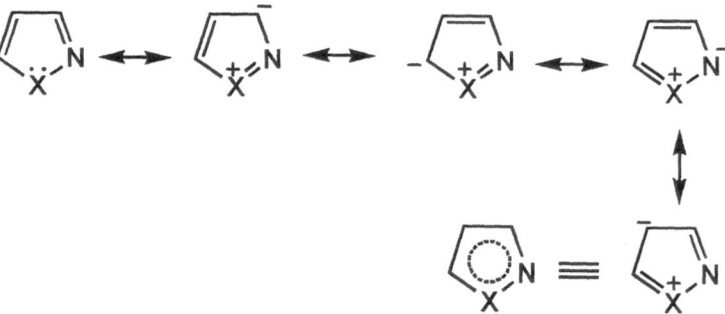

Fig. 28. Resonating structures in 1,2-azoles

All the resonating structures for azoles are not equivalent in their contribution to the resonance hybrid. The resonating structures, anion with α C=N and C=X^{+}, are stabilized and contributing more to the resonance hybrid.

1,3-Azoles :

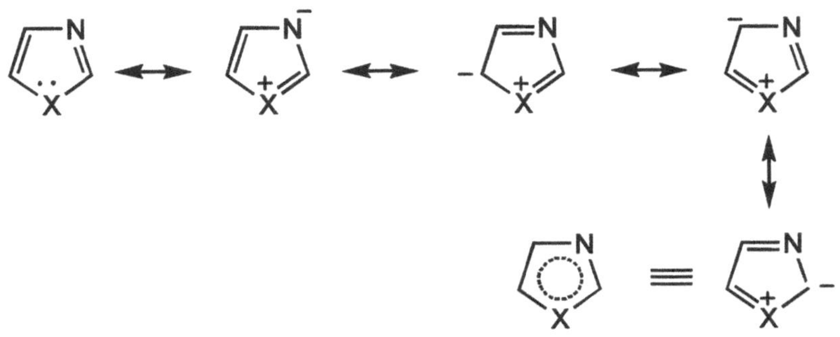

Fig 29. Resonating structures in 1,3-azoles

The azoles are more basic than their monoheteroaromatic analogs because the lone pair on the additional pyridine-type nitrogen is not involved in maintaining aromaticity, but is available for the protonation. The direct linking of two hetero-atoms decreases basicity and thus 1,3-azoles are more basic than 1,2-azoles.

The stability of the azoles decreases with increasing substitution of nitrogen for the carbon atoms with the stabilities of pyrazole and imidazole being comparable. Pyrazole and imidazole are aromatic with resonance energies; 123 kJ/mol and 56 kJ/mol, respectively[38] (pyrazole is more aromatic than imidazole).

The structure of thiazole is closely related to the structure of thiophene and has the same degree of aromatic character as in benzene and involves cyclic delocalization of the π-electrons.

Oxazole ring is planar and shows considerable bond fixation. Oxazole although possesses a sextet of π-electrons, but because of higher electronegativity of the oxygen atom, the delocalization of π-electrons is not extensive and causes localization of the π-electrons.

3.3.1.3 Benzo-fused Five-Membered Heteroaromatics

The aromaticity of benzo-fused five-membered aromatic heterocycles is comparable to that of naphthalene. The order of aromaticity depends on the electronegativity of the heteroatom and decreases with increasing electronegativity :

Benzo[b]thiophene > Benzo[b]pyrrole (indole) > Benzo[b]furan

(103.76 kJ/mol)   (99.58 kJ/mol)   (84.94 kJ/mol)

The resonating structures of these heterocycles do not involve the destruction of six π-electron system of the benzene ring (Fig. 30).

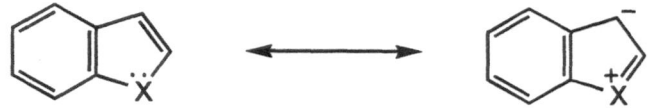

Fig. 30. Resonating structures involving five-membered ring

The fusion of five-membered heteroaromatic ring to the benzene ring across the carbon–carbon single bond decreases aromaticity of the system considerably and, therefore, different order of the aromaticity is observed :

benzo[c]pyrrole (isoindole) > benzo[c]thiophene > benzo[c]furan

(48.53 kJ/mol)   (38.91 kJ/mol)   (10.04 kJ/mol)

Isoindole **39** is considered to be contributed by the following resonating structures (Fig. 31) :

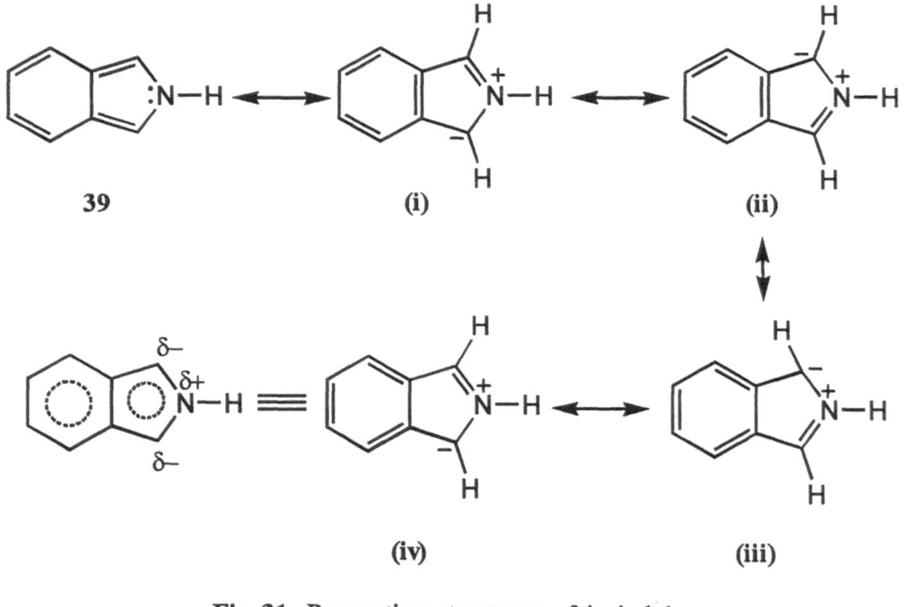

39   **(i)**   **(ii)**

(iv)   **(iii)**

Fig 31. Resonating structures of isoindole

Indolizine **46** is a 10π-electron system. The bridgehead nitrogen influences the properties of both the rings; five-membered ring with characteristic properties of electron rich pyrrole; and six-membered ring with characteristic properties of π-deficient pyridine. Indolizine is considered to be contributed by the following resonating structures in which five-membered ring has increased charge density (Fig. 32).

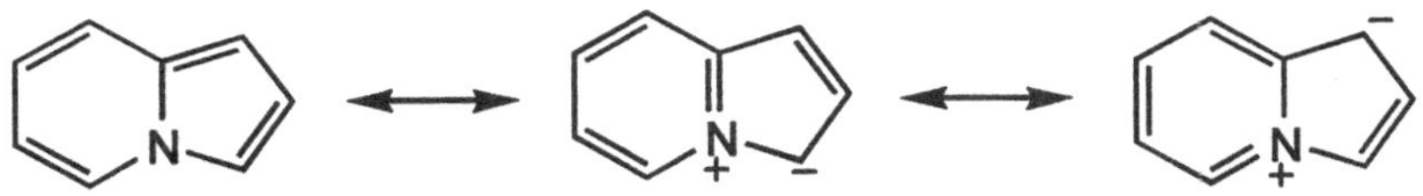

46   (i)   (ii)

Fig. 32. Resonating structures of indolizine

The fusion of the second benzene ring to the benzo-fused five-membered heterocyclic ring causes a further reduction in the aromaticity of the heterocyclic ring as is evidenced by the results of calculations of Dewar resonance energy of some five-membered heterocyclic rings fused with two benzene rings **53, 54** and **55**.

Benzene	Pyrrole	Thiophene	Furan
(94.56 kJ/mol)	(22.18 kJ/mol)	(27.20 kJ/mol)	(17.99 kJ/mol)

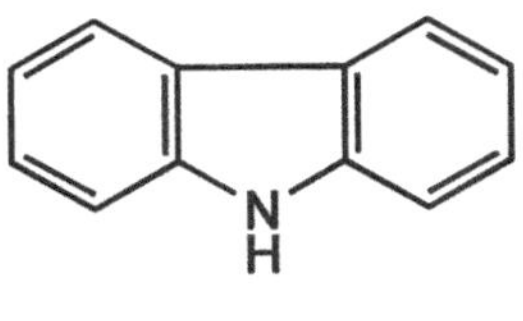

Carbazole	Dibenzothiophene	Dibenzofuran
(171.13 kJ/mol)	(186.61 kJ/mol)	(166.94 kJ/mol)
53	**54**	**55**

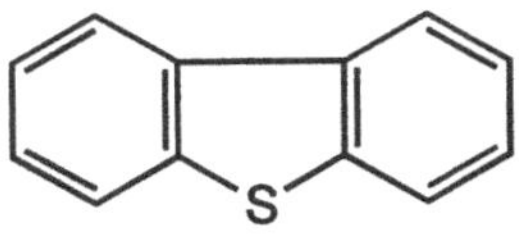

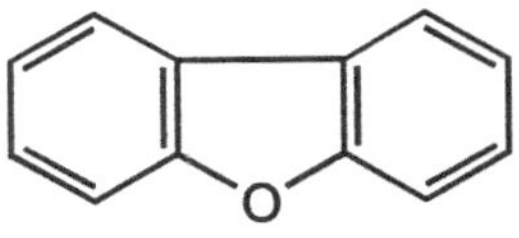

3.3.2 Six-Membered Heteroaromatic Ring Systems

3.3.2.1 Pyridine

Pyridine is a heteroaromatic system and is most similar to benzene. However, qualitatively substitution of one sp^2-carbon atom in the benzene ring by one sp^2-

nitrogen atom causes perturbation in the hexagonal structure because of the difference in the carbon–nitrogen and carbon–carbon bond lengths. The bond lengths in pyridine are intermediate between normal double and single bond lengths (C–C = 1.39Å and C–N = 1.34Å). The introduction of a nitrogen atom in the benzene ring makes it less symmetrical with the decrease in resonance energy and allows more resonating structures to contribute the resonance hybrid in which the negative charge is localized on the heteroatom (Fig. 33).

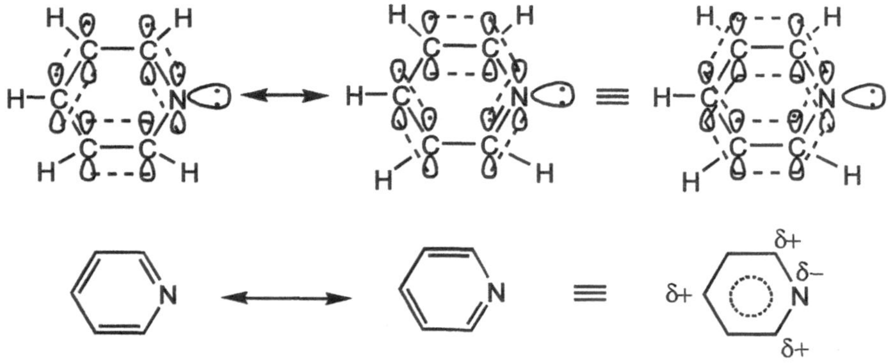

Fig. 33. Structure of pyridine

The effect of the nitrogen atom in pyridine is similar to the effect of an electron-withdrawing substituent in the benzene ring. The lone pair on nitrogen atom is not required to maintain aromatic sextet of π-electrons because of being orthogonal to the plane of the ring. The lower basicity of pyridine than the aliphatic amines is attributed to the sp^2-hybridized nitrogen. Thus, pyridine appears to have same or somewhat less aromatic character than in the benzene ring. The ring current of benzene and pyridine is reported to be close to unity, the energy estimations of pyridine, however, suggest it to be less aromatic than benzene. Conjugation energies of pyridine and benzene are 79.50 kJ/mol and 92.0 kJ/mol, respectively. Delocalization energy of pyridine is less than that of benzene and obeys the following eqation :

$$ERE = 15.8\,DE \hspace{4cm} .....(10)$$

3.3.2.2 Fused-Ring Derivatives of Pyridine

The fusion of a pyridine ring with a benzene ring in three different ways provides quinoline **27**, isoquinoline **28** and quinolizinium cation **45**.

These heterocycles are aromatic with 10π-electrons and the lone pair on the nitrogen atom is not involved in an aromatic sextet. These closely resemble naphthalene in the same way as pyridine to benzene. Quinoline **27** is highly aromatic with resonance energy of 47.3 kcal/mol (197.90 kJ/mol) and is considered to be contributed by the following resonating structures (Fig. 34) :

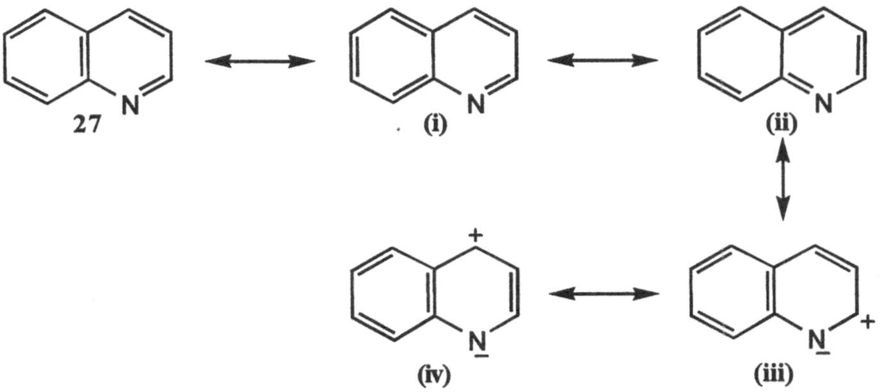

Fig. 34. Resonating structures of quinoline

These structures are similar to those of naphthalene, but the charged structures are also possible because of the electronegative nitrogen. The dipole moment of quinoline (2.10D) indicates charge-separation.

Isoquinoline **28** is also highly aromatic with high dipole moment (2.6D) and is contributed by the following resonating structures (Fig. 35) :

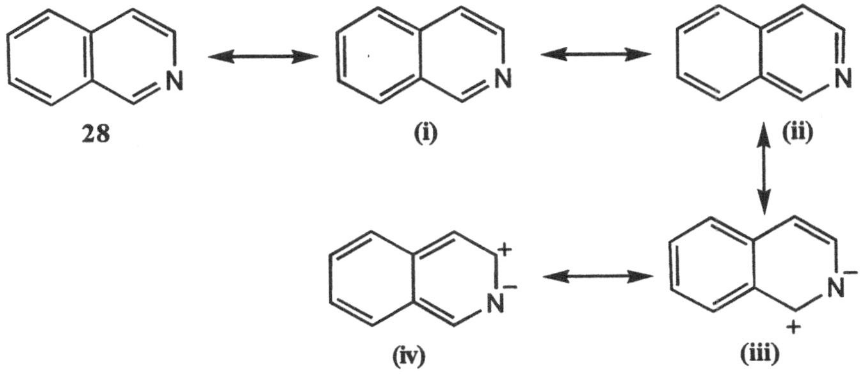

Fig. 35. Resonating structures of isoquinoline

The fused-ring derivatives of pyridine sustain ring current and the diamagnetic susceptibility exaltations for quinoline and isoquinoline are : -29.9×10^{-6} and -27.8×10^{-6} cm³/mol, respectively and are comparable with that of naphthalene $(-30.5 \times 10^{-6}$ cm³/mol$)$.

Bicyclic six-membered heteroaromatics with nitrogen atom in both the rings are called naphthyridines.

1,5-Naphthyridine   1,8-Naphthyridine

42   42a

The values of delocalization energy for the various members of naphthyridines are in the range of 3.76–3.85β and are comparable with that of naphthalene (3.68β).

3.3.2.3 Pyridones and Related Systems

2-Pyridones **56** and 4-pyridones **57** are potentially six π-electron systems and are considered to be aromatic with the charge-separated resonating forms (Fig. 36) :

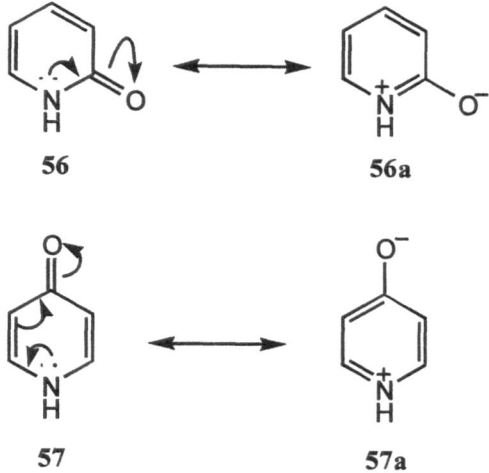

56   56a

57   57a

Fig. 36. Resonating structures of 2- and 4-pyridones

2-Pyridone is 6 kcal/mol (25.10 kJ/mol) less aromatic than 4-pyridone. Similarly, quinolones **58** and isoquinolones **59** retain much of the resonance energy of the quinoline and isoquinoline structures. 2- And 4-pyridones and 1- and 2-quinolones have substantial aromatic characters. NMR-chemical shift estimated aromaticity of 2-pyridone to be 35% of benzene and indicates relatively localized π-electron system.

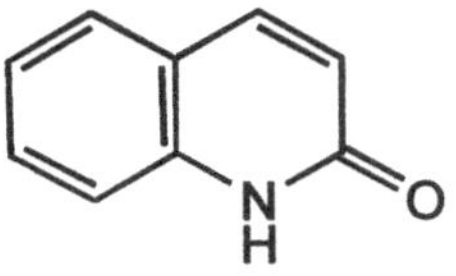

58 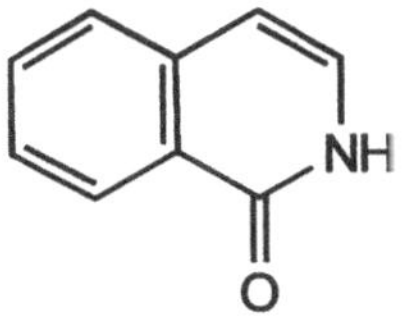   **59**

3.3.2.4 Pyrylium Salts

The oxygen atom can replace a carbon atom in the benzene ring only if the lone pair on oxygen atom is involved in the bonding and oxygen bears positive charge. Pyrylium cation **60** is a six π-electron system with an aromatic sextet and possesses aromatic character similar to that of benzene. Pyrylium cation **60** is considered to be contributed by the following resonating structures (Fig. 37). The high value of the aromaticity constant (+96) of pyrylium cation reflects its electrophilicity. The aromatic character of pyrylium ring is evidenced by the absorptions of ring protons in the range of δ 8.5–9.6 ppm and by the qualitative agreement between chemical shift and π-electron density.

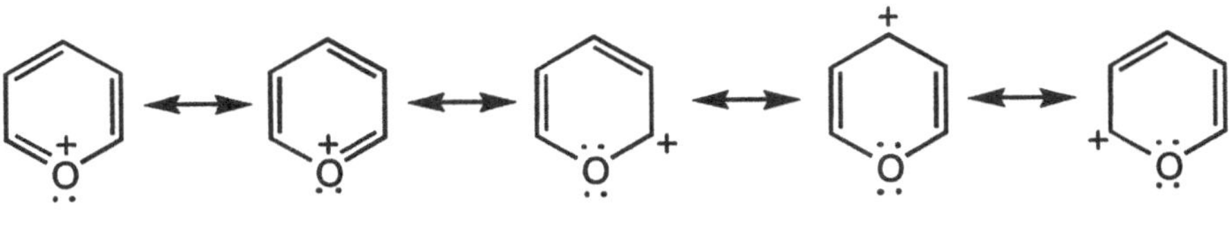

Fig. 37. Resonating structures of pyrylium cation

3.3.2.5 Pyrones

α-Pyrones **61** and γ-pyrones **62** are expected to have some aromatic character due to the contributing resonating structures involving an aromatic sextet (Fig. 38). Since oxygen is more electronegative than nitrogen, it causes localization of the electrons making α- and γ-pyrones less aromatic than 2- and 4-pyridones. The high value of the dipole moment (3.7D) supports the contribution of charge-separated resonating structures. The delocalization energies of α-pyrone and γ-pyrone are very close (2.896 β and 2.863 β).

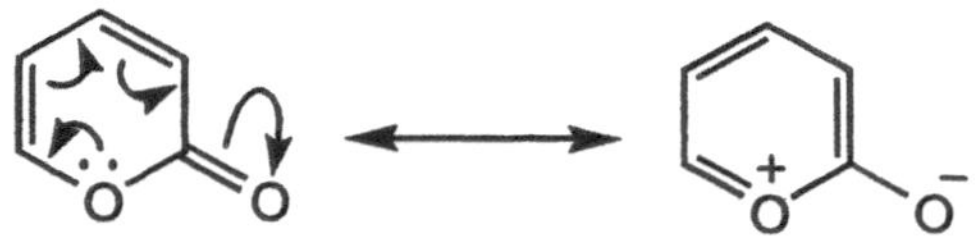

Fig. 38. Resonating structures of α- and γ-pyrones

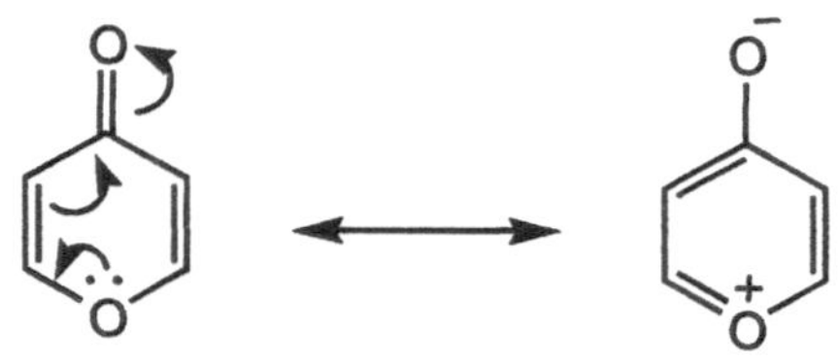

4 HETEROAROMATIC REACTIVITY[39,40]

Aromatic compounds are characterized by their extra stability due to delocalized π-electron system, and substitution reactions rather than addition reactions with the retention of aromatic character. The aromaticity is related with the stability which, in turn, may be considered to be related with the reactivity. However, the reactivity may not be considered to be a criterion of aromaticity.

The presence of heteroatom in the heteroaromatic system usually affects the reactivity. The basic principles governing the degree and the type of reactivity in the heteroaromatic systems are :

(i) Nucleophilic attack : The heteroatom (oxygen, nitrogen or sulfur) bonded to the carbon atom with multiple bond can accept shared pair of π-electrons and permits the attack of nucleophilic reagent. The attack of nucleophile is facilitated when heteroatom is positively charged (Fig. 39).

Fig. 39. Nucleophilic attack in heteroaromatic ring system

(ii) Electrophilic attack : A pair of electrons on the heteroatom (oxygen, nitrogen or sulfur) attached to an unsaturated system can be made available for the electrophilic attack through the following system. This can also be operative when the heteroatom is negatively charged (Fig. 40).

Fig. 40. Electrophilic attack in heteroaromatic ring system

(iii) Retention of aromaticity : Aromatic heterocycles tend to retain their aromatic character. However, ring oxygen atom, increasing number of ring heteroatoms and benzannelation reduce the aromaticity.

4.1 Selectivity and Reactivity

The relative reactivity and the selectivity of the diferent positions in the heteroaromatic systems to be attacked by the reagents can be predicted by the reactivity indices.

4.1.1 Reactivity Indices

The reactivity indices method involves the correlation between activation free energy of the process and some intrinsic parameters i.e. reactivity indices, related to the electronic properties of the heteroaromatic system involved in the reaction. The introduction of heteroatom in an aromatic system causes considerable changes in the electronic properties and, therefore, the parameter relating to the heteroatom is used in the molecular orbital calculations in the heteroaromatic system. The reactivity indices are :

4.1.1.1 Electron Densities

The π-electron densities refer to the electron density at a given carbon atom obtained by summing up the contributions from all the filled molecular orbitals. The π-electron densities directly determine the orientation of electrophilic and nucleophilic substitutions. Electrophilic attack is considered to occur at the position where the electron density is highest, while nucleophilic attack takes place at the centre of low electron density. The π-electron densities in homolytic substitution are not appreciably affected by the electron density distribution.

4.1.1.2 Frontier Electron Densities[41]

The frontier electron densities are a property of an isolated molecule and their use as reactivity indices involves the assumption that the electron distribution in the transition state resembles that in the initial state. The frontier electron distribution is much more uneven and the relative order of the frontier electron densities is not changed when approached by the reagent. The frontier electron densities are associated with the electron exchange rather than with electrostatic interaction and if electron exchange (charge transfer) is significant in stabilizing transition state, the frontier electron densities influence the reactivity.

In electrophilic substitution, the frontier electron density is the density in the highest filled molecular orbital (HOMO) and these electrons are considered to be analogous to the valence electrons of an atom. The electrophilic substitution reaction involves the interaction of highest occupied molecular orbital (frontier orbital) of heteroaromatic system with the lowest unoccupied molecular orbital of electrophile. The activation energy depends on the extent of mixing which, in turn, depends on the difference in the energy between these two types of orbitals. If smaller is the difference in the energy, lower will be the activation energy and faster will be the reaction with more efficient mixing of the orbitals.

In nucleophilic substitution, the frontier molecular orbital is the lowest unoccupied molecular orbital (LUMO). Nucleophilic substitution reaction involves the mixing of lowest unoccupied molecular orbital of the heteroaromatic system with a filled molecular orbital of nucleophile. The heteroaromatic compound tends to accept the electron pair in the transition state and the frontier electron density at the carbon atom is then the electron density in this molecular orbital as if it were occupied by two electrons. The electrophilic and nucleophilic reactions occur at the carbon with the greatest appropriate frontier electron density (Fig. 41).

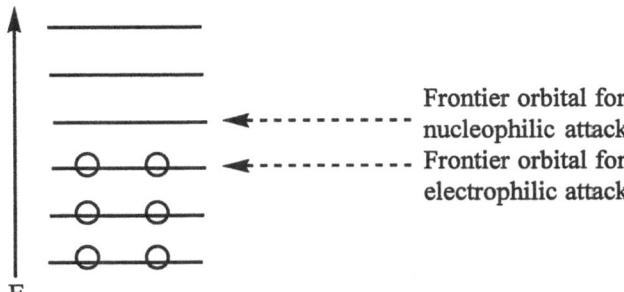

Fig. 41. Diagrammatic representation of frontier orbitals for electrophilic and nucleophilic substitutions

4.1.1.3 Localization Energy[42]

The localization energy refers to the energy difference between the π-electron energy in the isolated molecule and the transition state complex. The relative stabilities of the different transition states in the same molecule are in the same order as the localization energies. The localization energy in the heteroaromatic system reflects the relative stability of the transition state complex and therefore predicts the orientation in substitution reactions.

Thus, the following inferences on orientation can be drawn on the basis of reactivity indices in the heteroaromatic systems :

(i) The π-electron distribution is the main factor controlling the orientation. The transition states very closely resemble to the unpurturbed molecule and the electrostatic interactions contribute significantly to the orientation.

(ii) The physical picture for correlation of reactivity with frontier electron densities is less clear, but probably involves electron transfer interaction between the reagent and the slightly purturbed aromatic system in which the cyclic conjugation is retained.

(iii) The correlationship between orientation and localization energies implies that the transition state is very different from the initial state, since transition state reflects the structure of σ- complex in which loss of cyclic conjugation is partly compensated by the formation of a new σ-bond.

4.2 Selectivity and Reactivity in Heteroaromatic Rings

4.2.1 Six-Membered Heteroaromatic Rings

The relative reactivity of the ring atoms in six-membered heteroaromatic compounds can be predicted on the basis of reactivity indices which predict that the ring-nitrogen atom of π-deficient heteroaromatic ring is the most reactive site for the electrophiles and the electrophilic attack at the ring carbon atoms is exceptionally difficult. Under acidic conditions, nitrogen of pyridine is protonated and further decreases reactivity of the ring carbon atoms toward electrophiles (Table 20). The comparison of reactivity indices for both pyridine and pyridinium ion accounts for the less reactivity of carbon atoms of azines. The preferential electrophilic and nucleophilic attacks in the azines are represented (Fig. 42).

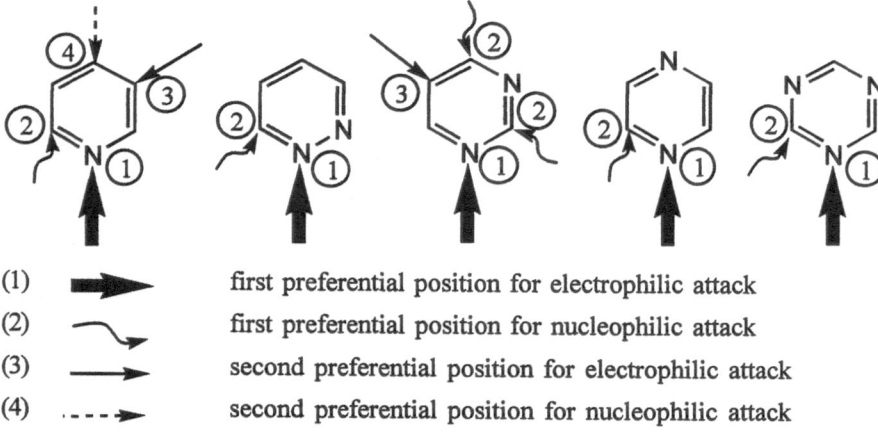

(1) ➤   first preferential position for electrophilic attack
(2) ↴   first preferential position for nucleophilic attack
(3) →   second preferential position for electrophilic attack
(4) ----➤   second preferential position for nucleophilic attack

Fig. 42. Selective positions for electrophilic and nucleophilic substitutions

These effects can also be applied to the benzo analogs of the azines where electrophilic attack preferentially occurs at the carbon atoms of the benzo-fused ring (Fig. 43).

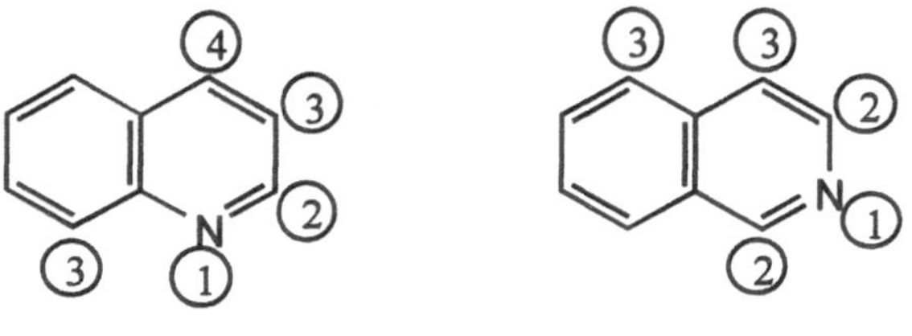

Fig. 43

Table 20. Reactivity indices of pyridine and pyridinium ion

Compound	Atoms	Net total charge	π-charge	π-electron density
	N	−0.06	−0.04	1.19
	C-1	−0.02	+0.02	0.92
	C-2	−0.09	−0.03	−1.00
	C-3	−0.04	+0.04	0.95
	N	0.02	−0.35	1.62
	C-1	0.05	+0.10	0.76
	C-2	−0.08	−0.02	1.01
	C-3	+0.04	+0.19	0.83

4.2.2 Five-Membered Heteroaromatic Rings

Five-membered heteroaromatic rings with one heteroatom possess partial positive charge on the heteroatom which restricts the reaction with electrophilic reagents. While the presence of partial negative charge on the carbon atoms facilitates the electrophilic attack at the ring carbons. This charge distribution follows from the valence bond approach as a contribution to the resonance hybrid of the resonating structures (Fig. 44).

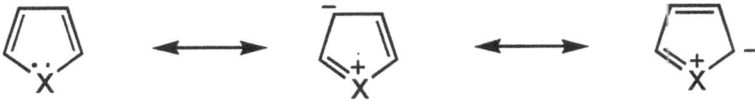

Fig. 44. Charge distribution in five-membered heteroaromatic rings

The valence bond approach provides a general method of predicting relative reactivities of the different positions of a particular heteroaromatic compound. Molecular orbital calculations also lead to the similar predictions.

The preferential attack of electrophile at α-position rather than at β-position can be rationalized in terms of more effective delocalization of charge in the intermediates or σ-complexes. The relative stabilities of the σ-complexes are correlated with the localization energies and thus predict the relative reactivity of the positions in the aromatic heterocycles (Fig. 45).

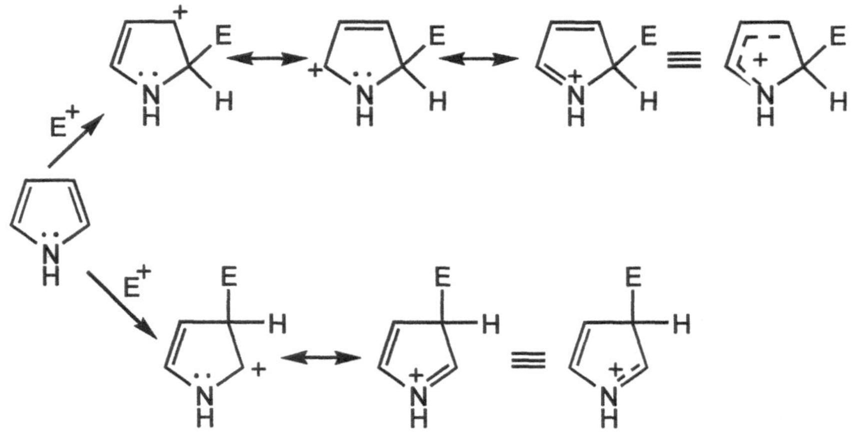

Fig. 45. Relative stabilities of σ-complexes

Benzo[b] heterocycles are expected to favour β-substitution in the heterocyclic ring over α-substitution which is based on the σ-complex stability. However, benzo[b]furan undergoes mainly α-substitution due to the strong directing effect of the oxygen atom to the α-position (Fig. 46).

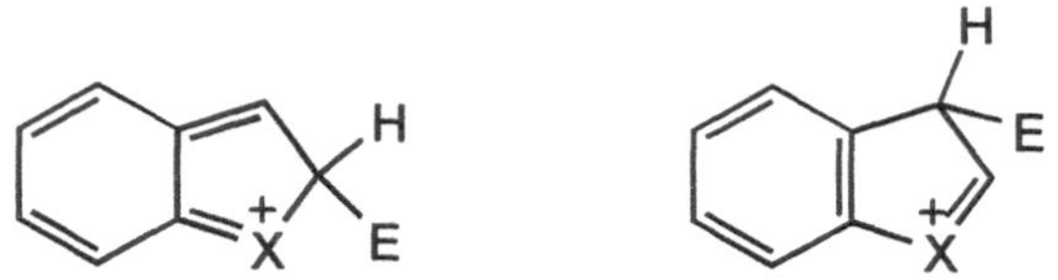

α-substitution   β-substitution

Fig. 46. α- And β-electrophilic substitutions in benzo-fused heterocycles

REFERENCES

1.   M. J. Cook, A. R. Katritzky and P. Linda in A. R. Katritzky (Ed.), *Adv. Heterocycl. Chem.* Vol. **17**, Wiley-Interscience, New York, 1974, pp. 255 : C. W. Bird and G. W. H. Cheeseman in A. R. Katritzky and C. W. Rees (Eds.), *Comprehensive Heterocyclic Chemistry* Vol. **4**, Pergamon Press, Oxford, 1984, pp. 1.

2. A. R. Katritzky, P. Barczynski, G. Musumarra, D. Pisano and M. Szafran, *J. Am. Chem. Soc.* **111**, 7 (1989); A. R. Katritzky, V. Frygelman, G. Musumara, P. Barczynski and M. Szafran, *J. Prakt. Chem.* **332**, 853 and 870 (1990); A. R. Katritzky and P. Barczynski, *J. Prakt. Chem.* **332**, 885 (1990); A. R. Katritzky et al., *Heterocycles* **32**, 127 (1991); V.G.S. Box, *Heterocycles*, **32**, 2023 (1991).

3. M. Burke–Laing and M. Laing, *Acta Crystallogr.* (B) **32**, 3216 (1976).

4. S. F. Mason in A. R. Katritzky (Ed.), *Physical Methods in Heterocyclic Chemistry* Vol. **2**, Academic Press, New York, 1963, pp. 1.

5. A. R. Katritzky, *Handbook of Heterocyclic Chemistry*, Pergamon Press, Oxford, 1985, pp. 64.

6. R. Bonnett, R. F. C. Brown and R. G. Smith, *J. Chem. Soc. Perkin Trans.* 1, 1432 (1973).

7. D. Wege, *Tetrahedron Lett.* 2337 (1971).

8. A. R. Katritzky and A. P. Ambler in A. R. Katritzky (Ed.), *Physical Methods in Heterocyclic Chemistry* Vol. **2**, Academic Press, New York, 1963, pp. 61.

9. A. R. Katritzky and P. J. Taylor in A. R. Katritzky (Ed.), *Physical Methods in Heterocyclic Chemistry* Vol. **4**, Academic Press, New York, 1971, pp. 265.

10. A. T. Balaban, S. Badilescu and I. I. Badilescu in R. R. Gupta (Ed.), *Physical Methods in Heterocyclic Chemistry*, Wiley-Interscience, New York, 1984, pp. 1.

11. H. Bock, *Pure Appl. Chem.* **44**, 343 (1975).

12. C. N. R. Rao and P. K. Basu in R. R. Gupta (Ed.), *Physical Methods in Heterocyclic Chemistry*, Wiley-Interscience, New York, 1984, pp. 231.

13. R. Zahradnik and J. Koutecky in A. R. Katritzky (Ed.), *Adv. Heterocycl. Chem.* Vol. **5**, Academic Press, New York, 1965, pp. 69.

14. T. J. Barton, R. W. Roth and J. G. Verkade, *J. Am. Chem. Soc.* **94**, 8854 (1972).

15. K. Schofield, *Heteroaromatic Nitrogen Compounds : Pyrroles and Pyridines*, Butterworths, London, 1967, pp. 124.

16. P. George, *Chem. Rev.* **75**, 85 (1975).

17. A. Streitwieser, Jr., *Molecular Orbital Theory for Organic Chemists*, Wiley-Interscience, New York, 1961.

18. N. C. Baird, *J. Chem. Ed.* **48**, 509 (1971).

19. M. J. S. Dewar, *The Molecular Orbital Theory of Organic Chemistry*, McGraw-Hill, New York, 1969.

20. M. J. S. Dewar and A. J. Holder, *Heterocycles* **28**, 1135 (1989).

21. R. K. Harris, *Nuclear Magnetic Resonance Spectroscopy,* Pitman, London, 1983.

22. R. C. Haddon, V. R. Haddon and L. M. Jackman, *Top. Curr. Chem.* **16**, 103 (1971).

23. J. Aihara, *J. Am. Chem. Soc.* **103**, 5704 (1981); I. Hasan and F. W. Fowler, *J. Am. Chem. Soc.* **100**, 6696 (1978).

24. J. A. Elvidge and L. M. Jackman, *J. Chem. Soc.* 859 (1961).

25. W. B. Smith, W. H. Watson and S. Chiranjeevi, *J. Am. Chem. Soc.* **89**, 1438 (1967).

26. N. Jonathan, S. Gordon and B. P. Dailey, *J. Chem. Phys.* **36**, 2443 (1962).

27. H. Gunther, *Tetrahedron Lett.* 2967 (1967).

28. B. A. Hess, Jr. and L. J. Schaad, *Tetrahedron Lett.* 535 (1977).

29. G. C. Levy and G. L. Nelson, ^{13}C *Nuclear Magnetic Resonance for Organic Chemists,* Wiley-Interscience, New York, 1972.

30. R. J. Radel, B. T. Keen, C. Wong and W. W. Paudler, *J. Org. Chem.* **42**, 546 (1977) and references therein.

31. H. J. Dauben, J. D. Wilson and J. L. Laity in J. P. Snyder (Ed.), *Non-benzenoid Aromatics* Vol. **2**, Academic Press, New York, 1971, pp. 167.

32. E. A. Boudreux and R. R. Gupta in R. R. Gupta (Ed.), *Physical Methods in Heterocyclic Chemistry,* Wiley-Interscience, New York, 1984, pp. 281.

33. J. F. Labarre and F. Crasnier, *J. Chem. Phys.* **64**, 1664 (1967).

34. P. J. Garratt, *Aromaticity,* Wiley-Interscience, New York, 1986.

35. P. Jutzi, *Angew. Chem. Int. Edn. Engl.* **14**, 232 (1975).

36. A. J. Ashe, *Accounts Chem. Res.* **11**, 153 (1978).

37. G. Marino in A. R. Katritzky (Ed.), *Adv. Heterocycl. Chem.* Vol. **13**, Academic Press, New York, 1971, pp. 235.

38. M. Roche and L. Pujol, *J. Chem. Phys.* **68**, 465 (1971).

39. J. Ridd in A. R. Katritzky and A. J. Boulton (Eds.), *Physical Methods in Heterocyclic Chemistry* Vol. **1**, Academic Press, New York, 1963, pp. 109.

40. M. Speranza in A. R. Katritzky (Ed.), *Adv. Heterocycl. Chem.* Vol. **40**, Academic Press, New York, 1986, pp. 25.

41. I. Fleming, *Frontier Orbitals and Organic Chemical Reactions,* Wiley-Interscience, London, 1976.

42. A. R. Katritzky and R. Taylor in A. R. Katritzky (Ed.), *Adv. Heterocycl. Chem.* **47**, Academic Press, New York, 1990, pp. 467.

NONAROMATIC HETEROCYCLES

CONTENTS

1 GENERAL

Saturated and partially unsaturated alicyclic compounds closely resemble acyclic analogs in most of the chemical and physical properties. However, relatively small, but significant differences exist in their properties due to the conformational effects, the size of the ring and overall shape of the molecule. Similarly, the heterocycles (without cyclic delocalization) are the cyclic analogs of amines, ethers, amides, enamines, sulfides *etc.* and possess many properties common with their acyclic analogs. However, when the heteroatom is constrained in a ring the properties relating to the relative ring size and the steric requirement of the lone pair on the heteroatom are considerably changed. The ring constrain causes considerable enhancement in the reactivity of the three-membered heterocycles and to a lesser extent, but still significant, in the four-membered heterocycles.

The heteroatom in the nonaromatic hetrocycles exhibits characteristic properties. Trivalent nitrogen inserts conformational properties in the ring due to its pyramidal inversion and lone pair of electrons. The introduction of an oxygen atom into the ring produces strong dipolar effects (anomeric effect) and sulfur causes effects due to the expansion of its valence shell. In addition to the characteristic atomic property, the heteroatom alters geometry of the ring with respect to the alicyclic (carbocyclic) analog by changing the bond distances and bond angles adjacent to the heteroatom. The alteration may be small or large depending on the nature or the number of the heteroatoms and their relative positions. The striking differences that exist between the forces controlling the carbocyclic and heterocyclic conformations are due to :

(i) Torsional interactions : torsional interactions along the carbon–heteroatom bonds differ from those along the carbon–carbon bonds.

(ii) Non-bonded interactions : non-bonded interactions are different in the heterocyclic and carbocyclic systems.

(iii) Dipolar interactions : the presence of the heteroatoms increases dipolar interactions.

(iv) Force constants : the force constants for the bond angle deformations of the heteroatoms are different from those of carbon.

(v) Hydrogen bonding : internal hydrogen bonding between hydroxyl group and heteroatoms can influence the molecular conformations.

2 STRAIN

The molecule preferentially exists in the most stable energetically favourable staggered conformations in which bonding attractive interactions are maximized and the non-bonding repulsive interactions are minimized. The molecule cannot adopt the energetically unfavourable eclipsed conformations without the distortion of the normal bond angles and the bond distances because of the increased internal energy of the system. When the bonds are forced to form a cyclic structure of a molecule, causing deviation in the internal angle from the normal bond angle, the molecule cannot attain the features of the staggered conformer. Thus, there exists strain in the molecule and the molecule is said to be strained[1]. The deviation of bond angle from the normal valence bond angle causing strain in the molecule is associated with the internal energy of the molecule. The greater bond angle deviation is associated with the higher energy and the lower stability of the molecule.

Baeyer has proposed a theory of strain that the deviation of an internal bond angle from the normal bond angle causes internal strain on the ring and the amount of the deviation depends on the size of the ring. Baeyer strain theory, however, is not consistent with the experimental results because it is based on the wrong assumption that all the rings are planar. The strain theory is applicable to the small rings, but not to the six-membered and large-membered rings because such rings are puckered and exist in a completely strain free chair form. Thus, the concept of non-planar structure of the rings exists and the distortion of the bond angle (angle strain) is one of the contributing factors to the ring strain.

3 FACTORS

The ring strain that exists in the cyclic molecule can be considered to be contributed by the following factors which increase the internal energy of the molecule affecting molecular conformation :

(i)   Bond angle strain (angle strain or classical strain or Baeyer strain) : deviation from the normal bond angle.

(ii) Bond strain : the alteration of interatomic distances i.e. the stretching or compression of the chemical bonds.

(iii) Torsional strain (eclipsing strain or Pitzer strain) : forced deviation from the most favourable staggered conformations.

(iv) Non-bonded interactions (steric strain) : the intramolecular van der Waals forces or the steric repulsion between closely spaced atoms.

(v)   Dipole-dipole interactions : the positions of the non-bonded atoms that minimize dipole-dipole repulsions or maximize dipole-dipole attractions.

These factors are interdependent and the change in any one factor affects the other and ultimately molecular conformation. The molecule will adopt such conformation in which sum of all these strains is at a minimum.

3.1 Bond Angle Strain

The deviation of bond angle from the normal value of the tetrahedral angle (109°28') causes strain within the ring which, in turn, brings instability to the ring. The energy required to deviate the bond angle from the normal value is known as bond angle strain.

The deviation or distortion in the bond angle (α) can be calculated from (Eq. 1) :

$$\alpha = \frac{1}{2} \ (109°28' - \text{actual bond angle}) \qquad .....(1)$$

or in the generalized form it can be represented as (Eq. 2) :

$$\alpha = \frac{1}{2} \left[109°28' - \frac{2\,(n-2)}{2} \times 90 \right] \qquad .....(2)$$

where n is the number of members in the rings.

The deviation of bond angles is maximum in the three-membered rings and decreases with increasing ring size (minimum with five-membered rings). Bond angle strain is particularly significant in the small rings (three-and four-membered), while less significant or negligible in the five-and six-membered rings.

The bond angle strain is related to the bond angle deviation or bond distortion. The bond angle strain (energy required to distort the bond angle from its normal value, E) is proportional to the square of the angle of deviation (α), and is represented as (Eq. 3) :

$$E \propto \alpha^2 \qquad ......(3)$$

$$E \approx \frac{K\theta}{2} \ \alpha^2 \ \text{kJ/mol} \qquad ......(4)$$

where $K\theta \approx 0.0840$ (bond bending force constant)

or $\qquad E \approx \frac{0.084}{2} \ \alpha^2 \ \text{kJ/mol} \qquad ......(5)$

$$E \approx 0.042 \ \alpha^2 \ \text{kJ/mol} \qquad ......(6)$$

The bond angle deviations of 5° and 10° require energy of 1 kJ/mol and 4 kJ/mol, respectively.

3.1.1 Bond Angle Strain in Small Ring Heterocycles

The replacement of a $-CH_2$ group in the small ring cycloalkanes by a heteroatom leads to the saturated small ring heterocycles and causes changes in the structural parameters. The carbon–heteroatom bond distances are often appreciably different from the carbon–carbon bond distance. Thus, carbon–oxygen bond (1.436Å) and

carbon–nitrogen bond (1.475Å) are shorter than the carbon–carbon bond (1.54Å).
The variation in the bond angles is less important for oxygen and nitrogen, but
C–S–C bond angle is substantially different.

In the small ring heterocycles (three-and four-membered), bond angle strain is
large and the ring strain is mostly due to the bond angle deviation. The ring strain
in three-and four-membered heterocycles is approximately of the same magnitude
and depends on the nature of the heteroatom than on the size of the ring
(cyclopropane=115 kJ/mol; cyclobutane=112 kJ/mol). The introduction of
unsaturation in the small ring heterocycles further increases angle strain. Some
molecular dimensions in the saturated three-membered rings with their strain
energies[2] are summarized in Table 1.

Table 1. Molecular dimensions of saturated three-membered rings and their strain
energies

Compound	Cyclic compound			Acyclic analog		Strain
	C–C (Å)	C–X (Å)	C–X–C	C–C (Å)	C–X (Å)	kJ/mol
	1.510	1.510	60°	1.54	–	115
	1.481	1.475	60°	1.54	1.470	113
	1.472	1.436	61°	1.54	1.430	114
	1.49	1.820	48.5°	1.54	1.810	83

3.1.2 Bonding in Small Rings

Small ring heterocycles have relatively similar bonding pattern as in small ring cycloalkanes. It is essentially suitable to mention bonding in the small ring cycloalkanes (cyclopropane and cyclobutane) in order to establish their relationship with the small ring heterocycles and to analyze conformational changes caused by introduction of the heteroatom.

3.1.2.1 Bonding in Three-Membered Rings

3.1.2.1.1 Bonding in Cyclopropane

In cyclopropane ring, carbon–carbon bond is not formed by maximum overlapping of sp^3-hybrid orbitals along the axes required for the strong carbon–carbon σ-bond with the bond angle of 109°28', but formed by the oblique overlapping of sp^3-hybrid orbitals being neither head-to-head (along the axes as in normal C–C bond) nor side-to-side (parallel overlapping as in normal C=C π–bond) (Fig. 1). The carbon–carbon bonds in cyclopropane are called bent or banana bonds and

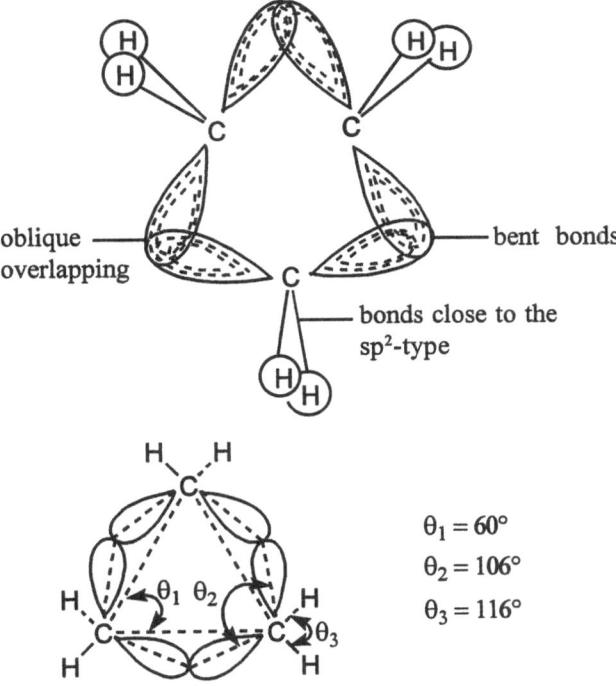

Fig. 1. Bonding in cyclopropane (orbital overlapping)

possess the charecters of both the σ- and π-bonds. Therefore, carbon–carbon bond in cyclopropane is weak and behaves as a carbon–carbon double bond[3]. The electron density along the bond axis is therefore low, but is directed away from the ring.

In cyclopropane, four sp^3-hybrid orbitals on a carbon atom are not equivalent. The two sp^3-hybrid orbitals which are involved in the formation of ring bonds have greater p-character than those forming a normal sp^3-hybrid σ-bond. Therefore, each of the three carbon–carbon bonds is considered to be formed by sp^4 to sp^5-hybrid orbitals and the bond angle is somewhere between the tetrahedral angle (109°28') for sp^3-carbon orbitals and the angle for p-orbitals (90°). The two orbitals directed to the out-side bonds have more s-character than a normal sp^3-orbital because greater p-character of the ring bonds are balanced by the increased s-character of the orbitals forming C–H bonds. The external orbitals have about 33% s-character and are expected to be approximately sp^2-hybrid ($sp^{2.8}$). Carbon uses sp^2-hybrid orbitals for carbon–hydrogen bonds which are shorter and strong, while sp^4 to sp^5 ($sp^{4.2}$) hybrid orbitals are used for the ring carbon–carbon bonds (approximately with 17% s-character and 83% p-character).

Thus, this type of arrangement of the bonding orbitals is usually considered to raise the total energy of the molecule which renders the molecule strained and reactive. The strain has effect on the physical and chemical properties of the molecule.

3.1.2.1.2 Bonding in Three-Membered Heterocycles

The bonding in saturated three-membered heterocycles is similar to that in cyclopropane because these are formally derived by replacing –CH_2 group by heteroatom. The orbitals associated with the ring bonds are not sp^3-hybridized but have greater p-character and are considered to be sp^4–sp^5 hybridized. These orbitals are directed outside the axes joining the nuclei of the ring atoms and involve oblique overlapping resulting in the formation of bent or banana bond. The change in the hybridization of the orbitals forming ring bonds (with greater p-character) is balanced by the increasing s-character of the hybrid orbitals forming external bonds (to the substituents). Thus, the internal bond angles in the saturated three-membered heterocycles, similar to cyclopropane, reflect more p-character than in sp^3-hybrid orbitals and this is compensated by the external bond orbitals having greater s-character and less p-character than the normal sp^3-hybrid orbitals. The increase in the p-character of the hybrid orbitals forming ring bonds results in much less distortion in the interorbital angle than in the internuclear angle and the bond angle is reduced from 109°28' to 106° to form banana-shaped bonds (Fig. 2).

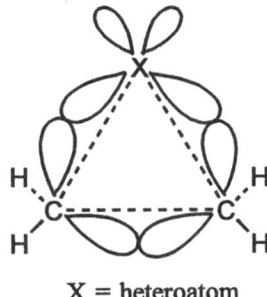

X = heteroatom

Fig. 2. Bonding in three-membered heterocyclic rings

The bond angles in three-membered heterocycles alongwith cyclopropane are summarized in Table 2.

Table 2. Bond angles in three-membered heterocycles

Compound	Internal bond angle C–X–C	External bond angle H–C–H
H₂C (cyclopropane)	60°	116°
NH (aziridine)	60°	116.6°
O (oxirane)	61°	116.4°
S (thiirane)	48.5°	116.0°

The bonding system in three-membered heterocycles[4] can be considered to be constructed from the interaction of π and π^* orbitals of two structural groups; basal group and apical group (Figs. 3 and 4). Basal group is consisting of C–C bond, while heteroatom forms apical group. The basal group (C–C bond behaving as C=C bond in three-membered rings) forms π-complex with an apical group involving electron donation from the filled olefinic π-orbitals to the unfilled orbitals of apical group. This electron donation, in turn, is balanced by the electron donation in the reverse direction. Apical group (acceptor) in π-complex has filled *p*-orbitals, that can be used to form a reverse dative bond in which π^* orbital of olefin (basal group) acts as acceptor. Three-membered saturated rings can be considered as π-complexes with back-coordination in which the basal group and acceptor are doubly linked by two opposite dative bonds. The saturated three-membered heterocycles involve two types of interactions :

(i) interactions involving transfer of electron density from the ethylene fragment to the heteroatom (π-orbital of ethylene with unfilled orbital on heteroatom)

(ii) interactions involving transfer of the electrons from the heteroatom to the ethylene fragment (π^*-orbital of ethylene with filled orbital on heteroatom)

If heteroatom is extremely electronegative, the second type of interactions are negligible and first type of interactions dominate. The distribution of the electrons in the ring are determined by the relative strength of both the interactions.

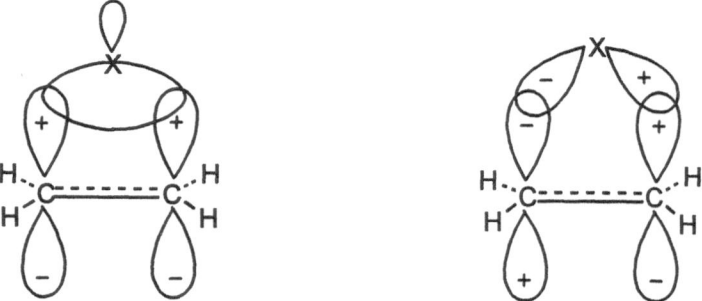

Figs. 3 and 4. Bonding system in three-membered saturated heterocycles (π-complex with back-coordination)

3.1.2.2 Bonding in Four-Membered Rings

3.1.2.2.1 Bonding in Cyclobutane[5]

In cyclobutane, similarly in cyclopropane, the orbitals are not in a condition of regular *sp³*-tetrahedral overlapping, but are bent outward from the hypothetical internuclear straight line. This effect is not so pronounced in cyclobutane as in cyclopropane, although exists to the extent of about 20% as compared to in cyclopropane. This arrangement results in the overlapping with the increased, although small, p-character of the ring bonds. The increased p-character of sp^3-hybrid orbitals involved in the ring bonds is compensated for by the decrease in the p-character or increase in the s-character of the sp^3-hybrid orbitals involved in the external (C–H) bonds. The overlapping of the orbitals is depicted in Fig. 5.

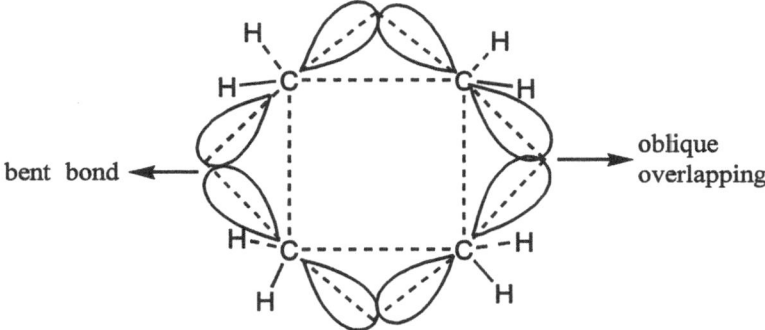

Fig. 5. Bonding in cyclobutane

The molecule of cyclobutane may be represented by two conformations: planar and puckered. If cyclobutane molecule is considered to be planar, there will be angle strain as well as bond eclipsing strain due to four pairs of C–H bonds thus causing increased strain in the molecule (Fig. 6). The puckering of the ring with one carbon atom either below or above the plane of other three carbon atoms (Fig. 7) causes additional strain, but reduces considerably the eclipsing strain (eclipsed hydrogen interactions) and torsional strain. The repulsion between the pairs of non-bonded carbon atoms probably accounts for some what longer C–C bonds in cyclobutane than in cyclopropane. Thus, four-membered rings are less strained than the three-membered rings.

Fig. 6. Planar conformation of cyclobutane

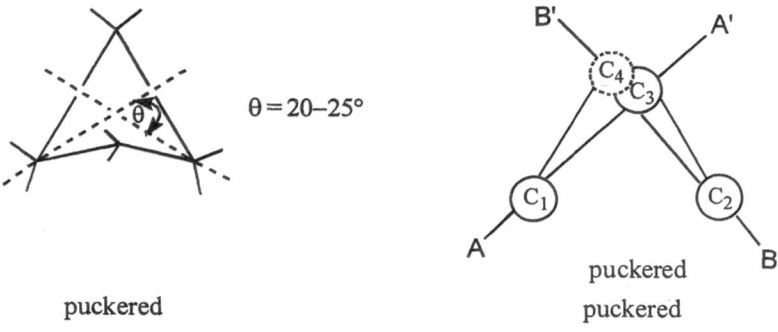

Fig. 7. Puckered conformation of cyclobutane

3.1.2.2.2 Bonding in Four-Membered Heterocycles

The bonding in four-membered heterocycles can be considered very similar to that
in cyclobutane involving oblique overlapping of the sp^3-hybrid orbitals. The
azetidine molecule **1** is puckered with the puckering angle of 33°, but oxetane **2**
and thietane **3** are essentially planar molecules with low energy barrier because the
replacement of $-CH_2$ in cyclobutane by a divalent heteroatom causes reduction in
the number of non-bonding interactions. However, the oxidation of thietane to
thietane sulfone **4** causes reintroduction of non-bonding interactions making the
molecule puckered.

Azetidine (X = NH)

1a

1b

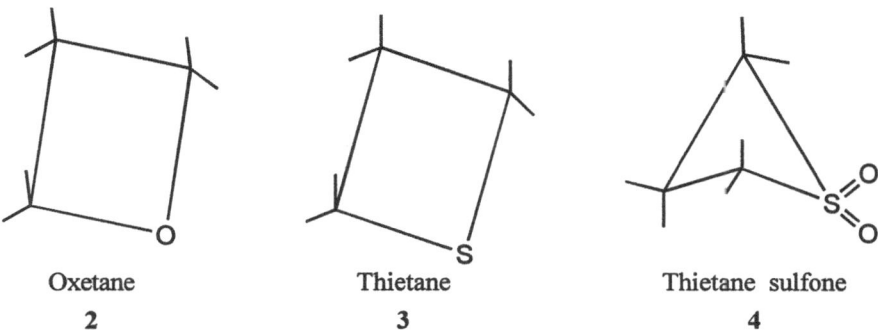

| Oxetane | Thietane | Thietane sulfone |
| 2 | 3 | 4 |

3.1.3 Consequences of Bond Angle Strain in Small Ring Heterocycles

Bond angle strain in small ring heterocycles accounts for their chemical reactivity and, therefore, small ring heterocycles undergo ring opening reactions more readily than the normal rings. The effect of bond angle strain can be visualized well in their structures.

3.1.3.1 IR Spectra[6,7]

Bond angle strain affects the frequencies of vibrations and causes relatively large shift to higher frequency. In small ring heterocycles, C–H stretching vibrations move to higher frequency (3000–3080 cm^{-1}) as compared in the unstrained molecules and the C–H stretching frequency decreases with increasing ring size (aziridine-3047 cm^{-1} and azetidine-2966 cm^{-1}). Analogous changes have been observed in the saturated oxygen heterocycles (oxirane-3052 cm^{-1} and oxetane-2928 cm^{-1}) and in sulfur heterocycles (thiirane-3047 cm^{-1} and thietane-2959 cm^{-1}). The increase in the stretching frequency is attributed to the bond angle strain in the small rings. The bond angle in small rings is substantially contracted and results in the increased *s*-character in C–H bonds. Therefore, with the increase in s-character the bond becomes shorter and the frequency increases. The higher C=N stretching frequency (1800 cm^{-1}) in 2*H*- azirines as compared in unstrained acyclic imines (1650 cm^{-1}) is also attributed to the bond angle strain. Relatively higher frequencies for exocyclic C=O bond in α-lactams (aziridinones-1830–1850 cm^{-1}) and β-lactones (azetidinones-1745–1765 cm^{-1}) as compared in unstrained amides (1680 cm^{-1}) are associated with the increased s-character in C=O bond due to the bond angle strain. The exocyclic C=O bond is shortened and strengthened and, therefore, C=O bond frequency is increased.

3.1.3.2 NMR Spectra

The effect of bond angle strain in small ring heterocycles can be evidenced by ^{13}C–H coupling constants in their NMR-spectra. The coupling constants increase with increasing *s*-character in the C–H bonds within a series of small ring heterocycles. The order of J (^{13}C–H) coupling constants in three membered heterocyclic rings[8] is as : oxirane (176) > thiirane (170) > aziridine (168) > cyclopropane (164) (Fig. 8).

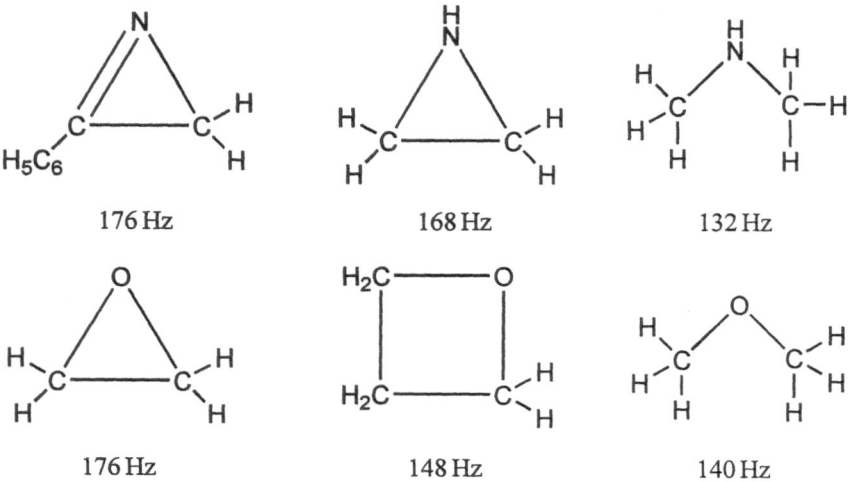

176 Hz   168 Hz   132 Hz

176 Hz   148 Hz   140 Hz

Fig. 8. Coupling constants for some small ring heterocycles

The higher value of ^{13}C–H coupling constant for methylene hydrogen in 2-phenyl-1-azirine (176 Hz) as compared to that in aziridine (168 Hz) is attributed to the greater *s*-character of the orbitals at carbon involved in the C–H bond[9].

3.1.3.3 Conjugative Effect

Bond angle strain in small ring heterocycles results in rehybridization of the ring atoms giving additional *p*-character to the orbitals forming ring bonds. Thus, the ring bonds behave in some respects like double bonds. If the small ring is conjugated with a π-electron system (containing double bond), the conjugation is transmitted through the small ring due to overlapping of the double bond π-orbital with *p*-like orbitals of the small ring[10].

The transmission of conjugation through the small heterocyclic ring can be evidenced as the λ_{max} appears at high wavelength in the ultraviolet absorption spectra[11] of substituted aziridine 5.

3.1.3.4 Basicity

Aziridine itself is less basic than the acyclic secondary amines. The low basicity of aziridine is attributed to the bond angle strain and can be explained by

$$R$$ over $$N$$

H$_5$C$_6$ — (triangle ring) — C — C$_6$H$_5$, with O double-bonded to C

5

considering the hybridization of lone pair on nitrogen. The bond angle strain in aziridine results in an increase in *s*-character of the lone pair and the hybridization of lone pair on nitrogen will be between sp^2 and sp^3. The increased *s*-character of the hybrid orbitals brings the lone pair closer to the nucleus and therefore, will be less available for the protonation than that in acyclic amines. Thus, the lone pair, due to increased *s*-character, interacts less effectively with π-electron system to which aziridine is conjugated. The order of basicity in small ring heterocycles is as follows :

four-membered nitrogen heterocycles > three-membered nitrogen heterocycles.

3.1.3.5 Pyramidal Inversion at Nitrogen

Pyramidal inversion[12] is a property of nitrogen which distinguishes it from the tetravalent carbon. The trivalent nitrogen in a flexible ring inserts another conformational property to the ring. Hydrogen or a substituent at the nitrogen atom in a ring can attain two equilibrating configurations by virtue of inversion of the nitrogen atom. The rate of atomic inversion is affected by the size of the ring in which nitrogen is inserted and the substituent(s) attached to nitrogen. But when nitrogen atom is involved in three-membered ring i.e. aziridine, the conformational changes arise mainly from the pyramidal inversion of nitrogen because the aziridine ring is planar and rigid. The pyramidal inversion in aziridine is reduced because of much higher energy barrier in aziridine ($\Delta G^* = 72.0$ kJ/mol) as compared to the energy barrier in acyclic unstrained compounds which have very low energy barrier ($NH_3 = 24$–25 kJ/mol and $(CH_3)_3N = 34.4$ kJ/mol) (Fig. 9).

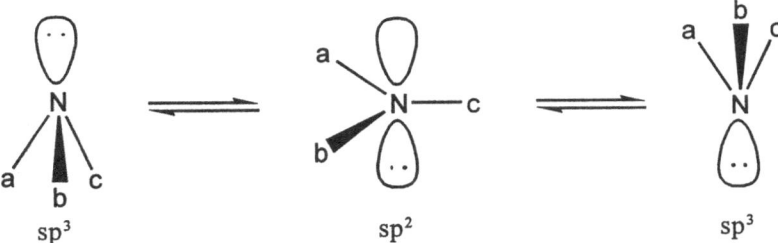

Fig. 9. Pyramidal inversion in unstrained acyclic compound with low energy barrier

The relatively high energy barrier in aziridine is attributed to the highly strained transition state involving *sp²*-hybridized nitrogen (for which normal valence angle is 120°) constrained to the angle of approximately 60° (Fig. 10).

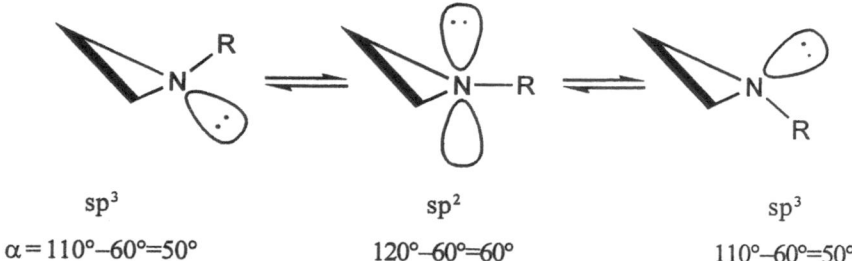

$$sp^3 \qquad\qquad\qquad sp^2 \qquad\qquad\qquad sp^3$$
$$\alpha = 110°-60°=50° \qquad 120°-60°=60° \qquad 110°-60°=50°$$

Fig. 10. Pyramidal inversion in aziridines

The energy barrier to pyramidal inversion in aziridines depends on the nature of the substituent attached to the nitrogen atom :

(i)   Electron delocalizing substituents on small ring nitrogen lower the inversion barrier by lowering the energy of the transitonal 'flat' geometry in which three substituents of nitrogen are in the same plane (with acyl, aryl or carboalkoxy-energy barrier reduced)[13].

(ii) The substituents bearing unshared electron pairs (NH_2, Cl, OCH_3) increase the energy barrier considerably and the enantiomers can be isolated[14]. The increased energy barrier is probably due to the unfavourable lone pair–lone pair interactions in the planar transition state .

$$\triangle\!\!-N\!\!-R$$

6	R = C_2H_5	81.0 kJ/mol.
7	R = $C(CH_3)_3$	71.0 kJ/mol.
8	R = C_6H_5	49.0 kJ/mol.
9	R = $CON(CH_3)_2$	41.0 kJ/mol.
10	R = $COOCH_3$	30.0 kJ/mol.
11	R = NH_2, Cl, OCH_3	90.0 kJ/mol.

Energy barriers to N-inversion in substituted aziridines

(iii) Steric effect destabilizes pyramidal form and lowers the energy barrier[15].

(iv) The solvents which stabilize ground state conformer through the hydrogen bonding or by solvation raise the energy barrier.

3.1.4 Angle Strain in Larger Rings

Larger ring heterocycles exhibit large angle strain arising out of the angle deviation due to the larger values of the bond angles than that of the normal bond angles. 1-Azabicyclo[3.3.3] undecane **12**, commonly named manaxine[16,17] exhibits interesting physical features (lower pK_a value = 9.9 than the related tert-amine, in ultraviolet spectra λ_{max} = 240 nm (ε = 2935) at exceptionally longer wave length than that in tert-amine) which are attributed to the flattening of the bridgehead regions which affects the hybridization of nitrogen atom. The chemical shift of the bridgehead H-5 at relatively low field (δ = 2.57 ppm) in ^1H NMR spectra is ascribed to the unusual hybridization around H-5 due to the proximity of nitrogen atom. Thus, the flattening of the bridgehead regions of the molecule is accompanied by the widening of the angles in bridges and affects the hybridization of the bridgehead atoms with the result of strain in the molecule.

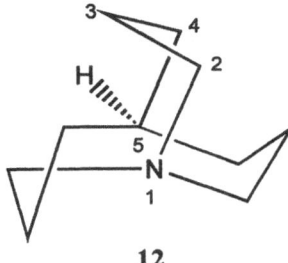

12

The constraining effect is more pronounced in out-6*H*-1-azabicyclo[4.4.4] tetradecane **13** and causes the molecule to adopt preferential conformation in which molecule behaves as an exceptionally weak base. The weak basicity of the molecule is ascribed to the adoption of preferential conformation with an inwardly pyramidalized nitrogen with a hiden lone pair[18].

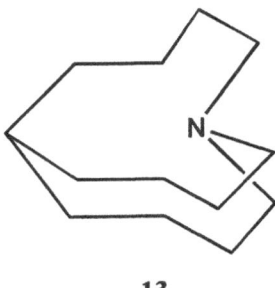

13

Cadged structure

3.2 Torsional Strain

The forced deviation from the most favourable conformation causing relative instability to the molecule is known as bond torsion and the energy that creates bond torsion is called torsional strain. Torsional energy is a measure of the torsional strain for a molecule and changes with the change in dihedral angle. The torsional energy E is expressed as (Eq. 7) :

$$E = 5.87 \ (1 - \cos 3\omega) \qquad \qquad ......(7)$$

where ω is the deviation of dihedral angle from the most favourable conformation.

Torsional strain may be called as Pitzer strain or eclipsing strain or bond opposing strain and is considered to be repulsive force between electrons of the bonding orbitals (bond pair–bond pair repulsion). The repulsive force will be highest in the eclipsed conformation in which bonds are closest to each other, while it will be lowest in the most favourable staggered conformation in which bonds tend to remain as far as possible. The energy corresponding to the eclipsed conformation serves as the energy barrier between the conformers.

The rotation around the single bond is not completely free as the potential energy of the molecule is changed with the change in dihedral angle. The rotation is, therefore, some what restricted by the energy barrier. The energy barrier to rotation of single bond in ethane (the energy difference between the most stable staggered conformation with lowest energy and the least stable eclipsed conformation with highest energy) is very low (12-12.5 kJ/mol) and, therefore, the interconversion of conformers is fast and even takes place at room temperature. The energy barrier (torsional strain in eclipsed form) is contributed by (i) steric repulsion and (ii) bond pair-bond pair repulsion. The steric contribution due to the non-bonded interaction between hydrogens at the adjacent carbon atoms in ethane is negligible because the internuclear distance (0.23nm) is almost equal to twice of the van der Walls atomic radius of hydrogen (0.12 nm).

Torsional strain, referred to as bond eclipsing strain, can be relieved in the acyclic compounds by the rotation of single bond to give staggered conformation. But in cyclic compounds the rotation of single bond is not possible and the strain resulting from the eclipsing of the bonds makes substantial contribution to the strain. Cyclohexane assumes two conformations; chair form (most stable with minimum potential energy) known as staggered form which is without eclipsing strain, and boat form (least stable with maximum potential energy) known as flexible (eclipsed) form with eclipsed strain. The energy barrier (energy difference between two forms) is 21–25 kJ/mol. The increased energy of the boat form is attributed to the eclipsing strain due to four pairs of eclipsed C–H bonds and to the steric repulsion or van der Waals strain due to crowding between 'flag pole' hydrogens.

The replacement of carbon atom(s) by heteroatom(s); oxygen, sulfur and nitrogen, causes considerable change in the rotational energy barrier. The presence of lone pairs on the heteroatoms considerably influence the preferred conformations.

3.2.1 Single Bonds

The rotation around the carbon–heteroatom bond is not quite free and, therefore, leads to the conformers of different energies. The lone pairs on the heteroatom serve as the missing substituents. The rotational energy barriers for carbon–heteroatom bonds are relatively lower than for carbon–carbon bond[19]. The lower rotational energy barrier for carbon–heteroatom bond may be attributed to the reduction in the non-bonding interaction due to H–H eclipsing. The rotational energy barriers for carbon–heteroatom bonds are summarized in Table 3.

Table 3. Rotational energy barriers for single bonds

Compound	Energy barrier
$CH_3–CH_3$	12.2 kJ/mol
$CH_3–NH_2$	8.3 kJ/mol
$CH_3–OH$	4.6 kJ/mol
$CH_3–SH$	5.3 kJ/mol

However, the rotational energy barrier in carbon–nitrogen bond is increased considerably with increasing substituents than in carbon–carbon bond (Table 4).

Table 4. Effect of substitution on rotational energy barriers

Compound	Energy barrier
$CH_3–CH_2–CH_3$	13.80 kJ/mol
$CH_3–NH–CH_3$	13.72 kJ/mol
$CH_3–\underset{\underset{CH_3}{\vert}}{N}–CH_3$	18.40 kJ/mol
$CH_3–O–CH_3$	10.46 kJ/mol
$CH_3–S–CH_3$	8.91 kJ/mol

The increase in rotational energy barrier is attributed to the adoption of preferential conformation by the molecule in which a lone pair is placed gauche to a bulky group.

The rotational energy barriers to rotation about heteroatom–heteroatom bonds are higher than that for carbon–carbon bond (NH_2–NH_2 = 49.7 kJ/mol). The higher energy barrier to rotation about heteroatom–heteroatom bond is ascribed to the lone pair–lone pair interaction on the adjacent heteroatoms. The compound with two heteroatoms at the adjacent positions exists in the conformation in which two lone pairs are gauche to eact other. Thus, the lone pairs on two adjacent heteroatoms tend to occupy preferentially skew position or gauche to each other and this is known as gauche effect[20] (Fig. 11).

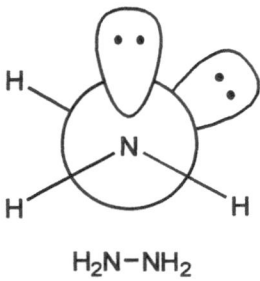

H_2N–NH_2

Fig. 11. Gauche effect

The consequences of the torsional strain (energy barrier) can be evident in the heterocyclic compounds containing heteroatom–heteroatom bond. These compounds exist in the conformation in which lone pairs on the adjacent heteroatoms are perpendicular to each other. The conformation in which the lone pairs on adjacent heteroatoms are eclipsed is destabilized (Fig. 12).

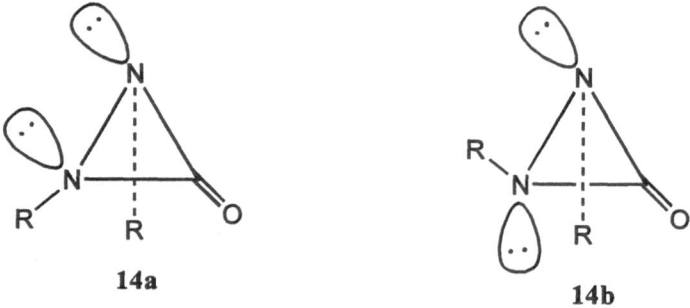

14a

14b

Fig. 12. Conformers of diaziridinone

The structure of diaziridinones provides information on the conformational effect of the lone pairs on the adjacent nitrogen atoms. The properties of diaziridinones are changed with the change in the conformation of lone pairs. Diaziridinone exists in two conformations[21] (Fig. 12) : the conformer with eclipsed arrangement of the lone pairs on the heteroatoms **14a** and the conformer with the lone pairs perpendicular to each other (lone pair–lone pair dihedral angle-120°) **14b**. The conformer **14b** with perpendicular arrangement of the lone pairs on heteroatoms is more stable than the conformer with eclipsed arrangement of the lone pairs[22].

The bicyclic diaziridinone **15** exhibits different properties from those of monocyclic diaziridinone **14** because bicyclic diaziridinone is forced to adopt the conformation in which the lone pairs on the adjacent heteroatoms (nitrogen atoms) are eclipsed. The eclipsed arrangement of the lone pairs (Fig. 13) makes bicyclic diaziridinone less stable and undergoes readily decarbonylation (scheme-1).

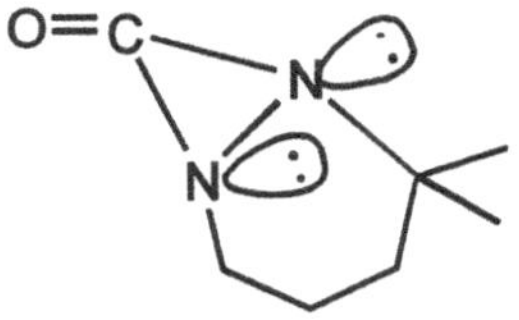

15

Fig. 13. Conformation of bicyclic diaziridinone

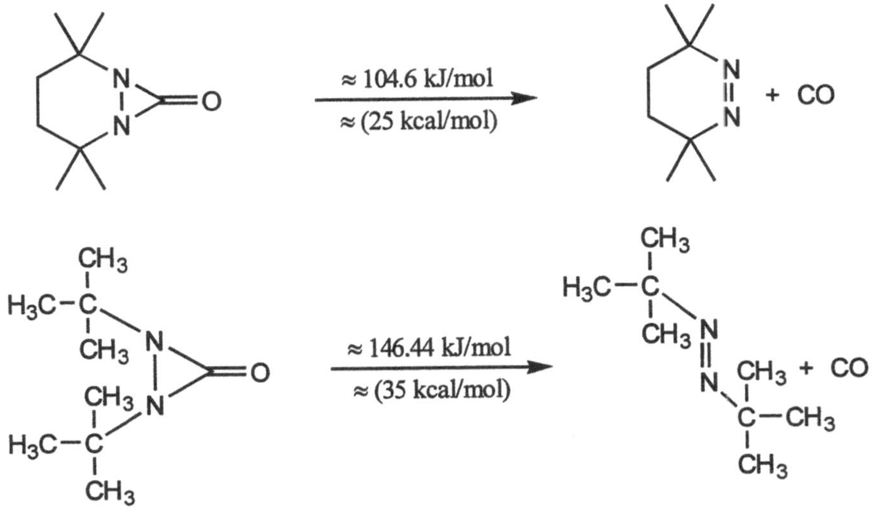

Scheme-1

3.2.2 Double Bonds

The rotation around the double bond is highly restricted due to high energy barrier to rotation, and the molecule is considered to be rigid. Moreover, the incorporation of the double bond into cyclic structure compels the π-bonding system to be twisted causing strain in the molecule.

The twisting of π-bond about the axis joining two olefinic atoms causes four bonding atoms involved not to be in the same plane and prevents optimum overlapping. The distortion of π-bond is accompanied by the change in hybridization with admixture of *s*-character into p-orbitals with the result of increasing p-character in σ-bond. Two olefinic atoms neither achieve the usual coplanar arrangement with four atoms bonded to them nor they have normal bond angle of 120°.

Small bicyclic compounds with the double bond at a bridgehead position are highly strained and, according to the Bredt's rule, cannot exist. The instability of such bicyclic systems is a result of increased ring strain. However, this rule is violated in larger rings[23,24] because the planarity of the π-bond is accommodated by puckering of larger rings **16**.

The instability criterion of Bredt's rule is also applicable to the unsaturated heterocycles **17** containing C=N bond at the bridgehead which is highly reactive and cannot be isolated, but may be trapped with methanol **18**. The instability is because of the structural similarities to carbocyclic bridgehead olefin and the planarity required by a π-bond cannot be maintained in the system[25,26].

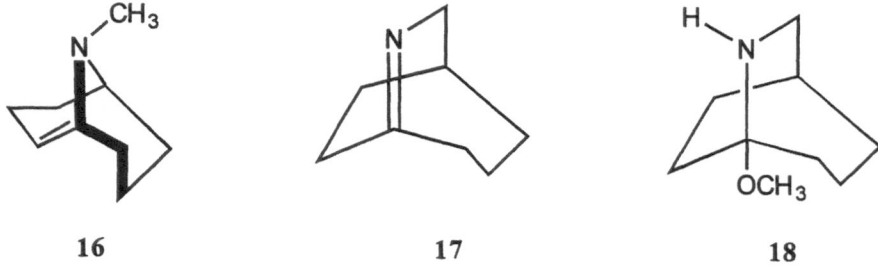

16   **17**   **18**

3.2.3 Partial Double Bonds

The rotation about the single bond in molecules with conjugated double bonds or double bond conjugated with an atom having lone pair of electrons is restricted because of acquiring partial double bond character. The higher rotational energy barrier (~ 75 kJ/mol) to carbon–nitrogen bond in amide is attributed to the partial double bond character of carbon–nitrogen bond due to delocalization of nitrogen lone pair (Fig. 14). The carbon–nitrogen bond in amide is shorter (1.32 Å) than the normal carbon–nitrogen bond, while the carbon–oxygen bond, on the contrary, is longer than the normal carbon oxygen double bond. The partial double bond character in the carbon–nitrogen bond creates certain barrier to rotation and thus leads to two isomers.

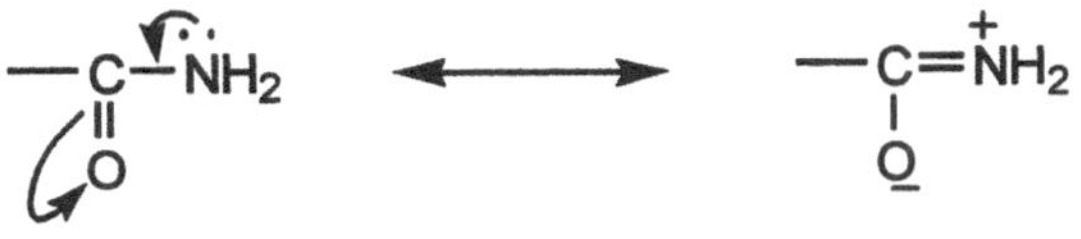

Fig. 14. Delocalization in planar amides

The incorporation of an amide linkage into four-membered ring forms planar β-lactam (nitrogen with exocyclic C=O bond) 19 and causes change in hybridization of the atoms with a corresponding change in the preferred ring geometry[21].

C_3–N–C_1	=	95 ± 1 Å	N–C_1	=	1.49 ± 1 Å
N–C_3–O	=	131 ± 1 Å	C_1–C_2	=	1.58 ± 2 Å
C_2–C_3–N	=	93 ± 1 Å	C_2–C_3	=	1.52 ± 1 Å
N–C_1–C_2	=	86 ± 1 Å	C_3–N	=	1.36 ± 1 Å
C_1–C_2–C_3	=	86 ± 1 Å	C_3–O	=	1.21 ± 1 Å

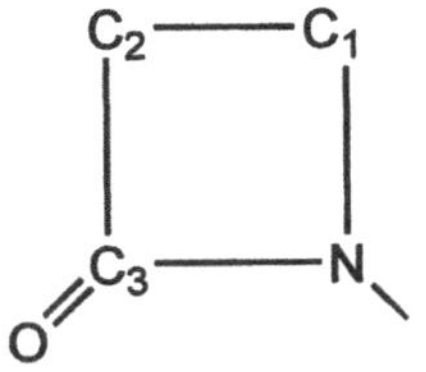

19

β-Lactam system shows difference from the simple azetidine non-planar ring in dihedral angle (azitidine–171° and β-lactam–180°) and in bond distances and bond angles. In β-lactam the angles around the nitrogen (95°) and carbonyl carbon (93°) are larger than in azetidine ring and the other two ring angles are consistently small (86°). This change can be explained in terms of change in hybridization from sp^3 to sp^2. The nitrogen–carbon bond (carbonyl carbon) is shorter than the carbon–nitrogen bond in azetidine which can be ascribed to the resonance due to delocalization of nitrogen lone pair giving partial double bond character to the carbon–nitrogen bond (Fig. 15).

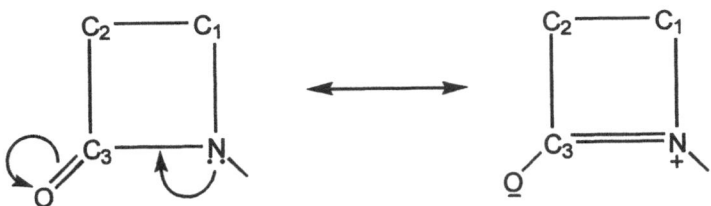

Fig. 15. Canonical forms of β-lactam ring

In bicyclic fused system, the fusion of four-membered heterocyclic ring to a larger ring causes strain and distortion. The amount of distortion depends on the size and nature of the fused ring as well as the substituents on both the rings. The distortion causes disruption of delocalized system (particularly with bridgehead nitrogen). The distortion is extremely significant in the biologically active molecules such as penicillins **20** and cephalosporins[21,27,28] **21**.

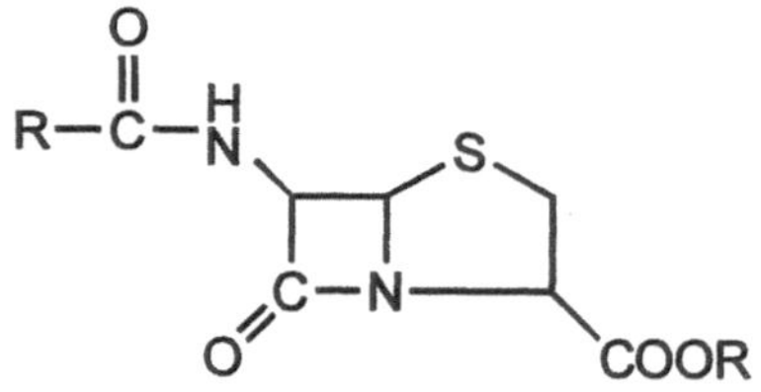

| **20** | **21** |
| Penicillins | Cephalosporins |

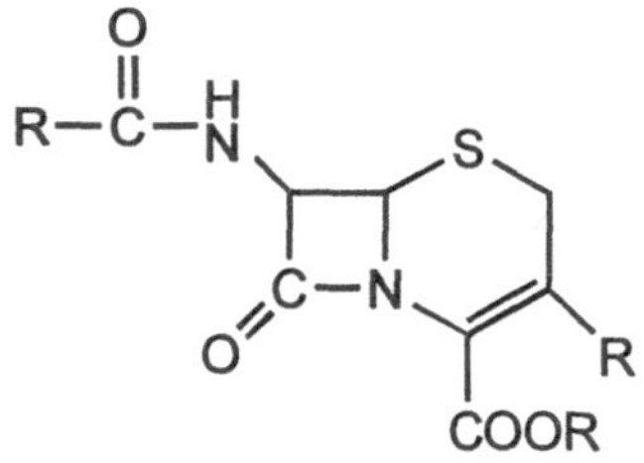

In bioactive penicillin molecule, the bridgehead nitrogen is bent (0.40 ± 0.01Å) out of the plane of the other three carbon atoms in the β-lactam ring. Ability of nitrogen lone pair to participate in π-bonding decreases and conform more to tetrahedral geometry favouring nucleophilic attack at the carbonyl carbon atom. The higher carbonyl stretching frequency (1780 cm^{-1}) in penicillin than in unfused β-lactam (1730 cm^{-1}) is attributed to the decrease in the planarity of bridgehead nitrogen. Distortion restricts delocalization and causes lengthening of the carbon–nitrogen bond and the shortening of the carbon-oxygen bond.

In cephalosporins **21**, the deviation of bridgehead nitrogen from the planarity is less than in penicillins **20** (0.21 ± 0.01). The carbonyl strectching frequency (1776 cm^{-1}) indicates the disruption of delocalization with the carbon–nitrogen bond distance (1.40Å) and carbon–oxygen bond distance (1.21Å). The reason, that the cephalosporins do not require much deviation from the planarity to be active, is that they involve second mechanism for isolating of carbon–oxygen double bond in which the lone pair on the nitrogen atom interacts with the six-membered ring involving some degree of enamine resonance and restricts amide resonance (Fig. 16).

Fig. 16. Resonance in cephalosporins

4 CONFORMATIONS OF FLEXIBLE HETEROCYCLES

The conformations of the flexible heterocycles[19,29,31] are qualitatively very similar to those of the alicyclic compounds, but differ quantitatively due to the changes in the structural parameters caused by the replacement of one or more CH_2 groups by heteroatom(s). In addition to the torsional strain, other factors also affect considerably to the conformations and determine preferred ones in flexible heterocycles. The structural factors which affect preferred conformations in flexible heterocycles are :

(i) **Bond distances :** The carbon–heteroatom bond distances are appreciably different from the carbon–carbon bond distance. Carbon–oxygen bond (1.43Å) and carbon–nitrogen bond (1.47Å) are shorter than the carbon–carbon bond (1.54Å), whereas the carbon–sulfur bond (1.82Å) is longer.

(ii) **Bond angles :** The variation in the bond angles for oxygen and nitrogen is less significant, but C–S–C bond angle is substantially different in heterocycles from the near tetrahedral angle in alicyclic compounds.

(iii) **Dipole interactions :** Heteroatoms produce appreciable dipole moment. If there is only one heteroatom in the ring, the dipole moment has no effect on the conformations, but when two or more heteroatoms are present in the ring or when there is hetero substituent (OH or OR) on the heterocyclic ring, dipole-dipole interactions considerably affect the conformations. However, these interactions are solvent dependent and are inversely proportional to the dielectric constant of the solvent. Thus, the conformations vary with the solvent.

(iv) **van der Waals radii :** The van der Waals radii of oxygen, nitrogen and sulfur are different from that of carbon. The van der Waals radii follow the order :

$$CH_2 \ (200 \ pm) > S \ (185 \ pm) > NH \ (150 \ pm) > O \ (140 \ pm)$$

Thus, other non-bonded atoms are able to approach closer to oxygen than amino group and closer to an amino group than to sulfur which is closer than to a CH_2 group. Moreover, the presence of a lone pair of electrons on the heteroatom which is considered absent ligand reduces non-bonded repulsive interactions because of the smaller volume (size) of the lone pair than the size of hydrogen atom.

(v) **Torsional interactions and Force constants :** The torsional interactions along the carbon–heteroatom bonds differ from those along the carbon–carbon bonds. Torsional energy about carbon–heteroatom bond is generally lower than that about carbon–carbon bond which, in turn, is lower than that about heteroatom–heteroatom bond.

The bending force constants for the bond angles C–X–C and X–C–C (X = heteroatom) are different from those of carbon atoms.

(vi) Hydrogen bonding (attractive interactions through space) : The intramolecular hydrogen bonding between hydroxyl group and heteroatoms can influence the conformations. The intermolecular hydrogen bonding between heteroatoms and hydrogen-donor solvent may also affect the conformations in flexible heterocycles.

The attractive interactions caused by the attraction of nucleophilic centre with electrophilic centre (space interaction) in the same molecule also determine the preferred conformation.

(vii) Pyramidal inversion : Pyramidal inversion at heteroatom especially nitrogen atom has an effective role in the conformational analysis of the nitrogen heterocycles.

4.1 Conformations of Five-Membered Rings

The replacement of $-CH_2$ group(s) by heteroatom(s) affects conformations in five-membered rings, however the conformations of five-membered heterocyclic rings are very similar to those of five-membered carbocyclic rings. The conformational analysis of cyclopentane is included in order to determine the preferred conformations in five-membered heterocycles.

4.1.1 Conformations of Cyclopentane

Cyclopentane may be expected to be planar. In planar structure the bond angle deviation from the normal bond angle is less than 1° (+ 0° 44') as shown by the

$$\left[\left(\frac{1}{2}(109°.28' - 108°)\right) = + 0°44'\right]$$

expression and thus experiences minimum of angle strain. However, the presence of five pairs of eclipsed C–H bonds should have caused an additional strain of about 63 kJ/mol, bringing instability to the ring. These interactions, however, can be minimized by puckering of the ring. Puckered conformation of cyclopentane involves some angle strain, which is compensated by the reduction in C–H bonds eclipsing effect, and the strain energy falls down to 27 kJ/mol due to the change into non-planar conformation. The puckering is not fixed but rotates around the ring in constant wave like motion. This effect is termed as pseudorotation (energy barrier = 17 kJ/mol). Thus, cyclopentane exists in two non-planar conformations : envelope form **17a** and half-chair form **17b** (Fig. 17).

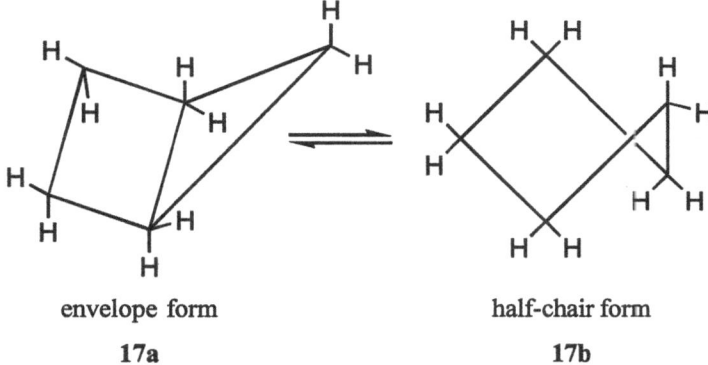

envelope form   half-chair form

17a   **17b**

Fig. 17. Conformations in cyclopentane

In the envelope conformation, four carbon atoms are in a plane and fifth carbon atom is displaced outside the plane. The half-chair conformation has three carbon atoms in a plane and the other two carbon atoms are placed one above and one below the plane. In the forms **18a** and **18b** (Fig. 18) the numbers indicate (in Å) twisting of the carbon atoms above (+ive) or below (-ive) the plane. The interconversion between two conformers does not involve substantial amount of the potential energy and thus exists equilibrium between them.

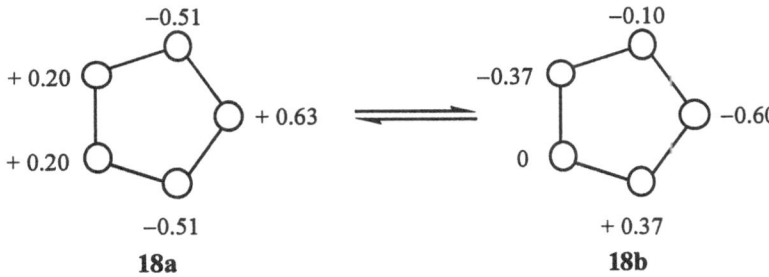

18a   **18b**

Fig. 18. Numbers (in Å) indicate out of plane twisting of carbon atoms

The carbon–carbon single bonds in cyclopentane are some what longer (1.546Å) than the normal carbon–carbon single bond in alkanes (1.54Å), which is accounted for by the transannular type repulsions, since the distance between non-bonded carbon atoms is smaller in cyclopentane (2.44Å) than in alkane (2.545Å)[32].

4.1.2 Conformations of Five-membered Heterocycles

The replacement of ring carbon of cyclopentane by heteroatom such as oxygen, sulfur and nitrogen leading to five-membered heterocycles results in the change in bond angles, bond distances and diminishing of C–H bond eclipsing interactions. Five-membered heterocycles are non-planar and exist in two conformers similar to cyclopentane; envelope and half-chair form. Five-membered heterocycles are considered to prefer half-chair conformation with the heteroatom in the middle of three atoms plane (three adjacent atoms coplanar), since this conformation possesses minimum of non-bonded repulsive interactions between remaining hydrogen atoms.

Tetrahydrofuran[33,34] and pyrrolidine[34] are freely pseudorotating systems (or slightly restricted in pyrrolidine) with pseudorotation among various half-chair and envelope conformations. The half-chair conformations are slightly more stable than the envelope conformations (barrier to pseudorotation is about 0.7 kJ/mol in tetrahydrofuran). The appreciably increased size of the heteroatom in tetrahydrothiophene results in the restricted rotation with higher energy barrier to pseudorotation and preferentially adopts half-chair conformations.

2,3-Dihydrofuran and 2,3-dihydrothiophene are also non-planar in which C-2 methylene group is out of the plane of the other ring atoms with the energy barriers to ring inversion of about 1 kJ/mol and 4 kJ/mol, respectively[35,36]. The higher energy barrier for 2,3-dihydrothiophene is presumably due to a decrease in the ring strain (due to increased size of sulfur than oxygen). Torsional forces are comparable for both molecules and tend to overcome the lower ring strain forces of dihydrothiophene and pucker the ring to a larger degree.

4.2 Conformations of Six-Membered Rings

Saturated six-membered heterocycles have many common conformational characteristics with those of cyclohexane and its derivatives and as such conformations of cyclohexane are being discussed to correlate with the conformations of six-membered heterocycles and to determine the structural factors affecting molecular conformations.

4.2.1 Conformations of Cyclohexane

Cyclohexane ring is puckered and exists in chair, boat and twist-boat conformations. The chair conformation is the most stable and boat conformation is the least stable. The stability of the twist-boat conformation is more than that of the boat but less than that of the chair conformation.

4.2.1.1 Chair Conformation

Cyclohexane ring exists almost exclusively in chair conformation in which alternate carbon atoms occupy two separate planes: 1, 3, 5 in one plane and 2, 4, 6 in the other plane, with axial C–H bonds (*a*) and equatorial C–H bonds (*e*). The chair conformation is slightly flattened becauses of the increased C–C–C bond angle (111°), so that the dihedral angles between the adjacent bonds are 56° and C–H axial bonds are not exactly vertical but directed outwards by 7°. The interaction between 1*e* : 2*e* and 1*e* : 2*a* (skewed bonds) is known as 1,2-interaction, while 1*a* : 3*a* (parallel bonds) interaction is known as 1,3-interaction or synaxial interaction.

The interconversion of one chair conformation of cyclohexane into another chair conformation requires 46.0 kJ/mol energy, however both the forms are identical (Fig. 19). The chair conformation of cyclohexane is strainless and is (i) free of angle strain, (ii) free of eclipsing strain because all C–H bonds are staggered and each carbon atom exhibits tetrahedral arrangement, (iii) free of van der Waals strain as the distance of adjacent H-atoms (249–251 pm) is greater than the sum of van der Waals radii of two H-atoms (240 pm) and (iv) free of dipole-dipole interactions because of carbon and hydrogen atoms.

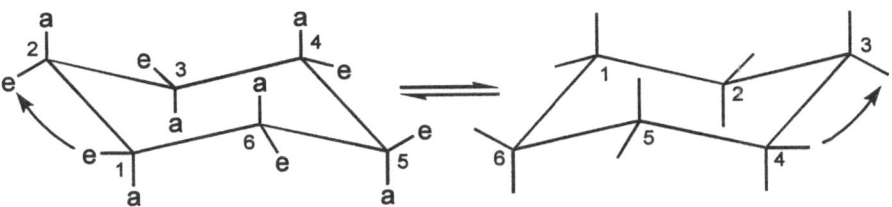

Fig. 19. Chair conformation of cyclohexane with axial and equatorial bonds

4.2.1.2 Flexible Conformation

4.2.1.2.1 Boat Conformation

Cyclohexane ring may also assume a boat-shaped conformation (Fig. 20). Boat form is free of bond angle strain because of the ring puckering: the C–C–C and H–C–C bond angles are 109°28'. Boat conformation has four eclipsed C–H bonds on C_2–C_3 and C_5–C_6. Thus, appreciable eclipsing effect (torsional strain) brings instability to the ring. In addition, van der Waals strain due to overcrowding between flag pole H-atoms which are 1.85 Å (185 pm) apart, closer than the sum of van der Waals radii 2.40 Å (240 pm), brings instability to the ring.

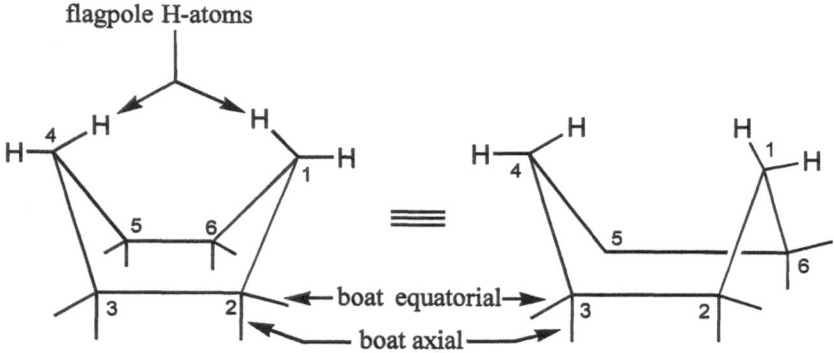

Fig. 20. Boat conformation of cyclohexane.

The boat form is sufficiently strained and contains 28.46 kJ/mol more energy than the chair conformation.

4.2.1.2.2 Twist-Boat Conformation

The C–C bonds of the boat conformation are rotated in such a way that flagpole hydrogen interaction and the eclipsing effect of four pairs of C–H bonds are reduced, the conformation is known as twist–boat form. This conformation is the most stable of the flexible conformations of cyclohexane (Fig. 21).

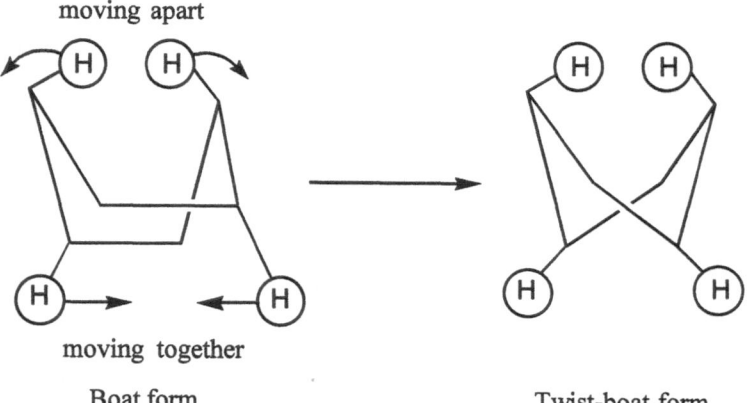

Fig. 21. Flexible forms of cyclohexane

The twist-boat conformation has eclipsing strain but less than that in the boat conformation. Flagpole hydrogen interactions are less in the twist-boat conformation than that in the boat conformation. Thus, twist-boat conformation is less strained than the boat conformation and its energy is 5.46 kJ/mol less than that of the boat conformation, but 23.0 kJ/mol more than that of the chair conformation.

4.2.2 Conformations of Six-Membered Heterocycles

Six-membered heterocycles also assume chair, twist-boat and boat conformations. The replacement of carbon atom(s) in the cyclohexane ring by heteroatom(s) although causes changes in the conformational properties, six-membered heterocycles exist almost exclusively in the chair conformation. The boat form is less favourable because of its higher energy. The conformational properties of the six-membered heterocycles are described in the light of the effects caused by the introduction of the heteroatom(s).

4.2.2.1 Molecular Geometry (Shape of the Ring)

The replacement of carbon atom(s) in the cyclohexane ring by heteroatom(s) to form six-membered heterocycles can alter the structural parameters because of the introduction of electronic interactions due to the lone pair of electrons on the heteroatom(s) and affects molecular geometry. The geometry (chair conformation) of tetrahydropyran and piperidine are expected to be slightly more puckered than then that of cyclohexane because of some what shorter C–O and C–N bonds.

The shape of six-membered rings[31] can be determined by vicinal NMR coupling constants. The coupling constant (J) between two protons depends on the number of separating bonds, nature of neighbouring substituents and the dihedral angle in vicinal protons. The coupling constant (J) is related to the dihedral angle by Karplus equation (Eq. 8):

$$J = A \cos^2\varphi + C \qquad \qquad .....(8)$$

or $\quad J = A \cos^2\varphi \qquad \qquad .....(9)$

(A is constant, C is very small and can be neglected)

However, this equation cannot be applied as such to the heterocyclic rings because it is not possible to evaluate constant A for vicinal protons attached to the atoms of different electronegativity (H–C–X–H or H–Y–X–H), as it depends on the electronegativity and the orientation of substituents. This method involves the use of two coupling constants that contain identical effect of the electronegativity and the other factors included in Karplus equation[37]. The ratio of both the coupling constants determines the shape of the heterocyclic ring.

$$\frac{J \text{ (trans)}}{J \text{ (cis)}} = \frac{A \cos^2\varphi \text{ (trans)}}{A \cos^2\varphi \text{ (cis)}} \qquad \qquad .....(10)$$

$$\frac{J \text{ (trans)}}{J \text{ (cis)}} = \frac{A \cos^2\varphi \text{ (trans)}}{A \cos^2\varphi \text{ (cis)}} = R \qquad\qquad .....(11)$$

$$\left[\quad \text{In } CH_2-CH_2 \text{ fragment}: \quad \begin{aligned} J \text{ (trans)} &= \frac{1}{2}(Jaa + Jee) \\[2mm] J \text{ (cis)} &= \frac{1}{2}(Jea + Jae) \end{aligned} \quad\right]$$

The ratio $R = \dfrac{J \text{ (trans)}}{J \text{ (cis)}}$ dose not depend on the electronegativity and the orientation of substituents, but on the conformation[38].

The molecules with R = (1.9–2.2) adopt conformation **22** similar to that of unsubstituted cyclohexane, *i.e.* chair form. A flattened distortion **23** in which the methylene groups are more nearly eclipsed results in a decrease in the R-value and a puckering distortion **24** causes an increase in R-value (Fig. 22). The ratio of the coupling constants, therefore, gives a direct measure of the deviation of fragment conformation from that of an undistorted cyclohexane (Table 5).

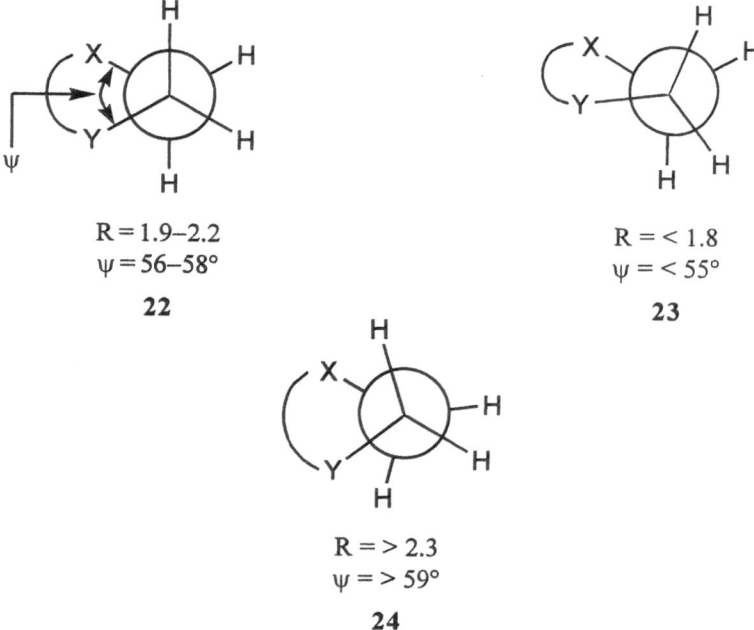

Fig. 22. Relationship of conformation with R-value

Table 5. Ring shapes determined by R-value

Hetero group	Segment	J(trans) (H_z)	J(cis) (H_z)	R	ψ
: NH	α,β	7.88	3.77	2.09	57
: N–CH$_3$	α,β	7.52	3.65	2.06	57
O	α,β	7.41	3.87	1.91	56
S	α,β	8.15	2.96	2.65	61
S	β,γ	8.47	3.28	2.58	60
: $\overset{+}{S}H$	α,β	8.5	3.9	2.2	58
$\overset{+}{S}CH_3 \bar{I}$	α,β	8.63	3.24	2.66	61

The R-value is related to the internal torsional angle ψ by (Eq. 12)[39] :

$$Cos\ \psi = [3/(2 + 4R)]^{1/2} \qquad\qquad .....(12)$$

Thus, the undistorted R-value 1.9–2.2 corresponds to a torsional angle of 56–58° (which is in agreement with the nontetrahedral geometry of cyclohexane). The flattened geometry (R < 1.8) corresponds to ψ = < 55°, and the puckered geometry (R > 2.3) corresponds to ψ = > 59°. The R-value of 1.91–2.2 supports the chair conformations of tetrahydropyran and piperidine.

4.2.2.2 Barriers to Ring Inversion

Six-axial and six-equatorial hydrogens of cyclohexane are chemically non-equivalent and have different chemical shifts in NMR spectra. The axial hydrogens appear at slightly higher field (0.4–0.5 ppm). However, at room temperature, NMR-spectra of cyclohexane shows sharp singlet due to the rapid ring reversal than NMR time scale and interchanging axial and equatorial hydrogens. At low temperature, the rate of the ring inversion is decreased with the non-equivalent hydrogens and the ring inversion is accompanied with the interchange of 1,3,5 and 2,4,6 carbon planes leading to the inverted chair conformation.

The ring inversion involves conformational change, but not in the configuration and proceeds via boat and twist-boat conformations (Fig. 23).

Bond angle deformation in the chair form is accompanied by the increased torsional energy and leads to the transition state (boat form) with high energy (42–43 kJ/mol). The transition state is then changed into twist-boat conformation which is without angle strain but with some torsional strain and corresponds to the energy of 23 kJ/mol. The twist-boat conformation is converted to the inverted chair conformation. Similarly, in six-membered heterocycles the process of chair-chair ring inversion converts the ring to its mirror image and interchanges the axial

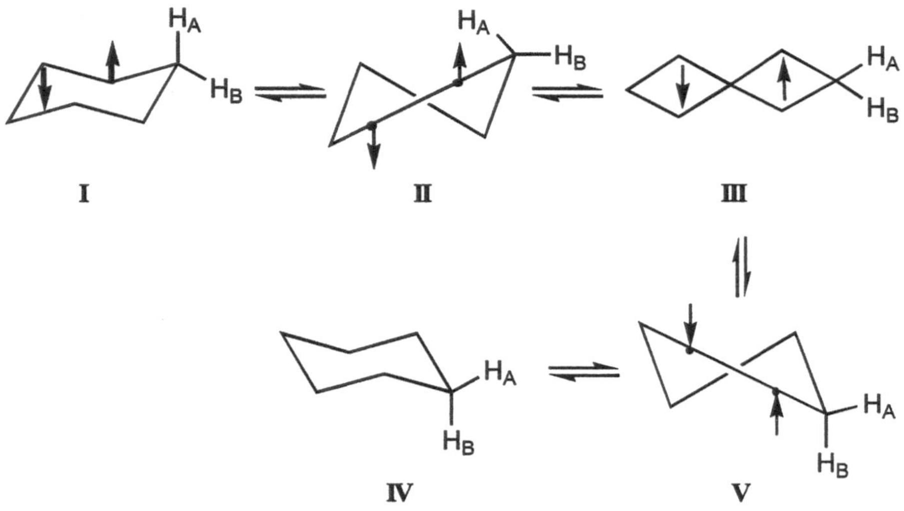

Fig. 23. Ring inversion in cyclohexane

or equatorial nature of all the substituents (Fig. 24). The barrier to ring inversion can be determined by examining NMR-spectrum as a function of temperature.

Fig. 24. Ring-inversion in six-membered heterocycles

The transition state of chair-chair ring inversion is generally considered to be the half-chair conformation in which four ring atoms are coplanar. Six-membered heterocycles with one heteroatom have the choice of three such conformations depending on the energy barriers (Fig. 25). The increased energy barrier in the transition state is attributed to the increased torsional interaction and the bond angle strain.

Torsional energy barrier of C–X bond (X = heteroatom) is different from the C–C bond and, therefore, the energy barrier for the ring inversion in six-membered heterocycles will be different from that in cyclohexane ring. The barriers to ring inversion depend directly on the magnitude of the carbon–heteroatom torsional

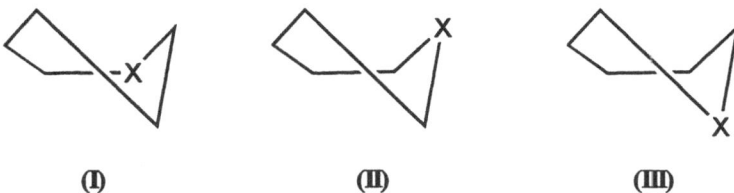

Fig. 25. Half-chair conformations of six-membered heterocycles

energy. If the C-X torsional barrier is lower than that of the carbon-carbon bond, the transition state (I) is preferred, since heteroatom relieves the angle strain. If the torsional barrier is higher, the transition state (III) is preferred because the heteroatom is in the least eclipsed position of the ring. However, if there is little difference between carbon–heteroatom and carbon–carbon torsional barriers, any of the transition states (I-III) may be involved in the ring inversion.

The rate of ring inversion K_c may be obtained at coalescence temperature T_c by (Eq. 13):

$$K_c = \pi \, \Delta v / \sqrt{2} \qquad \qquad .....(13)$$

where Δv is the chemical shift difference between two uncoupled nuclei at an exchange and Eq. 14 is used for the process between coupled nuclei.

$$K_c = \left(\frac{\pi}{2}\right)(\Delta v^2 + 6J^2)^{1/2} \qquad \qquad .....(14)$$

The free energy of activation $\Delta G^{\ddagger}$ can be obtained by (Eq. 15).

$$\Delta G^{\ddagger} = 2.3 \, RT_c \, (10.32 + \log T_c / K_c) \qquad \qquad ......(15)$$

Free-energy of activation for the ring inversion of some six-membered heterocycles is summarized in Table 6 :

Table 6. Free energy of activation for ring inversion of six-membered heterocycles

Compound	$\Delta G^{\ddagger}$ (kJ/mol)	Temperature (°C)
Cyclohexane (for comparison)	43.2	-67°C
Piperidine	43.6	-62.5°C
Tetrahydropyran	39.8	-80°C
Thiane	37.7	-93°C

4.2.2.3 Pyramidal Inversion

Pyramidal inversion[12,40] is a characteristic property which distinguishes heteroatoms, nitrogen, oxygen and sulfur, from the tetravalent carbon atom. Trivalent
nitrogen atom in a conformationally flexible ring inserts another conformational
property to the ring. The hydrogen atom or substituent attached to the nitrogen
atom and the lone pair on nitrogen attain two equilibrating conformations due to
the pyramidal inversion of nitrogen.

Nitrogen containing flexible heterocycles, i.e. piperidine, undergo two types of
conformational changes; ring inversion and pyramidal inversion. Both the
processes are rapid and go side by side and cannot be separated from the
complicating set of conformations. The pyramidal inversion is the lower energy
process (E = < 42 kJ/mol) than the ring inversion (E = > 42 kJ/mol) and the ring
inversion process is slower than the pyramidal inversion of nitrogen (Fig. 26).

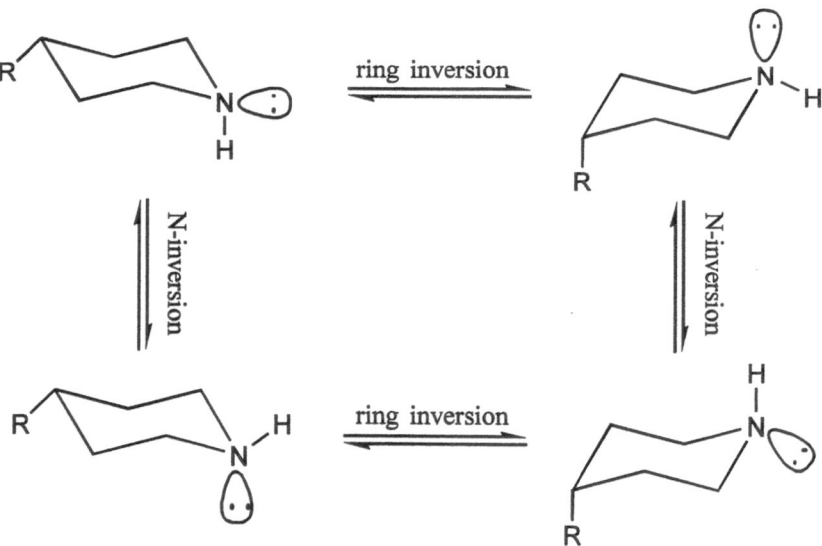

Fig. 26. Pyramidal inversion and ring inversion in piperidine

Two inversions bring out the same conformational changes i.e. N–H (axial) to
N–H (equatorial) and vice-versa. The N–H prefers equatorial position in gas phase
and in non-interacting solvents. The energy difference between N–H (equatorial)
and N–H (axial) is approximately 1.68 kJ/mol. The alkyl group attached to nitrogen
also preferentially occupies equatorial position and the energy difference between
axial and equatorial methyl groups is 11.30 kJ/mol.

The ring inversion leads to the change in conformation, whereas the nitrogen inversion (pyramidal inversion) results in the change in configuration. The difference can be evident in 1,3-dimethylpiperidine (Fig. 27) :

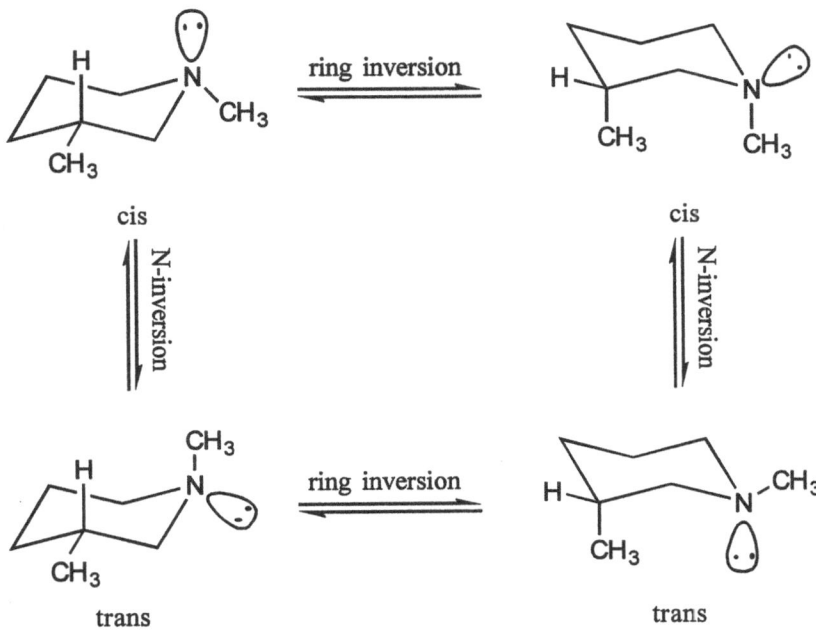

Fig. 27. Ring inversion and pyramidal inversion in 1,3-dimethylpiperidine

The rate of pyramidal inversion is affected by the substituent on nitrogen and by the attachment of nitrogen with other ring heteroatom and varies from one heteroatom to another[41,42]. An oxygen atom attached directly to nitrogen in the ring decreases inversion rate to an extent that the nitrogen atom is almost locked because the energy barrier to nitrogen inversion is raised. The energy barrier to nitrogen inversion in hexahydropyridazine **25** is 48.5 kJ/mol (Fig. 28).

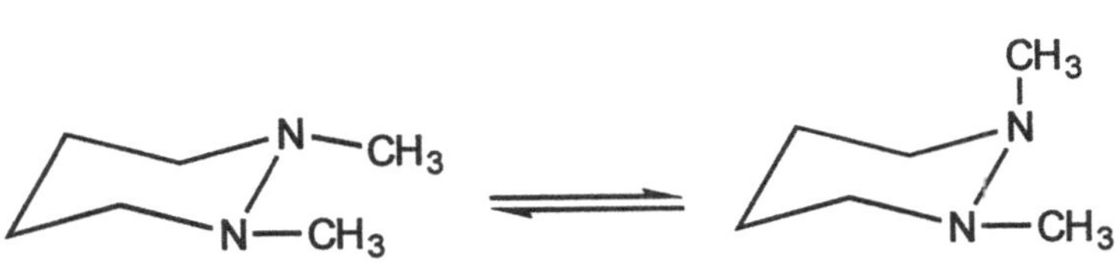

25a   **25b**

Fig. 28. Pyramidal inversion in hexahydropyridazine

The oxygen heteroatom also undergoes inversion, but it is not as interesting as of nitrogen. The oxygen-inversion does not cause appreciable change because two lone pairs of oxygen exchange their position during the atomic inversion. The sulfur heteroatom in sulfur heterocycles behaves similarly towards atomic inversion.

4.2.2.4 1,3-Diaxial Interactions

1,3-Diaxial interactions, referred to as synaxial interactions, are increased in six-membered heterocycles due to shorter bond distance than in cyclohexane because of increased puckering of the chair conformation in six-membered heterocycles.

2-Methyl-1,3-dioxane 26[30,43] assumes two conformations (26a and 26b) in which the conformer (26a) with equatorial methyl group is more favoured than in even methylcyclohexane. The shorter C–O bond distance causes larger 1,3-diaxial interactions in the conformer (26b) of 2-methyl-1,3-dioxane with axial methyl group than in methylcyclohexane.

26a   26b

However, 1,3-diaxial interactions are considerably reduced when hydrogen is replaced by a lone pair of electrons which are effectively smaller in the size than that of hydrogen. 5-Methyl-1,3-dioxane 27 shows only a slight preference for the conformer (27a) with equatorial methyl group. The conformer (27b) with axial methyl group is preferred because in the axial conformation the interactions of methyl group with lone pairs on oxygen are reduced. Bulky groups also adopt axial position at C-5 due to very small 1,3-diaxial interactions[43].

27a   27b

4.2.2.5 Conformational Preferences of Substituents on Heteroatom

In six-membered heterocycles bearing a substituent at the position-1, a simple equilibrium exists between axial and equatorial conformations (Fig. 29). The properties of the compounds exhibiting conformational isomerism cannot be determined on the basis of any single rigid structure, but depend on the nature of contributing conformers, their relative populations and their kinetic stability[31]. Different methods have been used to determine the conformational preferences of the substituents attached on the heteroatom in six-membered heterocycles.

Fig. 29. Axial and equatorial conformations

4.2.2.5.1 Thermodynamic Method

Conformational free energy ($\Delta G°$) provides valuable information on the relative stability of the conformers. The direct integration of resonances from both interconvertible conformers in slow exchange NMR-spectrum gives equilibrium constant K_e, from which $\Delta G°$ can be calculated (Eq. 16).

$$\Delta G° = -RT \ln K_e \qquad ......(16)$$

($\Delta G°$ is usally negative and represents the free energy difference between equatorial and axial conformers).

However, the failure to observe distinct resonances can arise beacause (a) only one conformer is predominantly present in equilibrium (a biased equilibrium), (b) slow exchange limit is not obtainable and (c) the spectra are fortuitously superimposed.

4.2.2.5.2 Spectroscopic Methods

4.2.2.5.2.1 Vibrational Spectra

The separate vibrational bands in the vibrational spectrum for axial and equatorial conformers also provide information on the conformational stability. The vibrational time scale is adjusted in such a way that two separate bands are obtained. The intensity ratio (of axial I_a and equatorial I_e) as a function of temperature, however, gives enthalpy difference ($\Delta H°$) between isomers (Eq. 17).

$$\frac{\Delta H^{\circ}}{R}\left[\frac{1}{T_{1}}-\frac{1}{T_{2}}\right]=l_{n}\frac{K_{2}}{K_{1}}=l_{n}\left[\frac{I_{e}}{I_{a}}\right]_{T_{2}}-\left[\frac{I_{e}}{I_{a}}\right]_{T_{1}} \quad ......(17)$$

where K_1 and K_2 are equilibrium constants and T_1 and T_2 are temperatures.

4.2.2.5.2.2 NMR Spectra

Since vicinal coupling constants depend on the dihedral angle (H–C–X–H) and as such coupling constants are used to identify whether a proton on the heteroatom is axial or equatorial. The larger coupling constant in **29** is to be due to the axial orientation of a proton on the heteroatom.

(J = 14.1 Hz)

29

In NMR-spectrum of thiane oxide **30** the axial isomer(**a**) invariably has the smaller chemical shift difference ($\Delta\upsilon = 0.48$ ppm) and larger coupling constant (J = 13.7 Hz), while the equatorial conformer (**b**) has larger chemical shift ($\Delta\upsilon = 1.87$ ppm) and smaller coupling constant (J = 11.7 Hz). However, the nature of the heteroatom and the substituents alter the absolute values of these spectral parameters and this criterion cannot be applied unless both the isomers are observed.

30a   **30b**

4.2.2.6 Conformational Preferences of lone Pair and Substituents on Nitrogen

Piperidine and its derivatives with substituents on nitrogen only exist in the chair conformation with R-value 2.06–2.09 and torsional angle 57°. Piperidine has been controversial for the orientation of lone pair and hydrogen atom on the nitrogen atom. However, considerable evidences have been given in the support of both N–H axial and N–H equatorial conformations.

It is accepted that the large groups have a greater preference for the equatorial conformation than the small group.Thus, the position of the equilibrium between two conformations may be determined by the relative size of the N-substituent and the lone pair (size of lone pair means its preference for equatorial position relative to the other N-substituent).

Molecular polarizability and Kerr constant measurements of piperidines lead to conclude that the lone pair is larger than a covalently bonded hydrogen atom and the volume of lone pair is nearly comparable to that of the methyl group. The order of smallest to largest is; NH : lone pair: $N-CH_3$,[44], and thus favours the preferred conformation with N–H axial and lone pair equatorial. Moreover, microwave studies also favour N–H axial conformation. Contrary to these studies, Katritzky et al.[45] have reported that a lone pair has smaller steric requirement than the covalently bonded hydrogen atom which, in turn, is smaller than the methyl group. The order of the size : $N-CH_3$ > N–H > lone pair, favours N–H equatorial conformation[46].

Evidences in favour of N–H equatorial :

(i) IR-spectrum of gaseous piperidine shows two bands in the N–H first overtone region which can be ascribed to the N–H equatorial and N–H axial conformations. The ratio of the band absorbances indicates an enthalpy difference of 2.1 kJ/mol favouring N–H equatorial conformation.

(ii) Dipole moment of 4-*p*-chloromethylpiperidine favours N–H equatorial conformation by 2.2 kJ/mol (0.4–0.5 kcal/mol).

(iii) The axial position of the lone pair on nitrogen (N–H equatorial) can be evidenced by the chemical shift difference criterion in NMR-spectra[31,47]. The chemical shift difference (δae) between the axial and equatorial positions of piperidine is larger (0.44 ppm similar to cyclohexane = 0.4–0.5 ppm) with the axial protons at higher field. The enhanced value of δae is attributed to the interaction between axial C–H group and a vicinal lone pair (axial position) on nitrogen. Thus, the larger δae value should occur with an equatorial N–H conformation.

Evidences in favour of N–alkyl equatorial :

In N-alkylpiperidines, the conformer with equatorial alkyl group and with axial lone pair is favoured. The evidences in favour of equatorial position of N-alkyl group are :

(i) Dipole moment studies suggest the piperidine conformation with alkyl group in equatorial position and lone pair on nitrogen in axial position.

(ii) Infrared spectral studies of the Bohlmann bands appearing in the region 2500–2820 cm^{-1} indicate that the lone pair on nitrogen is axially oriented, while N–alkyl group is in the equatorial position.

(iii) Chemical shift difference (δae) criterion in NMR-spectra suggests the equatorial position of the alkyl group and axial position of the lone pair on nitrogen. The larger chemical shift difference (δae) for N-methylpiperidine (δae = 0.94 ppm) is due to the interaction of vicinal axial C–H proton with axial lone pair on nitrogen because such interaction leads to the shielding of proton. The enhanced chemical shift difference (δae) occurs only with equatorial N–CH$_3$ and axial lone pair.

5 STEREOELECTRONIC EFFECTS IN SATURATED SIX-MEMBERED HETEROCYCLES (ANOMERIC AND RELATED EFFECTS)[48,49]

The presence of one or more heteroatoms in the ring affects the conformational preferences and hence conformational properties of the heterocycles due to the existing stereoelectronic effects.

5.1 Anomeric Effect

This effect was first studied in pyranose sugars and observed that the electronegative groups (halogen, acetoxy, methoxy, hydroxyl, amino *etc.*) situated on the carbon atom adjacent to oxygen in the pyranose ring prefer axial orientation.

2-Alkoxytetrahydropyran **31** assumes two possible conformations; with an equatorial **31a** and axial substituent **31b**. The conformation with axial alkoxy group is preferred over the equatorial orientation and attributed to the anomeric effect.

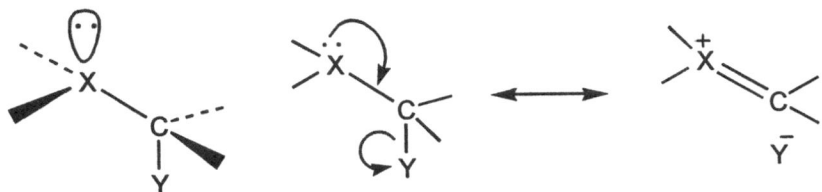

31a   **31b**

The anomeric effect is defined as the extra preference for axial orientation of an electronegative (polar) substituent attached at the anomaric carbon, i.e. α to the ring heteroatom. Quantitatively, anomeric effect may be defined as the free energy difference between the axial and equatorial conformers $(G_e^{\circ}-G_a^{\circ})$ plus the conformational free energy (ΔG°) of the substituent (which is assumed to be similar as in cyclohexane). Thus, the anomeric effect can be represented as :

Anomeric effect of axial substituent $= (G_e^{\circ}-G_a^{\circ}) - \Delta G^{\circ}$ axial substituent

Explanation :

The anomeric effect is caused by the interaction of an exocyclic polar substituent (electronegative group) with an endocyclic heteroatom and forces the molecule to adopt the stable conformation. The anomeric effect is considered to be contributed by :

(i) Non-bonding interactions and (ii) dipole-dipole interactions.

(i) Non-bonding interactions: The anomeric effect is contributed partly due to the non-bonding interactions of the non-bonding lone pair electrons on the heteroatom with the σ^* antibonding orbital of the antiparallel σ-bond of carbon with an electronegative group. The anomeric effect can be explained by considering X–C–Y fragment of cyclic system in which X is a heteroatom with lone pair electrons and Y is an electronegative substituent with antiperiplanar σ-bond and can be represented as (Fig. 30) :

Fig. 30. Anomeric effect : antiperiplanar interaction of lone pair and σ-bond

Thus, n–σ* interaction results in the transfer of electron density from the lone pair to the σ*orbitals with the strengthening of C–X bond due to the introduction of double bond character and the weakening of C–Y bond.

(ii) Dipole-dipole interactions : Dipole-dipole interactions also contribute to the anomeric effect by which an equatorial orientation of the substituent is destabilized with respect to an axial orientation of the polar group (Fig. 31). The instability of

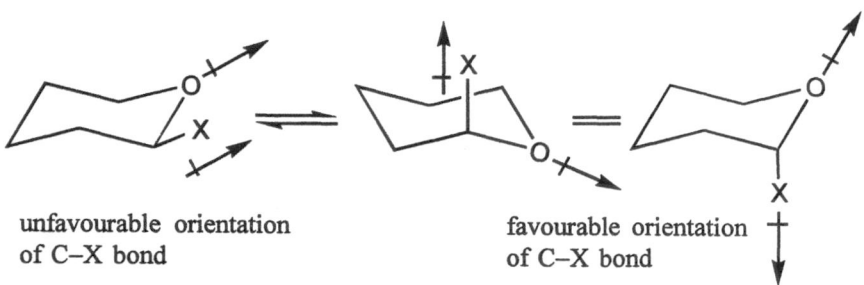

unfavourable orientation of C–X bond   favourable orientation of C–X bond

Fig. 31. Representation of dipolar interactions

the equatorial orientation is considered to be caused by the unfavourable orientation of the dipole of C–X bond and of the ring oxygen (orientation of free electron pairs) because two dipoles are parallel. The dipole interaction of ring oxygen and C–X bond (electronegative group) is more favourable when the orientation of substituent X is axial because two dipoles are divergent.

5.1.1 Factors Affecting Anomeric Effect

5.1.1.1 Effect of Substituents

The magnitude of the anomeric effect depends on the electron donor capacity of the heteroatom and the acceptor capacity of the electronegative substituent. It decreases in the following order :

$$\text{Halogens} > C_6H_5COO > CH_3COO > RO > RS > HO > NH_2 > \overset{+}{N}$$

The magnitude of the anomeric effect is not considerably affected with an increase in the size of R (methyl, ethyl, propyl, butyl, *etc.*) in alkoxy group (OR).

5.1.1.2 Solvent Effect

The equilibrium of the equatorial-axial conformers is influenced by the nature of the solvent. The proportion of axial conformer in solution decreases with decreasing dielectric constant of the solvent because the equatorial conformers are more polar. The equatorial form increases with increasing dielectric constant since the dipole-dipole repulsion diminishes.

5.1.2 Consequences of Anomeric Effect

5.1.2.1 Bond distance

The bond distances are considerably affected in the heterocycles exhibiting anomeric effect. *cis*-2,3-Dichloro-1,4-dioxane **32** with chloro group in both axial and equatorial positions exhibits abnormally long and short bond distances. The axial C–Cl bond is longer (1.82 Å) than the equatorial C–Cl bond (1.78Å), whereas the C–O bond, adjacent to the axial C–Cl bond is shorter (1.39Å) than the normal C–O bond (1.42 Å) (Fig. 32).

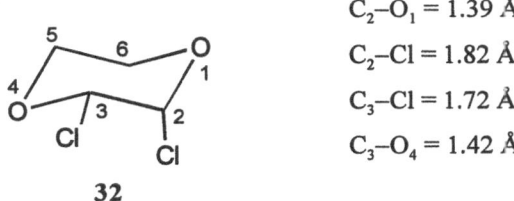

$C_2\text{–}O_1 = 1.39$ Å

$C_2\text{–}Cl = 1.82$ Å

$C_3\text{–}Cl = 1.72$ Å

$C_3\text{–}O_4 = 1.42$ Å

Fig. 32. Bond distances in cis-2,3-dichloro-1,4-dioxane

However, the bond distances in 2-alkoxytetrahydropyran **33** are not considerably changed because the exocyclic C–O bond (carbon-alkoxy group) can rotate in such a way that the lone pair on exocyclic oxygen becomes antiperiplanar to the ring carbon-oxygen bond. The anomeric effect is exerted in the opposite direction causing it less effective (Fig. 33).

33a   **33b**

Fig. 33. Anomeric effect in both directions

5.1.2.2 Chemical Reactivity

The anomeric effect results in the weakening of the bonds antiparallel to the lone pairs on heteroatom and thus affects chemical reactivity of the molecule. Cyclic acetal **34** with an axial substituent is much more reactive than with an equatorial substituent and is easily hydrolyzed. The reactivity is attributed to the bond weakening effect due to the anomeric effect[50].

equatorial   axial

34a   34b

The bond weakening effect due to anomeric effect is also exhibited by other ring systems. Tricyclic orthoamide **35** adopts the conformation in which the lone pairs of nitrogen atoms are antiperiplanar to the C–H bond and involves anomeric effect causing weakening of the C–H bond[51,52].

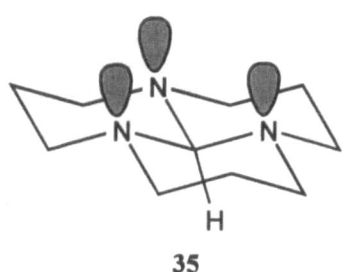

35

5.2 Other Related Effects

5.2.1 Reverse Anomeric Effect

The direction of the anomeric effect is reversed when the partial charge on a given substituent is to be positive instead of a negative (pyridinium type). The effect is known as reverse anomeric effect. The conformer with equatorial orientation is preferred (Fig. 34). The reverse anomeric effect is attributed to the attractive interaction of the dipoles which contribute to the conformational energy of the substituents, thus results in predominance of the equatorial conformation.

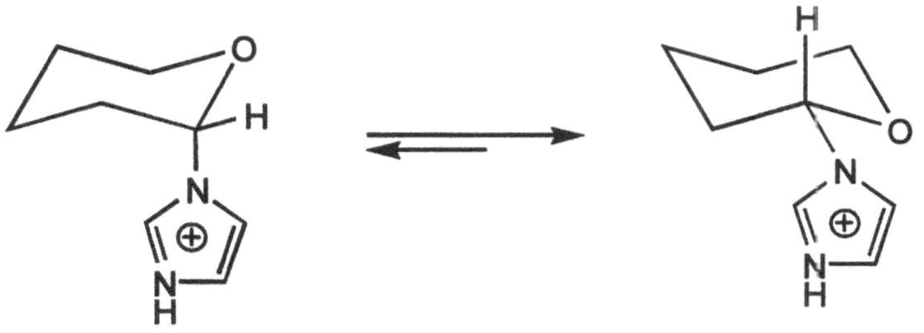

Fig. 34. Reverse anomeric effect

5.2.2 Double Anomeric Effect

1,3-Dioxanes substituted with an electronegative substituent at the position-2 **36** exhibit double anomeric effect. The conformer with the substituent in an axial orientation is preferred over the equatorial conformer. However, the synaxial interactions of an axial substituent with the axial hydrogen atoms at C_4 and C_6 exist due to the shortened carbon-oxygen bond by the anomeric effect.

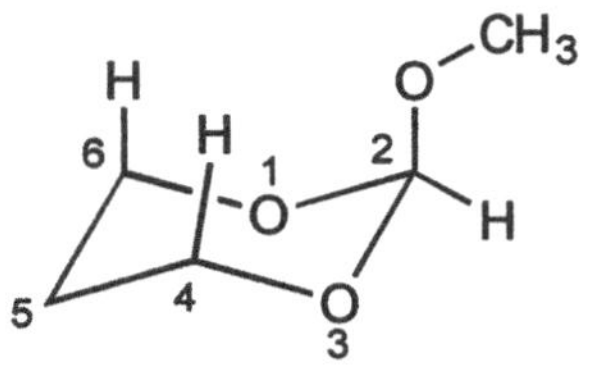

36

5.2.3 Rabbit-Ear Effect (Lone Pair-Lone Pair Interactions)

This effect involves the interaction of endocyclic heteroatoms in 1,3-position with each other. In the six-membered rings containing two heteroatoms in 1,3-positions, the conformation in which two lone pairs on two heteroatoms are synaxial is destabilized by the lone pair-lone pair repulsion (Fig. 35). This destabilizing effect of 1,3-synaxial lone pairs is known as rabbit-ear effect[53]. Thus, the most stable conformation will be with the least number of rabbit-ear effects.

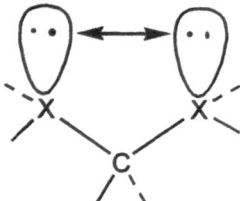

Fig. 35. 1,3-Synaxial lone pair-lone pair interactions (Rabbit-ear effect)

The preference of an axial orientation **37b** over an equatorial orientation **37a** of the substituent is attributed to the destabilizing effect of 1,3-synaxial lone pair-lone pair repulsion. The preferred axial orientation of an alkoxy substituent on C–2 in 2-alkoxy-1,3-dioxane can be explained similarly on the basis of rabbit-ear effect.

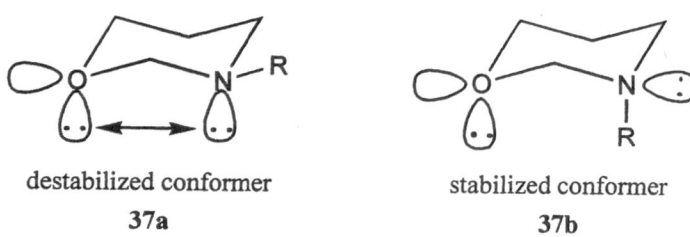

destabilized conformer   stabilized conformer

37a   **37b**

5.2.4 Repulsive-Gauche Effect (Hockey-Sticks Effect)

In 1,4-dioxanes, the axial orientation of an electronegative substituent (polar group) on the α-carbon **38** is preferred over the equatorial orientation, although to a lesser extent than in 1,3-dioxanes. This is attributed to the anomeric effect resulting from the interaction between endocyclic oxygen and exocyclic polar group.

38

In 1,4-oxathianes, the conformer with an equatorial orientation of the electro-negative substituent (polar group) on the α-carbon atom is much more preferred than that with an axial orientation. The destabilization of the conformer with axial orientation of substituent **39** is ascribed to the effect known as repulsive-gauche effect (hockey-sticks effect) (Fig. 36). The repulsive-gauche effect is contributed by the repulsive interactions between *p*-orbitals of the electronegative substituent and the lone pair orbital on the heteroatom, which are absent when interacting orbitals are antiparallel. The magnitude of the effect is increased with increasing the size of heteroatom and thus the effect in 1,4-oxathianes is much more pronounced than in 1,4-dioxanes because of the larger atomic size of sulfur.

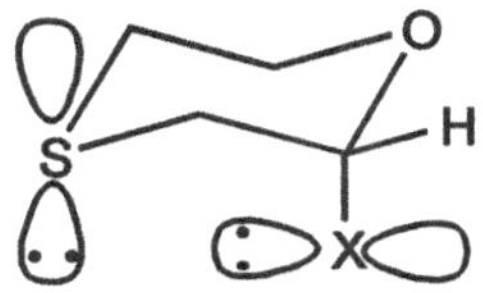

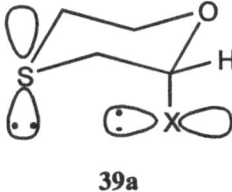

39

Fig. 36. Repulsive gauche effect (hockey-sticks effect) in 1,4-oxathianes

Thus, the molecule exhibiting hockey-sticks effect is forced to adopt the conformation with an equatorial orientation of the substituent to minimize the repulsive interactions (lone pair-lone pair interaction) (Fig. 37).

39a

Fig. 37. Preferred conformation in 1,4-oxathianes

5.3 Attractive Interactions 'Through Space'

5.3.1 Intramolecular Hydrogen Bonding

In certain cases, the ring heteroatoms are involved in the intramolecular hydrogen bonding formation with OH and NH_2 groups present in the heterocyclic molecule and affect the conformational preferences of the heterocycles.

5-Hydroxy-1,3-dioxane **40** exists in the preferred conformation with an axial hydroxyl group to form hydrogen bonding with the ring oxygen atom (Fig. 38).

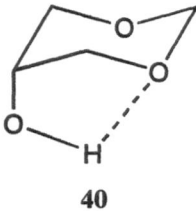

40

Fig. 38. Hydrogen bonding in 5-hydroxy-1,3-dioxane

3-Hydroxypiperidines **41** also exist preferantially in the conformation with an axial orientation of hydroxyl group (Fig. 39).

41a   **41b**

Fig. 39. Hydrogen bonding in 3-hydroxypiperidines

5.3.2 Intramolecular Nucleophilic-Electrophilic Attractive Interactions

The attractive interactions between nucleophilic centre and electrophilic centre within the same molecule affect conformational preferences in the heterocycles. 3-Phenyl-3-benzoyltropane **42** involves interactions between nucleophilic centre and electrophilic centre (transannular interactions) which force the molecule to adopt the boat conformation.

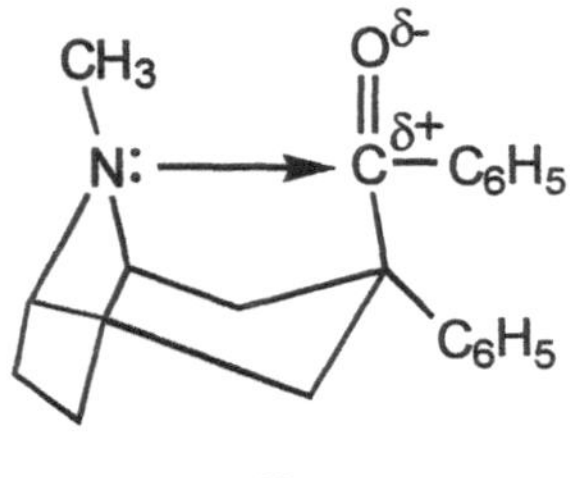

42

The comparison of IR and UV spectra of 3-phenyl-3-benzoyltropane **42** with 3-benzoyltropane **43** shows the normal bands of the carbonyl group in 3-benzoyltropane and not in 3-phenyl-3-benzoyltropane. This is considered to be a consequence of the flipping of the piperidine into the boat form and of the transannular interactions of the nucleophilic nitrogen with an electrophilic carbon of the carbonyl group, whereas such interactions are not present in 3-benzoyltropane[54].

43

REFERENCES

1. A. Greenberg and J. F. Liebman, *Strained Organic Molecules*, Academic Press, New York, 1978; J. F. Liebman and A. Greenberg, *Chem. Rev.* **76**, 311 (1976).

2. K. Pihlaja and E. Taskinen in A. R. Katritzky (Ed.), *Physical Methods in Heterocyclic chemistry* Vol. **6**, Academic Press, New York, 1974.

3. M. Charton in J. Zabicky (Ed.), *The Chemistry of Alkenes* Vol. **2**, Wiley-Interscience, New York, 1970, pp. 511–610.

4. M. J. S. Dewar and G. P. Ford, *J. Am. Chem. Soc.* **101**, 783 (1979)

5. A. Wilson and D. Goldhamer, *J. Chem. Edu.* **40**, 504 (1963).

6. A. R. Katritzky and A. P. Ambler in A. R. Katritzky (Ed.), *Physical Methods in Heterocyclic Chemistry* Vol. **2**, Academic Press, New York, 1963, pp. 161.

7. A. R. Katritzky and P. J. Taylor in A. R. Katritzky (Ed.), *Physical Methods in Heterocyclic Chemistry* Vol. **4**, Academic Press, New York, 1971, pp. 265.

8. C. V. Alsenoy, H. P. Figelys and P. Geerlings, *Theor. Chim. Acta* **55**, 87 (1980)

9. F. W. Fowler and A. Hassner, *J. Am. Chem. Soc.* **90**, 2875 (1968).

10. A. de Meijere, *Angew. Chem. Int. Edn. Engl.* **18**, 809 (1979).

11. N. H. Cromwell, R. E. Bambury and J. L. Adelfang, *J. Am. Chem. Soc.* **82**, 4241 (1960).

12. J. M. Lehn, *Fortschr. Chem. Forsch.* **15**, 312 (1970).

13. F. A. l. Anet and J. M. Osyany, *J. Am. Chem. Soc.* **89**, 352 (1967).

14. S. J. Brois, *J. Am. Chem. Soc.* **90**, 506 (1968).

15. J. B. Lambert, *Top. Stereochem.* **6**, 52 (1971).

16. N. J. Leonard and J. C. Coll, *J. Am. Chem. Soc.* **92**, 6685 (1970).

17. J. C. Coll, D. R. Crist, M. del C. G. Barrio and N. J. Leonard, *J. Am. Chem. Soc.* **94**, 7092 (1972).

18. R. W. Alder and R. J. Arrowsmith, *J. Chem. Res. (S)*, 163 (1980).

19. F. G. Riddell, *The Conformational Analysis of Heterocyclic Compounds*, Academic Press, London, 1980.

20. S. Wolfe, *Accounts Chem. Res.* **5**, 102 (1972)

21. L. M. Trefonas and R. J. Majeste in R. R. Gupta (Ed.), *Physical Methods in Heterocyclic Chemistry*, Wiley-Interscience, New York, 1984, pp. 314.

22. C. A. Renner and F. D. Greene, *J. Org. Chem.* **41**, 2813 (1976).

23. W. L. F. Armarego, *Stereochemistry of Heterocyclic Compounds* Part-I, Wiley-Interscience, New York, 1977.

24. G. Kobrich, *Angew, Chem. Int. Edn. Engl.* **12**, 464 (1973).

25. K. B. Becker and C. A. Gabutti, *Tetrahedron Lett.* 1883 (1982).

26. T. Sasaki, S. Eguchi, T. Okano and Y. Wakata, *J. Org. Chem.* **48**, 4067 (1983).

27. R. M. Sweet and L. F. Dahl, *J. Am. Chem. Soc.* **92**, 5489 (1970).

28. E. H. Flynn, *Cephalosporins and Penicillins in Chemistry and Biology*, Academic Press, New York, 1972.

29. F. G. Riddell, *Quart. Rev.* **21**, 364 (1967).

30. E. L. Eliel, *Accounts Chem. Res.* **3**, 1 (1970).

31. J. B. Lambert and S. I. Featherman, *Chem. Rev.* **75**, 611 (1975).

32. W. J. Adams, H. J. Geise and L. S. Bartell, *J. Am. Chem. Soc.* **92**, 5013 (1970).

33. D. Vremer and J. A. Pople, *J. Am. Chem. Soc.* **97**, 1358 (1975); E. Diez, A. L. Esteban and M. Rico, *J. Magn. Reson.* **16**, 136 (1974); G. G. Engerholm, A. C. Luntz, W. D. Gwinn and D. O. Harris, *J. Chem. Phys.* **50**, 2446 (1969).

34. J. P. Mc Cullough, *J. Chem. Phys.* **29**, 966 (1958); A. Kh. Mamleev and N. M. Pozdeev, *Zh. Strukt. Khim.* **20**, 1114 (1979); A. L. Esteban and E. Diez, *Can. J. Chem.* **58**, 2340 (1980); G. Barberella and P. Dembech, *Org. Magn. Reson.* **13**, 282 (1980).

35. J. R. Durig, Y. S. Li and C. K. Tong, *J. Chem. Phys.* **56**, 5692 (1972).

36. J. R. Durig, R. O. Carter and L. A. Carriere, *J. Chem. Phys.* **59**, 2249 (1973).

37. J. B. Lambert, *Accounts Chem. Res.* **4**, 87 (1971).

38. J. B. Lambert, *J. Am. Chem. Soc.* **89**, 1836 (1967).

39. H. R. Buys, *Recl. Trav. Chim. Pays-Bas.* **88**, 1003 (1969).

40. A. Rauk, L. C. Allen and K. Mislow, *Angew. Chem. Int. Edn.* **9**, 400 (1970); J. B. Lambert, *Top. Stereochem.* **6**, 19 (1971); W. L. F Armarego, *Stereochemistry of Heterocyclic Compounds* Part-I, Wiley-Interscience, New York, 1977, pp. 157.

41. S. F. Nelson, *Accounts Chem. Res.* **11**, 14 (1978).

42. A. R. Katritzky, R. C. Patel and F. G. Riddell, *Angew. Chem. Int. End. Engl.* **20**, 521 (1981).

43. E. L. Eliel, *Angew. Chem. Int. Edn. Engl.* **11**, 739 (1972).

44. M. J. Aroney, C. Y. Chen, R. J. W. Le Fevre and J. D. Sexby, *J. Chem. Soc.* 4269 (1964).

45. R. J. Bishop, L. E. Sutton, D. Dineen, R. A. Y. Jones and A. R. Katritzky, *Proc. Chem. Soc.* 257 (1964).

46. N. L. Allinger, J. G. D. Carpenter and F. M. Karkowski, *Tetrahedron Lett.* 3345 (1964); N. L. Allinger, J. G. D. Carpenter and F. M. Karkowski, *J. Am. Chem. Soc.* **87**, 1232 (1965).

47. J. B. Lambert and R. G. Keske, *J. Am. Chem. Soc.* **88**, 620 (1966); J. B. Lambert, R. G. Keske, R. E. Carhart, and A. P. Jovanovich, *J. Am. Chem. Soc.* **89**, 3761 (1967).

48. A. J. Kirby, *The Anomeric Effect and Related Stereoelectronic Effects at Oxygen*, Springer-Verlag, Heidelberg, 1983.

49. W. A. Szarek and D. Horton (Eds.), *The Anomeric Effect : Origin and Consequences*, ACS Symposium Series No. 87, Am. Chem. Soc., Washington, D.C., 1979.

50. J. Kirby and R. J. Martin, *J. Chem. Soc. Chem. Commun.* 803 (1978).

51. T. J. Atkins, *J. Am. Chem. Soc.* **102**, 6364 (1980).

52. J. M. Erhardt, E. R. Grover and J. D. Wuest, *J. Am. Chem. Soc.* **102**, 6365 (1980).

53. R. O. Hutchins, L. D. Kopp and E. L. Eliel, *J. Am. Chem. Soc.* **90**, 7174 (1968).

54. T. H. Chan and R. K. Hill, *J. Org. Chem.* **35**, 10 and 3519 (1970).

CHAPTER **5**

HETEROCYCLIC SYNTHESIS

CONTENTS

1 GENERAL

General methods of heterocyclic synthesis involve ring formation and the choice of method for the synthesis of a particular heterocycle depends on:

   (i)   size of the ring being synthesized,
   (ii) degree of unsaturation required and
   (iii) pattern of substituents required.

The nature of the ring closure reaction is mostly dependent upon the size of the ring being synthesized and its degree of unsaturation than upon the type(s) of heteroatom(s) present. A large number of methods of wide variety are available for the synthesis of heterocyclic compounds. Here an attempt is made to give a few general guidelines (for most common ring sizes and heteroatoms).

   (i)   In the synthesis of a monocyclic compound, the ring closure step often involves the formation of carbon–heteroatom bond.

   (ii) If the system contains two adjacent heteroatoms, it is unusual for the ring closure step to involve heteroatom–heteroatom bond formation (except in the cases where nitroso, nitro, nitrene or diazonium group acts as electrophilic moiety).

   (iii) A bicyclic ring system is formed almost invariably by the annelation of second ring on a monocyclic compound.

2 TYPES OF REACTIONS

The reactions involved in the ring synthesis can be broadly classified into two groups. In the first group of reactions, a single bond is formed in the ring closure step. Such reactions are known as cyclization reactions. The second group of the reactions involves the simultaneous formation of two bonds between usually two different molecules without elimination of small molecules. Such reactions are called cycloaddition reactions. However, in practice there is not always a clear-cut distinction between the two groups. This simple classification is followed here.

2.1 Cyclization Reactions

In the cyclization reactions, n-membered ring is formed by the cyclization of a chain of 'n' atoms. Cyclization reactions i.e. ring closure reactions comprise intramolecular variants of the reactions, the most common of which involve the interaction between electrophilic and nucleophilic centres.

2.1.1 Nucleophile-Electrophile Cyclization

Such kinds of reactions can occur either between a binucleophile and a bielectrophile or between two molecules each containing both nucleophilic and electrophilic centres. The various types of possible interactions are shown in Fig.1.

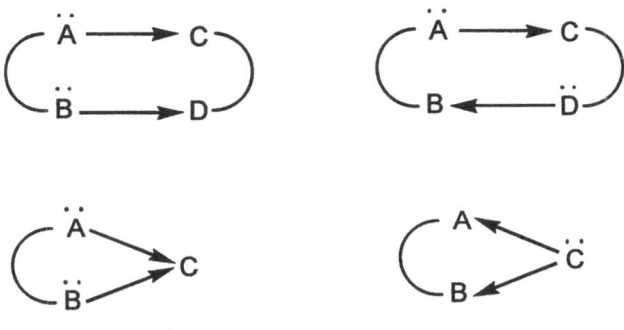

Fig. 1

Heterocyclic compounds of different ring sizes can be constructed by such interactions. Baldwin has suggested a systematic method for classifying the various types of nucleophilic-electrophilic cyclization processes[1]. One of the important factor which determines the feasibility of a ring closure is the geometry of approach of the nucleophile to the electrophile in the transition state. The geometry of these transition states has led to the formulation of Baldwin rules for ring closure[1]. If a transition state is obtained with a serious distortion of the normal bond angles or distances, it follows that the ring closure will occur with difficulty (or not at all) and such processes are called 'disfavoured'. A 'favoured' process occurs more readily than one which is 'disfavoured'. However, it does not mean that a 'favoured' process will necessarily occur readily in every case. It depends on the other factors also in addition to the angle strain (i.e. distortion of the normal bond angles), such as unfavourable steric interactions and distance factor. The efficiency of a cyclization reaction also depends upon the free energy of activation ΔG^{++}, which consists of enthalpy (ΔH^{++}) and entropy (T ΔS^{++}) terms (Eq. 1) :

$$\Delta G^{++} = \Delta H^{++} + T\Delta S^{++} \quad .....(1)$$

where T is the absolute temperature.

A nucleophilic attack at a tetrahedral carbon occurs at the side opposite to that occupied by the leaving group and the resulting transition state has an N–C–Y angle of 180° (scheme-1). The attack of a nucleophile on a trigonal carbon (as in

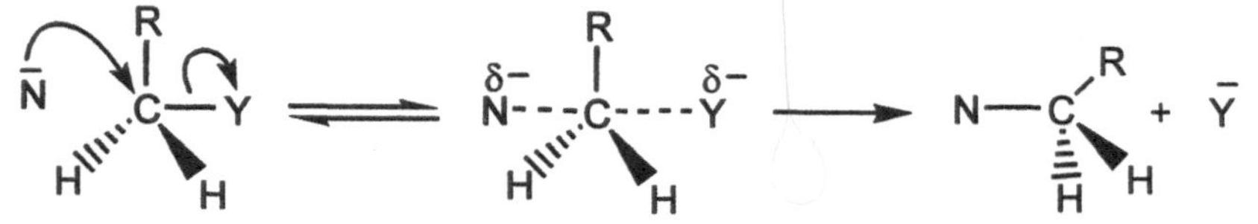

Scheme-1

carbonyl carbon) takes place preferentially from above or below the plane of the molecule at an angle of 109° to C=Y bond (scheme-2). The reactions in which the

Scheme-2

electrophilic carbon is diagonal (as in alkyne/cyano group), the approach of the nucleophile is at an angle of 60° to C≡Y bond (scheme-3).

Scheme-3

Baldwin's classification of cyclization i.e. ring closure reactions is based on three considerations:

(i)   size of the ring being synthesized,

(ii) whether the atom or group Y is a part of the ring system (endo) or lies outside the ring being synthesized (exo) (scheme-4) and

(iii) geometry of the electrophilic carbon (tetrahedral, trigonal or diagonal).

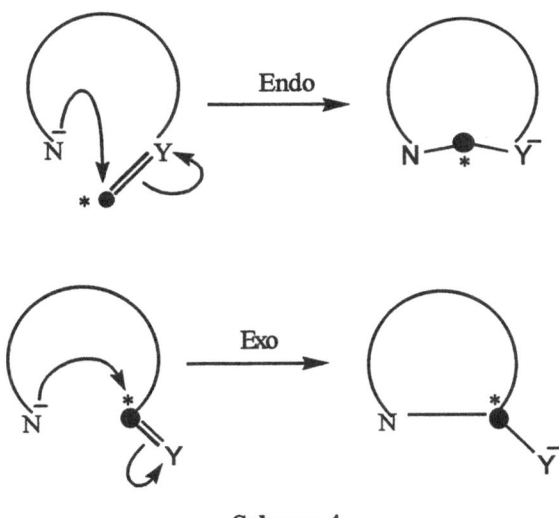

Scheme-4

Thus, a reaction of the type depicted in scheme-5 can be classified as 6-Exo-Tet (6-membered ring, Y outside the ring being formed, tetrahedral carbon undergonig substitution).

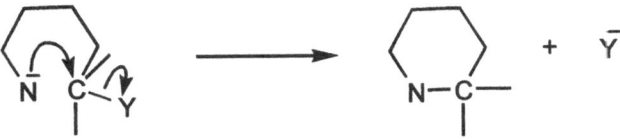

Scheme-5

Baldwin's rules for the cyclization reactions involving the nucleophilic atom of a first row element e.g. C, N or O are summarized as :

Rule No. 1 (for tetrahedral systems): (scheme-6)

      3 to 7-Exo-Tet are all favoured.

      5 to 6-Endo-Tet are disfavoured.

Rule No. 2 (for trigonal systems): (scheme-7)

      3 to 7-Exo-Trig are favoured.

      3 to 5-Endo-Trig are disfavoured.

      6 to 7-Endo-Trig are favoured.

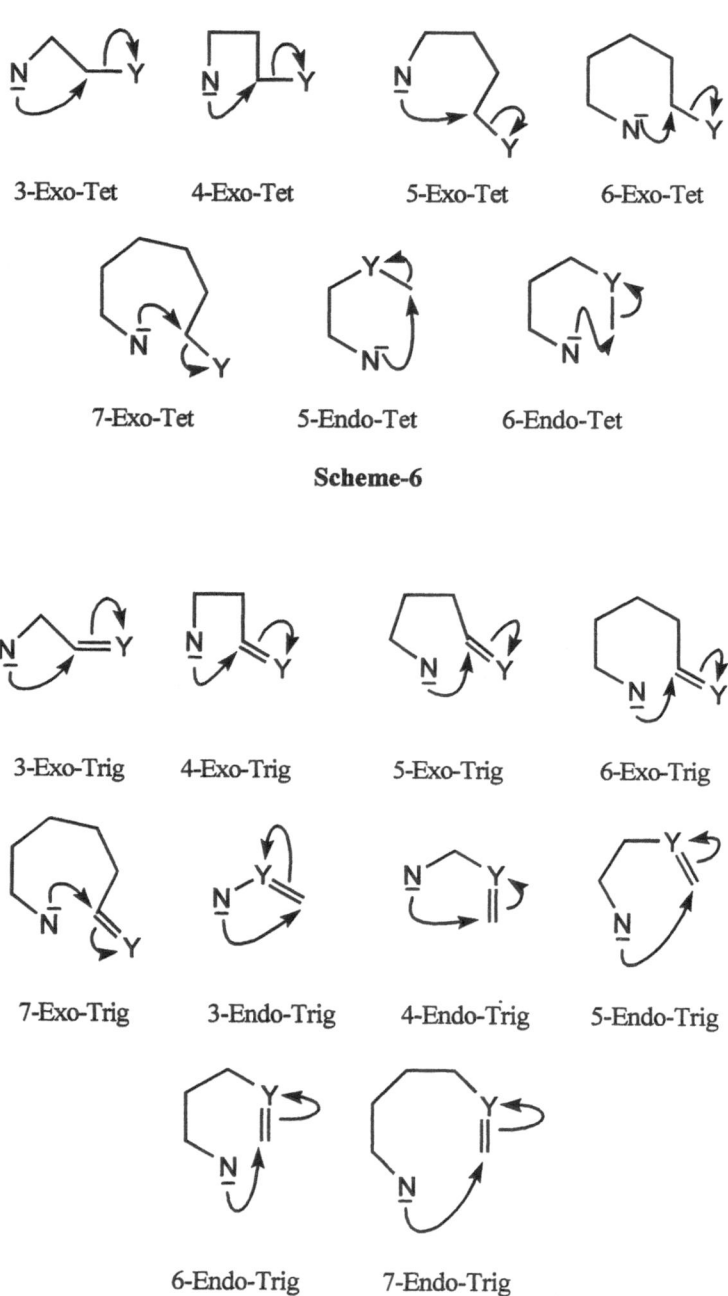

Scheme-6

Scheme-7

Rule No. 3 (for diagonal systems): (scheme-8)

    3 to 4-Exo-Dig are disfavoured.
    5 to 7-Exo-Dig are favoured.
    3 to 7-Endo-Dig are favoured.

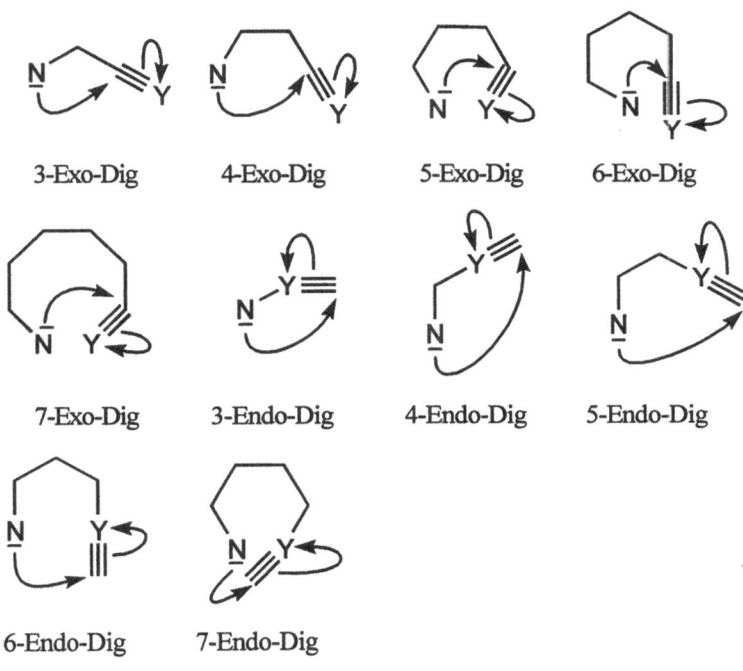

| 3-Exo-Dig | 4-Exo-Dig | 5-Exo-Dig | 6-Exo-Dig |

| 7-Exo-Dig | 3-Endo-Dig | 4-Endo-Dig | 5-Endo-Dig |

| 6-Endo-Dig | 7-Endo-Dig |

Scheme-8

An intramolecular nucleophilic substitution reaction is an exo-tet process. Nucleophilic addition and addition-elimination reactions of the carbonyl compounds are exo-trig processes. In addition to these, heterocycles can also be prepared by the intramolecular radical cyclization, carbene and nitrene cyclizations and by electrocyclization of the conjugated π-electron systems, all of which are included in the succeeding sections.

2.1.1.1 Intramolecular Nucleophilic Substitution

Synthesis of heterocyclic compounds by intramolecular nucleophilic displacement at a saturated carbon atom is an exo-tet process. Examples of such intramolecular cyclization reactions include the synthesis of oxiranes (epoxides), aziridines, lactones, furans, azetidines, *etc.* The convenient formation of the five-membered rings are due to the best balance between enthalpy and entropy forms.

Three-membered heterocycles can be conveniently prepared by intramolecular cyclization reactions, which occur by the back-side attack of heteroatom at the carbon atom bearing the leaving group with inversion of configuration (scheme-9). An important method of synthesizing an oxirane ring involves various

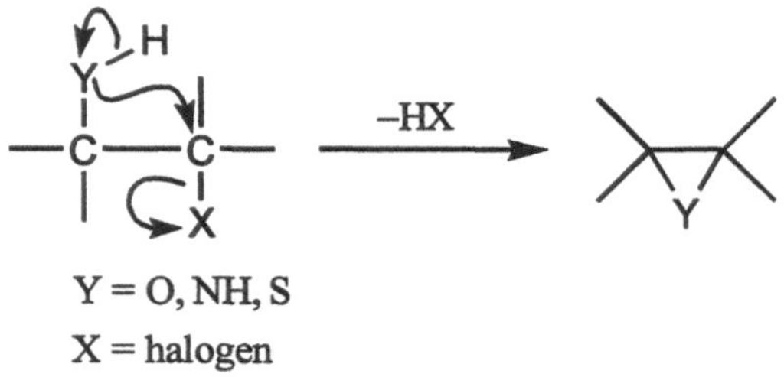

$Y = O, NH, S$

$X = $ halogen

Scheme-9

ring closure reactions. For example, a *trans*-chlorohydrin (obtained by the addition of hypochlorous acid to an alkene) undergoes a bimolecular dehydrohalogenation in the presence of a base resulting in cyclization to an oxirane ring (scheme-10)[2].

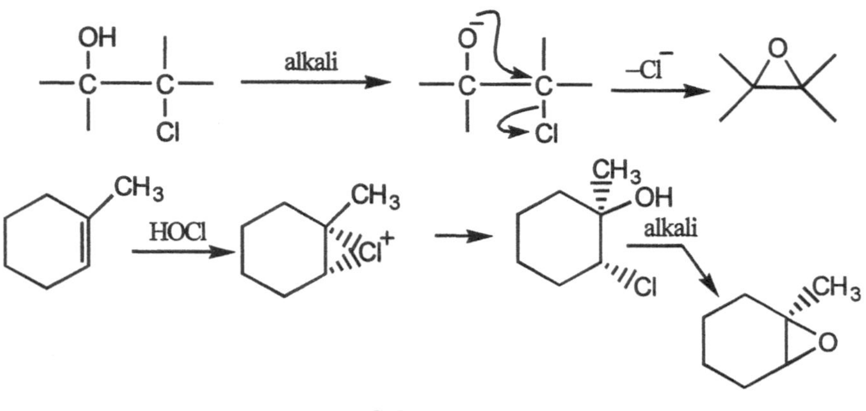

Scheme-10

The kinetic and products study of the basic decomposition of a series of methyl substituted 1,3-halohydrins in aqueous methanol shows that the fragmentation reaction competes with the ring closure[3] and the cyclic products are formed in poor yields (scheme-11).

The cyclization of *erythro*- and *threo*-3-bromo-2-hydroxybutanes yields *trans*- and *cis*-epoxides, respectively (scheme-12)[4]. The rate of ring closure is favoured by the presence of substituents at C-2, e.g., 1-chloro-2-methyl-2-propanol cyclizes eleven times faster than 1-chloro-2-propanol. Epoxide formation occurs several thousand times faster in diaxial **3b** than in diequatorial isomer **3a** (scheme-13)[5].

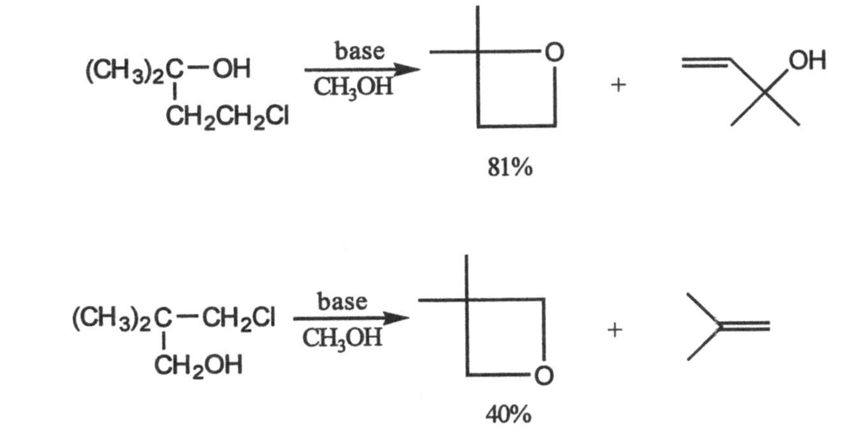

Scheme-11

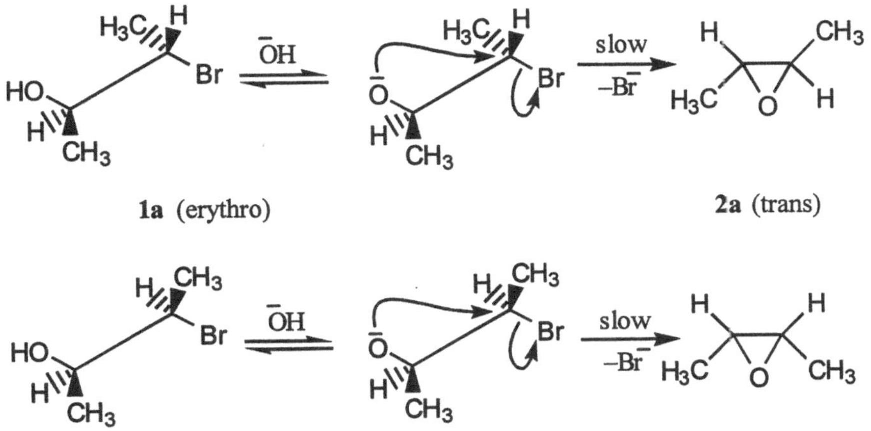

1a (erythro)   2a (trans)

1b (thero)   2b (cis)

Scheme-12

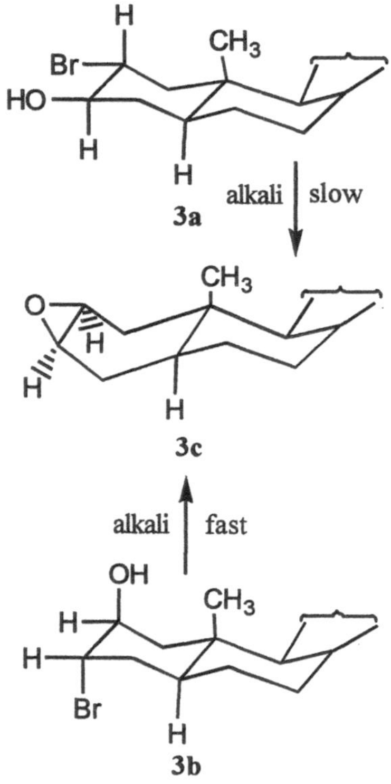

3a

alkali | slow

3c

alkali | fast

3b

Scheme-13

Another significant ring closure reaction is Darzens condensation[6]. It yields an intermediate **4b** by the reaction of a carbonyl compound with α-halo ketones or α-halo esters **4a** under the basic conditions which cyclizes by an internal nucleophilic displacement process (scheme-14). Like oxiranes, thiiranes **5b** have been prepared

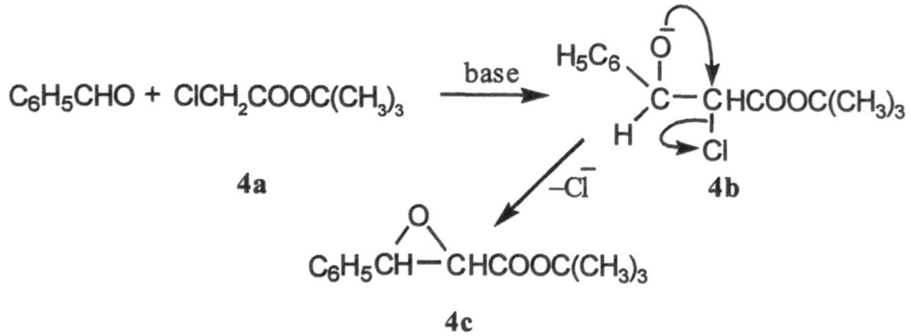

$C_6H_5CHO + ClCH_2COOC(CH_3)_3$

4a

base

H_5C_6 ... $C-CHCOOC(CH_3)_3$

4b

$-Cl^-$

$C_6H_5CH-CHCOOC(CH_3)_3$

4c

Scheme-14

by the cyclization of 2-halo mercaptans **5a** with an alkali (scheme-15)[7]. The

Scheme-15

cyclization of *trans*-2-chlorocyclohexanethiol **6a** in the presence of sodium bicarbonate provides cyclohexene sulfide **6b** (scheme-16)[8].

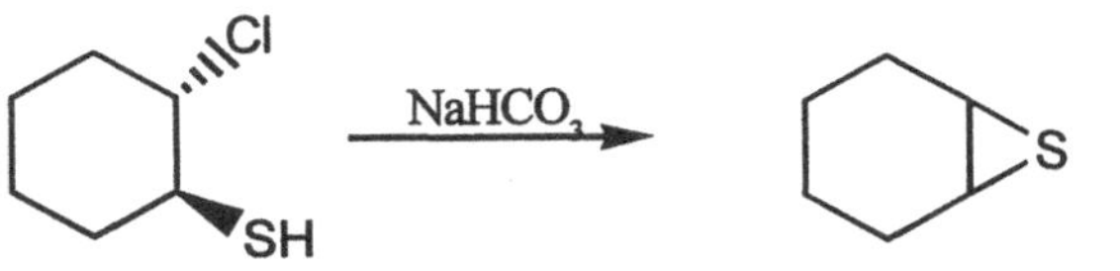

6a   **6b**

Scheme-16

The best preparative methods for the aziridines involve cyclization reactions. One of these is Gabriel ring closure method[9] which involves an intramolecular displacement of halogen atom by a free amino group. The cyclization is stereospecific and occurs with the inversion at a carbon bearing the leaving group (scheme-17).

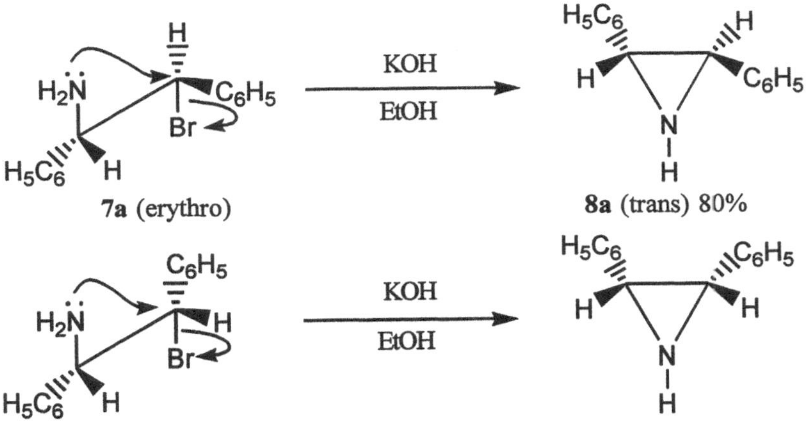

7a (erythro)   **8a** (trans) 80%

7b (thero)   **8b** (cis) 96%

Scheme-17

Wenker method also involves the conversion of β-amino alcohol **9a** to β-amino hydrogen sulfate **9b**, followed by cyclization in the presence of a base (scheme-18)[10].

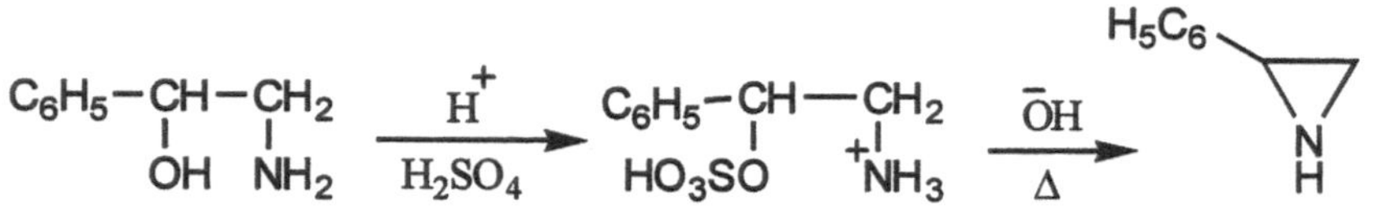

$$C_6H_5-\underset{\underset{\textbf{9a}}{OH}}{\overset{}{CH}}-\underset{NH_2}{\overset{}{CH_2}} \xrightarrow[H_2SO_4]{H^+} C_6H_5-\underset{\underset{\textbf{9b}}{HO_3SO}}{\overset{}{CH}}-\underset{\overset{+}{N}H_3}{\overset{}{CH_2}} \xrightarrow[\Delta]{\overline{OH}} \underset{\textbf{9c}}{\overset{H_5C_6}{\triangle}}$$

Scheme-18

Hassner synthesis involves stereospecific addition of iodoisocyanate to an alkene via *trans*-addition with the formation of iodoisocyanate **10c** which on treatment with methanol and subsequent cyclization under basic conditions yields an aziridine derivative **10f** (scheme-19)[11].

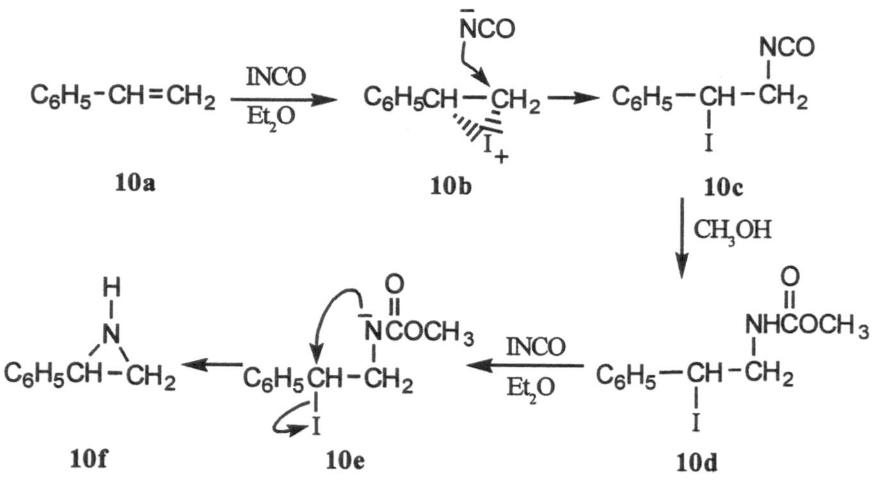

Scheme-19

α-Lactones **11b** have been prepared by the cyclization of ω-bromoalkanoic acids **11a** in the presence of a strong base (scheme-20)[12]. The rate constants for the lactonization in 99% dimethyl sulfoxide for ring sizes n = 3–16, 18 and 23 have shown that the formation of four-, five- and six-membered rings are especially favourable. α-Lactams **13a,b** are obtained by the cyclization of N-halo- or α-halo tert-butyl amides with a strong base (scheme-21)[13]. The cyclization of β-chlorohydroxomates **14a** yields β-lactams **14c** (scheme-22)[14].

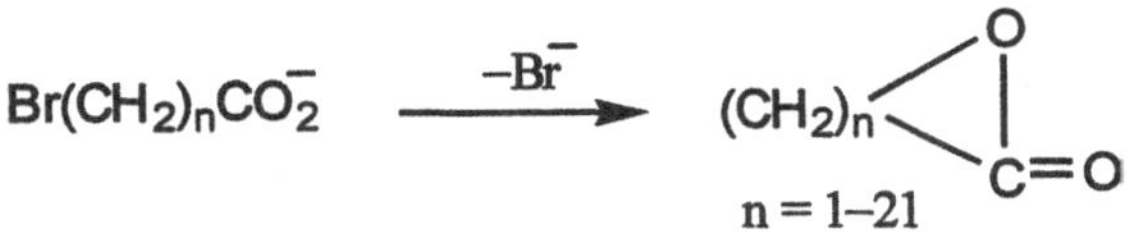

11a **11b**

Scheme-20

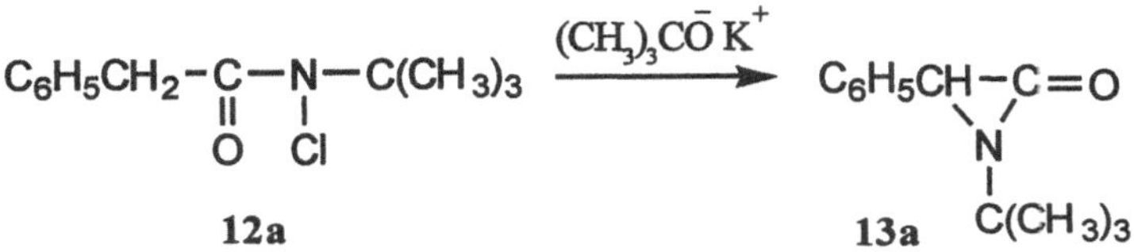

12a **13a**

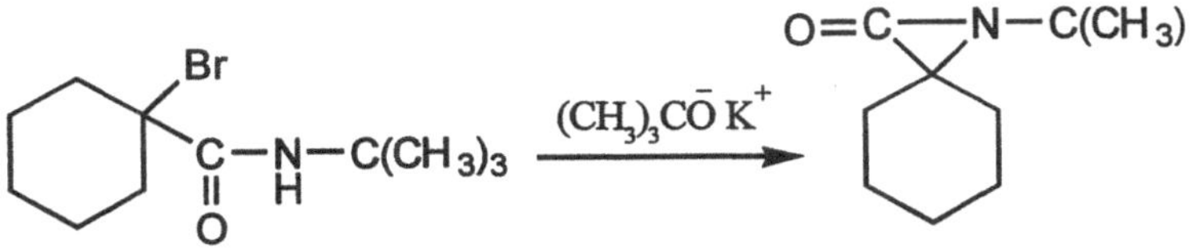

12b **13b**

Scheme-21

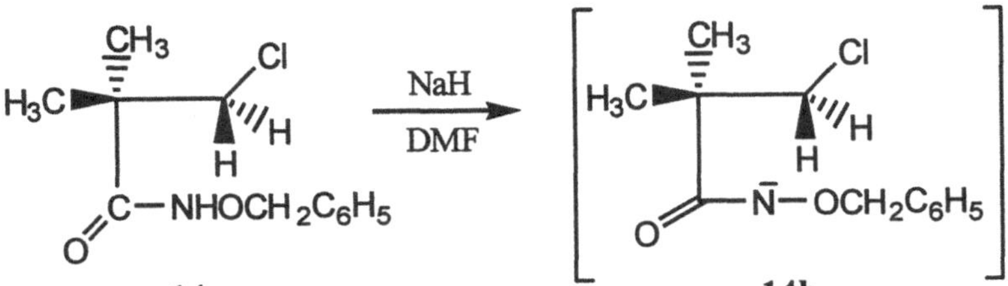

14a **14b**

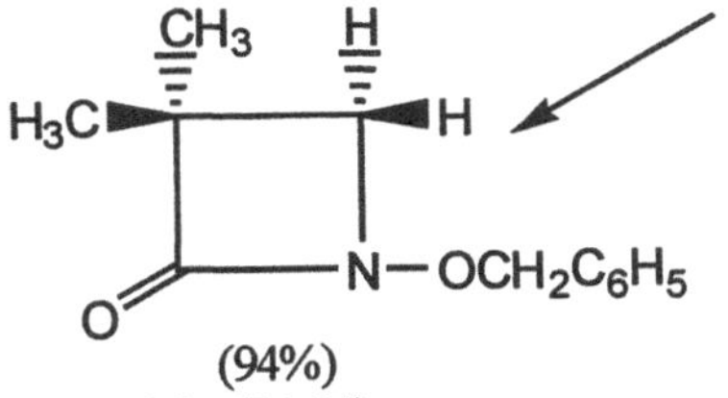

14c (94 %)

Scheme-22

The ring closure procedures have also been used to prepare furans. The reactions of β-keto esters and β-diketones with allenic salts **15a** produce furan derivatives in high yields[15]. The sulfonium group renders the 1,2-carbon–carbon double bond of allene extremely susceptible to nucleophilic addition and itself is readily displaced by the nucleophilic attack at the adjacent carbon atom (scheme-23).

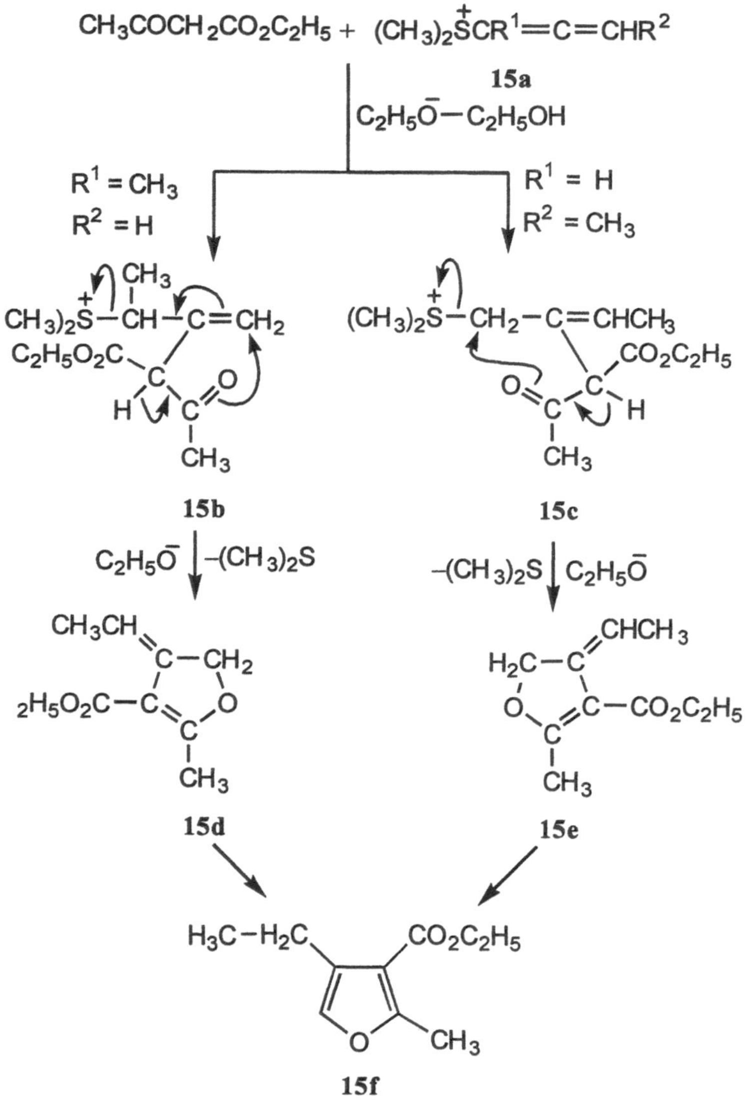

$$CH_3COCH_2CO_2C_2H_5 + (CH_3)_2\overset{+}{S}CR^1{=}C{=}CHR^2$$

15a

$$C_2H_5\overset{-}{O}-C_2H_5OH$$

$R^1 = CH_3$ $R^1 = H$
$R^2 = H$ $R^2 = CH_3$

15b   **15c**

$C_2H_5\overset{-}{O}$ ↓ $-(CH_3)_2S$ $-(CH_3)_2S$ ↓ $C_2H_5\overset{-}{O}$

15d   **15e**

15f

Scheme-23

Another cyclization reaction for the synthesis of furan derivatives is Fiest–Benary synthesis[16]. It consists in aldol condensation of α-halo ketone or α-halo aldehyde with β-keto ester or β-diketone in the presence of a base. The ester anion attacks the carbonyl group of α-halo ketone followed by the formation of an intermediate **16b** and cyclization by the intramolecular displacement of chloride ion with the loss of water molecule (scheme-24).

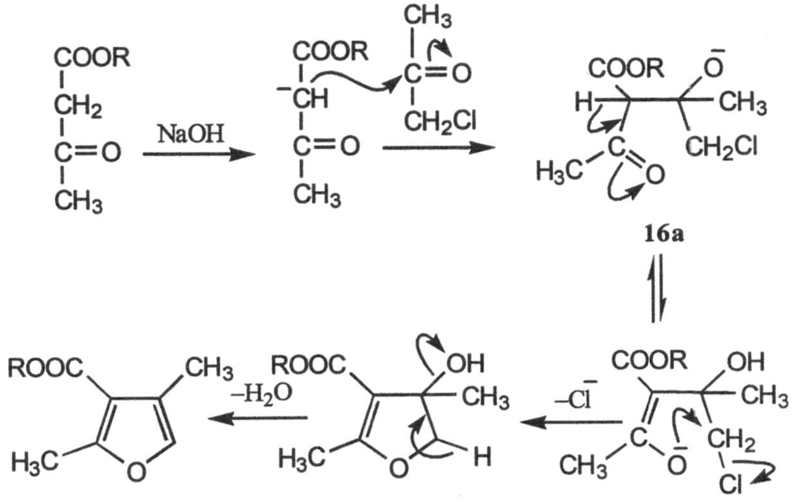

Scheme-24

2.1.1.2 Intramolecular Nucleophilic Addition to Double Bonds

The intramolecular nucleophilic attack at the carbonyl group of aldehydes, ketones, acid halides, esters, *etc.* comprises the most common method for the synthesis of some common heterocycles viz. pyrrole, thiophene, indole, quinoline, isoquinoline, *etc.* In some cases, cyclization involves the attack by nucleophilic carbon. The Madelung synthesis of indole involves the cyclic dehydration of N–acylated *o*-toluidine **17a** in the presence of a strong base at high temperature (scheme-25)[17]. Hinsberg thiophene synthesis involves the reaction between α-diketones and thiodiacetates **18b** (scheme-26)[18]. In Knorr pyrrole synthesis the cyclization step also involves the nucleophilic attack at the carbonyl carbon, followed by the loss of water molecule. It is an acid catalyzed reaction (scheme-27).

Some cyclization reactions occur through the nucleophilic heteroatoms e.g., Paal–Knorr pyrrole synthesis, Robinson–Gabriel oxazole synthesis, Reissert indole synthesis *etc.* Paal–Knorr method[19] for the synthesis of pyrrole (condensation of

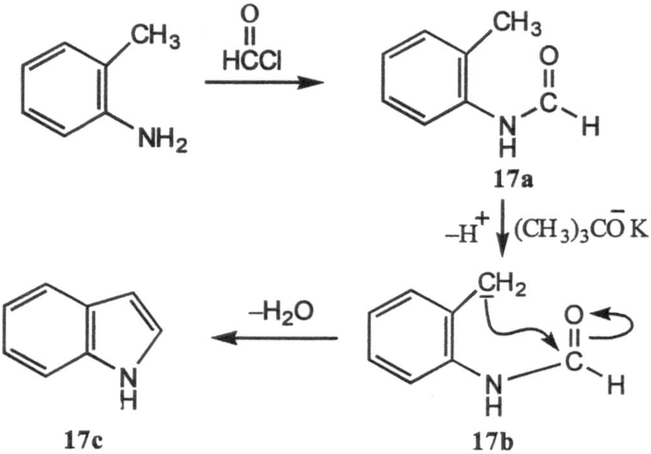

Scheme-25

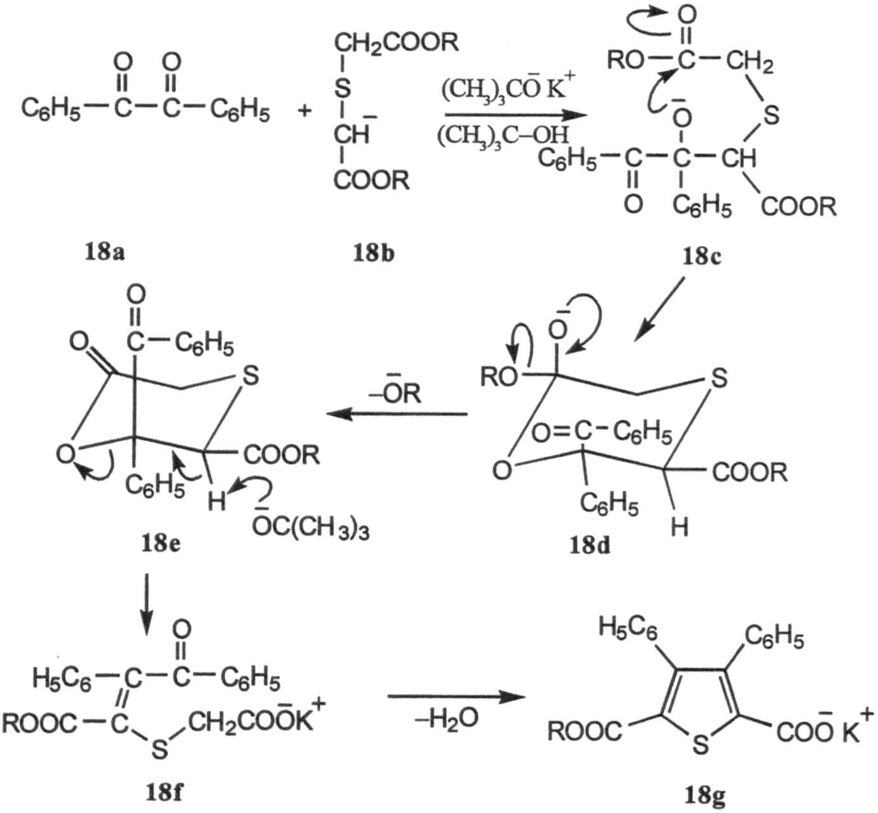

Scheme-26

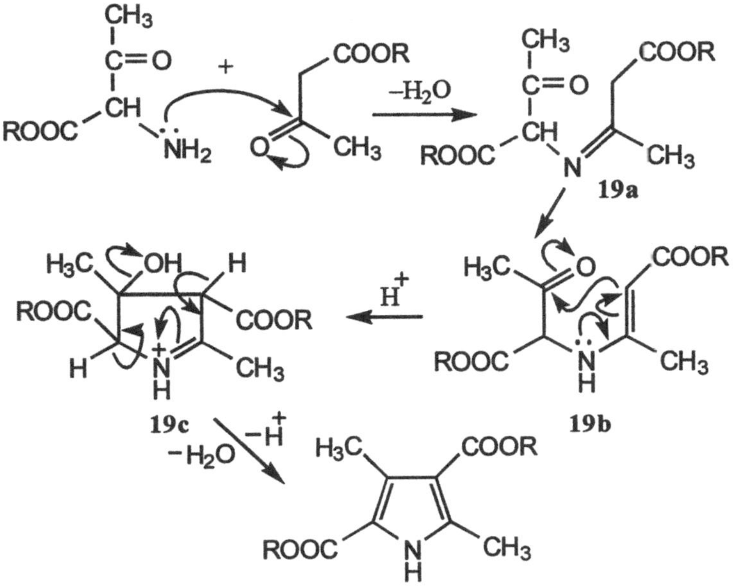

Scheme-27

1,4-diketone with ammonia or primary amine) involves the nucleophilic attack of an amino group at the carbonyl carbon followed by the cyclic dehydration (scheme-28).

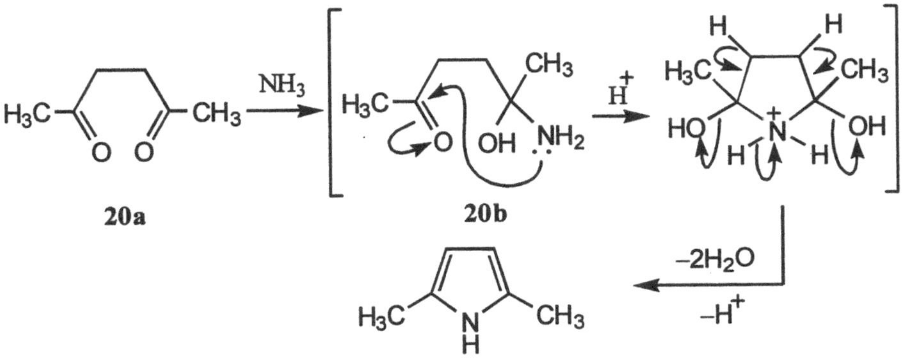

Scheme-28

In Reissert indole synthesis, *o*-nitrophenyl pyruvate **21a** obtained by the condensation of *o*-nitrotoluene with oxalic acid ester undergoes reductive cyclization involving nucleophilic attack at the carbonyl carbon to yield indole (scheme-29)[20]. Robinson–Gabriel oxazole synthesis presents an example where cyclization occurs by the nucleophilic attack at a carbonyl carbon by the acylamino group in α-acylamino ketones **22a**, followed by dehydration to yield oxazole **22c** (scheme-30)[21]. Synthesis of benzo-fused heterocycles involves the nucleophilic attack of an ortho carbon of the *o*-substituted benzenes on the carbonyl carbon. One such example is the Skraup synthesis[22] of quinoline (scheme-31). The vacant

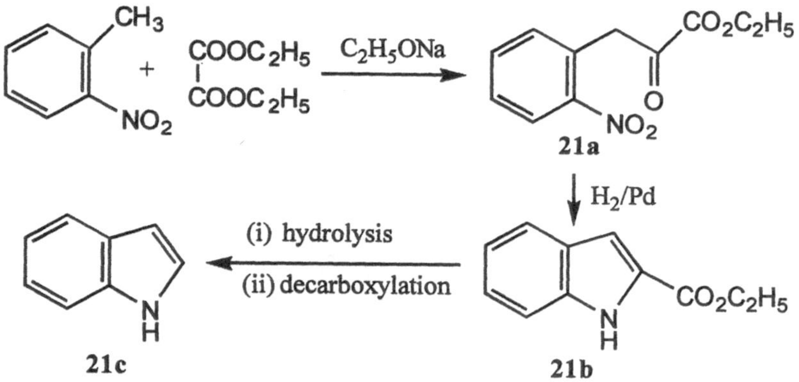

Scheme-29

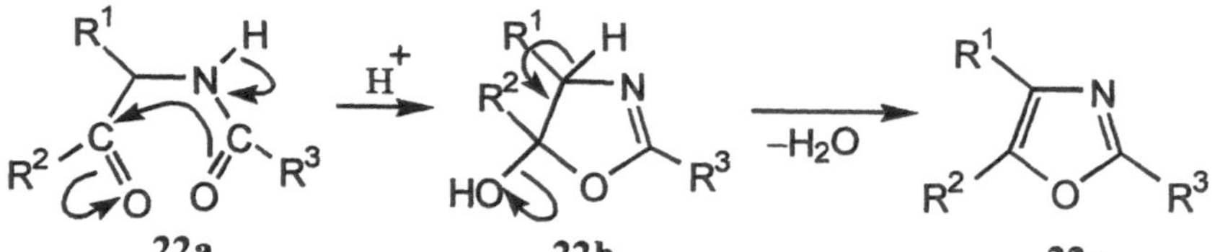

Scheme-30

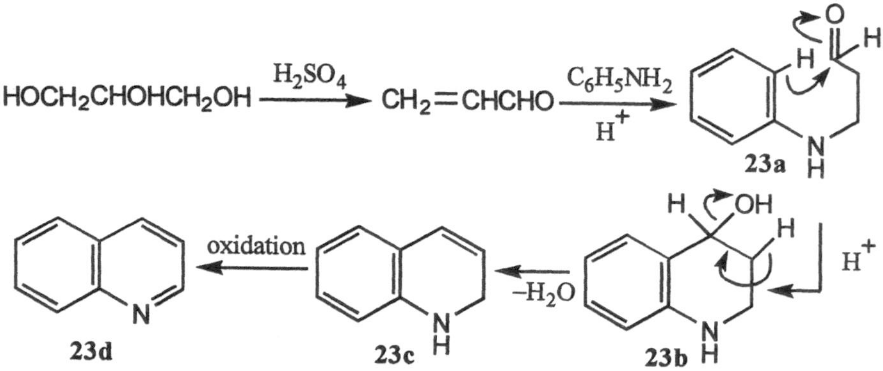

Scheme-31

ortho position of aniline derivative attacks the carbonyl carbon. The cyclization is followed by the dehydration and oxidation to yield quinoline **23d** . The cyclization of a Schiff base **24a** in the presence of an acid, followed by the dehydration yields quinoline derivative **24b** (scheme-32)[23].

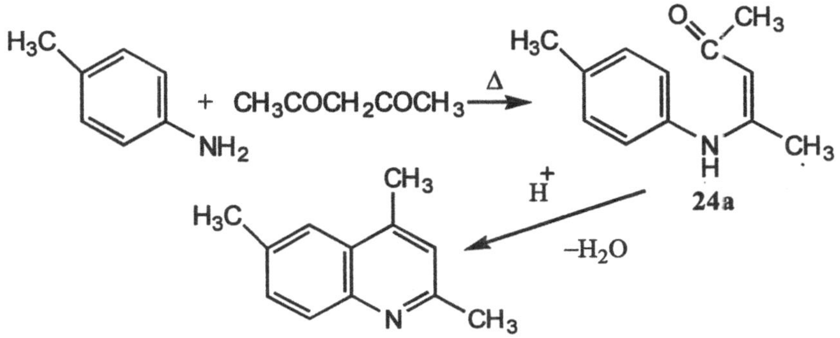

24b

Scheme-32

Bischler–Napieralski isoquinoline synthesis[24] involves the cyclodehydration of an acyl derivative of β-phenylethylamine **25a**. Cyclization occurs by the nucleophilic attack at the protonated carbonyl group. 3,4-Dihydroisoquinoline formed undergoes dehydrogenation to yield isoquinoline **25d** (scheme-33).

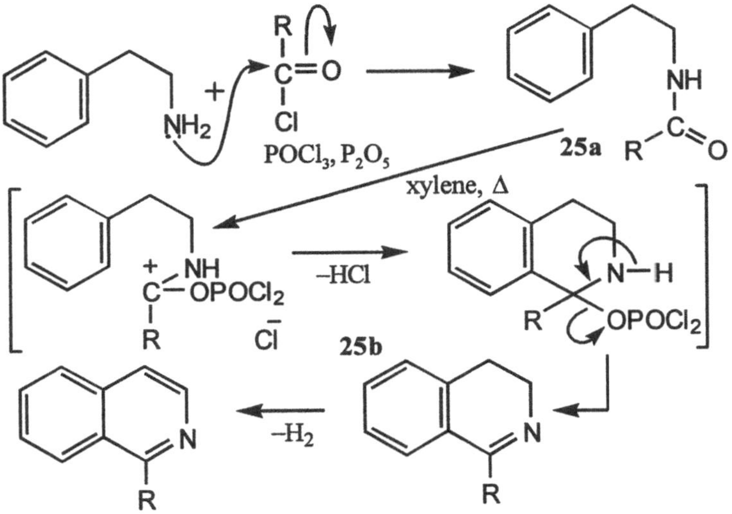

25d **25c** $R = CH_3/C_6H_5$

Scheme-33

Bischler synthesis[25] for indoles proceeds with the nucleophilic attack at the carbonyl group of an α-arylamino ketone or aldehyde **26a**. It is an acid catalyzed reaction (scheme-34).

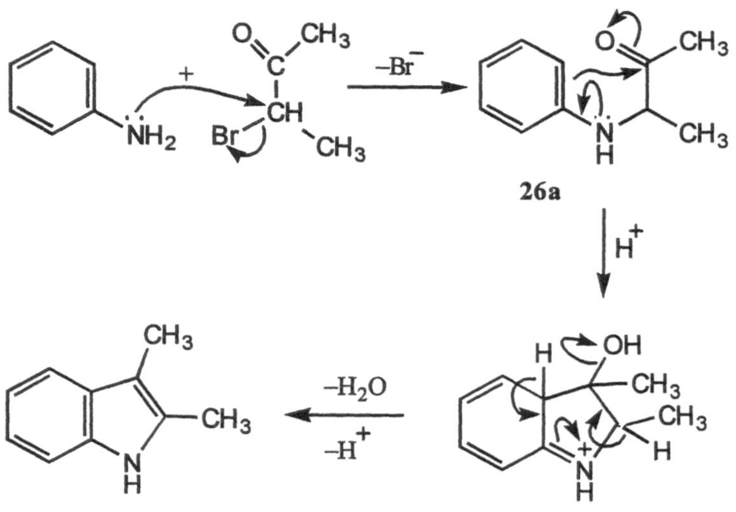

26a

26c

26b

Scheme-34

The construction of the heterocyclic compounds by the nucleophilic attack at the carbon–carbon double bond is the commonly used method. Organoselenium induced ring closure leads to the various cyclic systems often with a high degree of regio- and stereoselectivity[26]. The phenylseleno compounds **27c** formed in these cyclizations serve as excellent intermediates for the elaboration to a variety of products under mild conditions. Removal of the phenylseleno group oxidatively and reductively yields unsaturated and saturated systems respectively (scheme-35). The initial step of cyclization is presumed to be reversible electrophilic addition of the phenylselenonium ion to the double bond of the compound **27a** leading to the reactive intermediate **27b** which subsequently suffers intramolecular reaction providing phenylselenolactone **27c**. Based on a presumed SN[2] mode of capture of the selenonium ion and by analogy to the related halolactonization reaction, the stereochemistry of phenylselenolactones is assumed to be trans.

Organoselenium based techniques also provide sulfur herterocycles (scheme-36). This technique has also been extended to the macrolide synthesis (scheme-37).

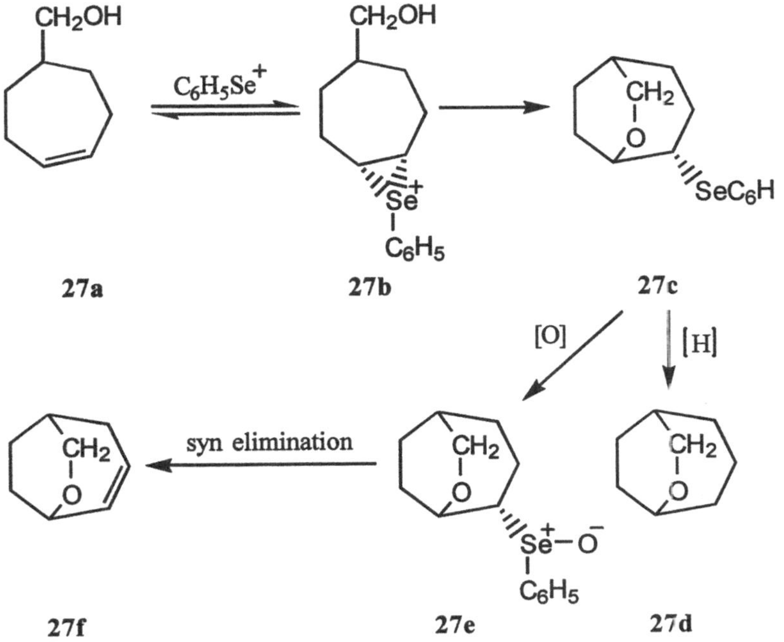

Scheme-35

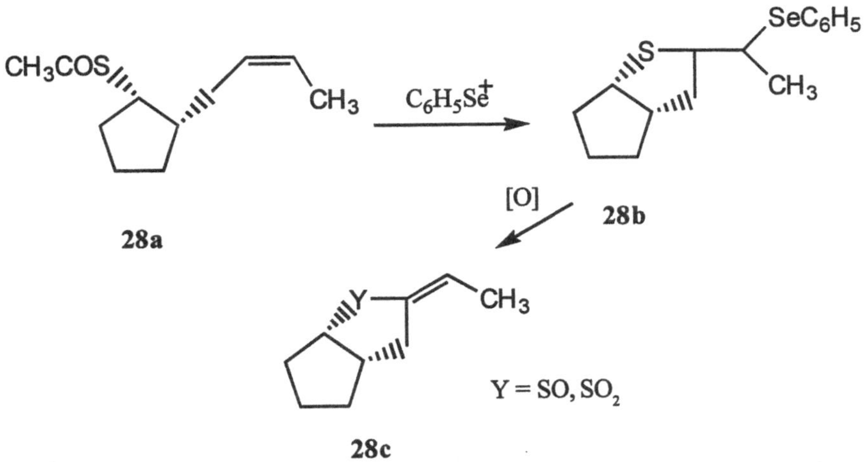

Scheme-36

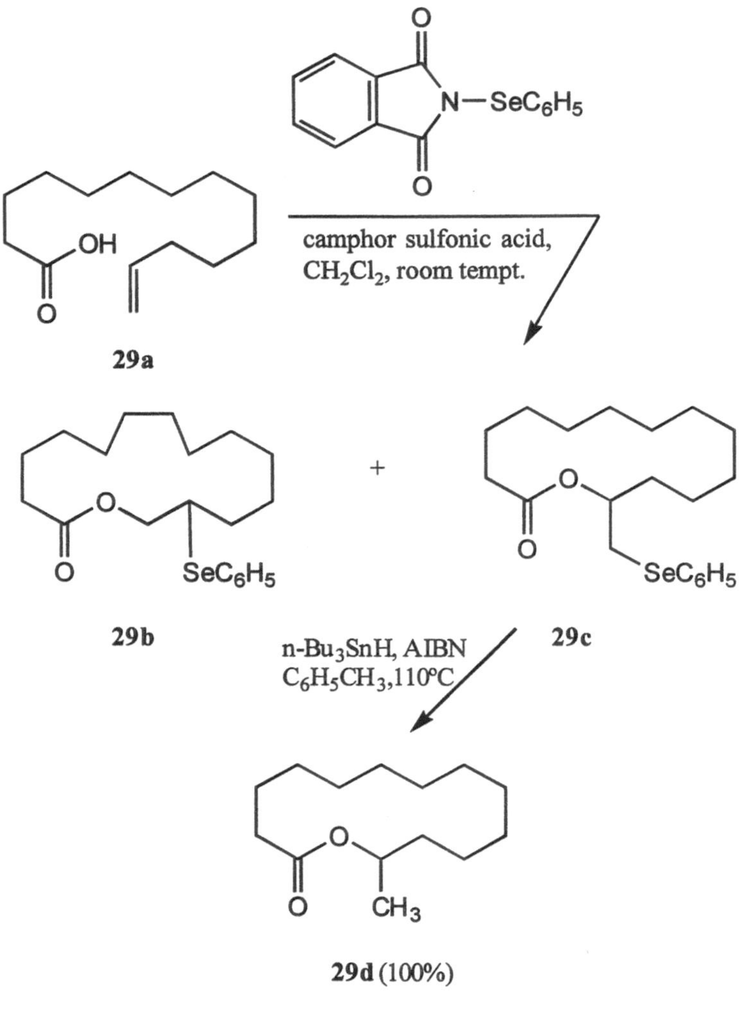

29a

camphor sulfonic acid,
CH$_2$Cl$_2$, room tempt.

29b   +   **29c**

n-Bu$_3$SnH, AIBN
C$_6$H$_5$CH$_3$,110°C

29d (100%)

Scheme-37

The usefulness of this organoselenium based technology is in the construction of advanced intermediates and in the synthesis of a series of O- and S-containing prostacyclins. The synthesis of O-containing prostacyclins is depicted in (scheme-38). Intramolecular ureidoselenenylation of the double bonds followed by the allylative deselenenylation which thus accomplishes a net ureidoallylation[27] has been found to be useful. Intramolecular ureidoselenenylation of the general system **31a** using N-phenylselenophthalimide (N-PSP) establishes the carbon–nitrogen bond. An allyl group is introduced by the reaction of **31b** with tri-n-butyl

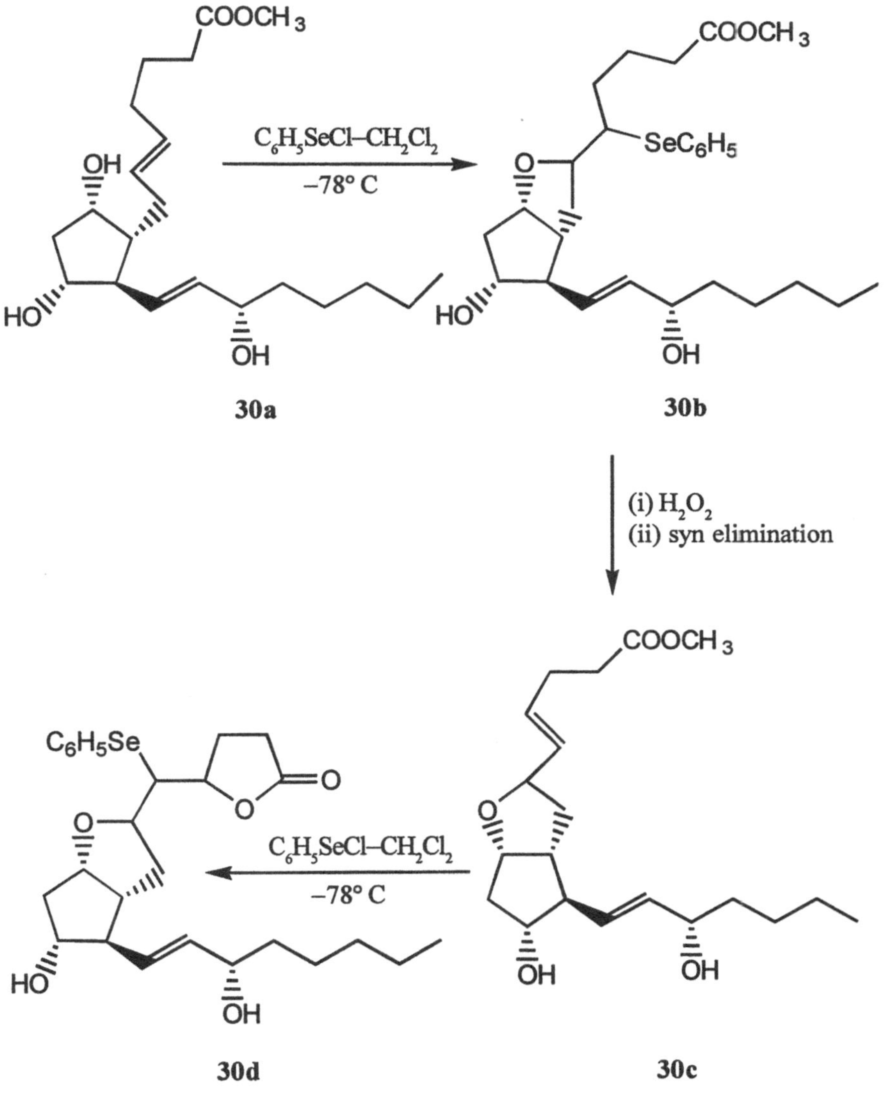

Scheme-38

allylstannane with an initiation by azobisisobutyronitrile (AIBN) (scheme-39). In a similar way, indoline **32b** has been prepared from N-Cbz-o-allylaniline **32a** (scheme-40).

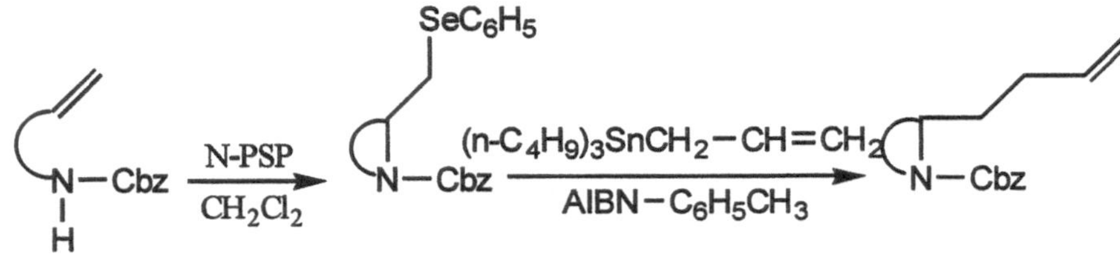

Scheme-39

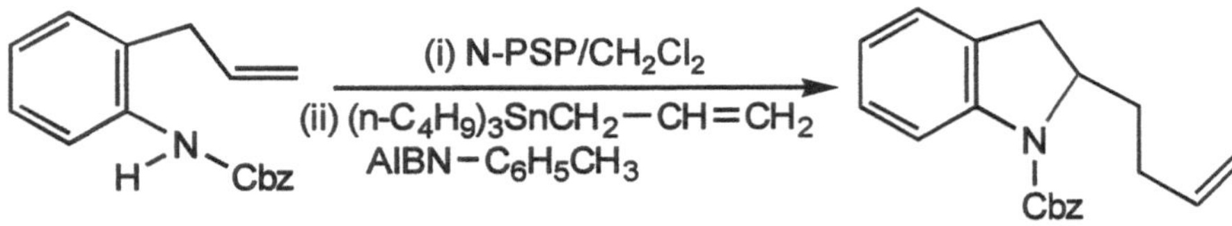

Scheme-40

Heterofunctionalized alkenes activated by mercuric salts undergo intramolecular addition to the double bonds and lead to the development of a general method for the synthesis of heterocycles[28]. A few examples are depicted in scheme-41. Organomercury intermediates 34a–c undergo heterogenolysis cleaving carbon–mercury bond with anhydride.

A synthesis of β-lactams 36b has been achieved by intramolecular Michael addition of a substituted acrylamide 36a (scheme-42)[29].

Cyclization under Vilsmeier conditions also yields hererocycles. Vilsmeier reaction is primarily concerned with the formylation of a wide variety of substrates. It involves the reaction of a Vilsmeier reagent derived from a tertiary amide and an acid chloride. In heterocyclic synthesis, the most commonly used Vilsmeier reagent (VR) is derived from dimethylformamide and phosphoryl chloride (Fig. 2). The activated C=N bond acts as an electrophile. Methyl group α or γ to an annular nitrogen in an N-heteroaromatic system easily undergo Vislmeier reaction. When an amino group is ortho to the methyl group, cyclization yields a fused pyrrole (scheme-43)[30]. The product of Vilsmeier reaction of aminopyrimidine acetic ester 38a depends upon the ratio of the reagent used (scheme-44)[31]. Furanoquinoxaline 39c is obtained by Vilsmeier reaction of quinoxalone 39a (scheme-45)[32].

Pyrazoles 41a-c can be obtained by diformylation and cyclization under Vilsmeier conditions from hydrazones, semicarbazones and azines (scheme-46)[33-35].

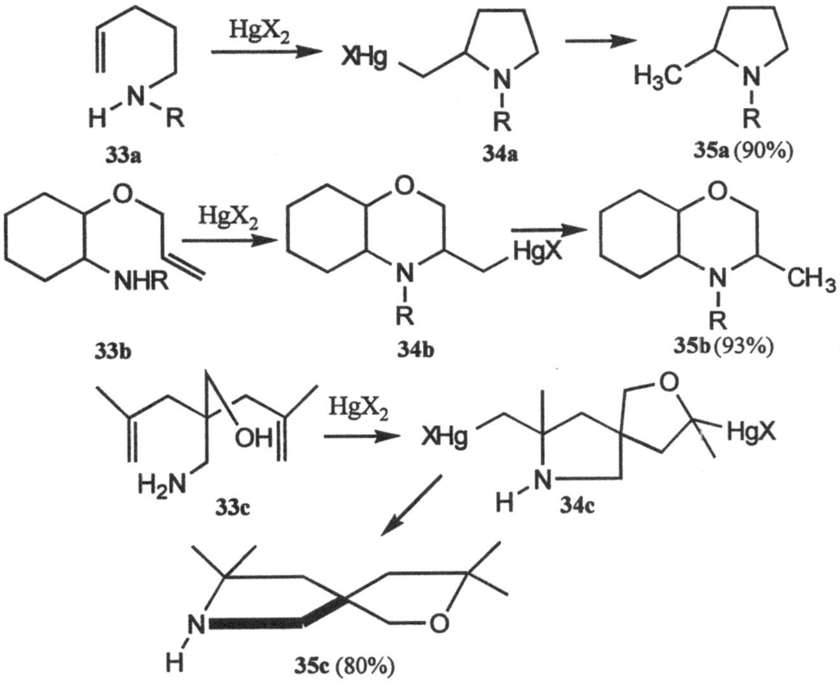

Scheme-41

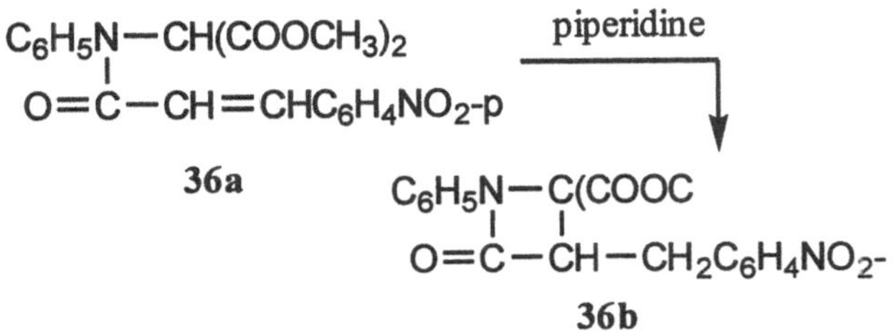

Scheme-42

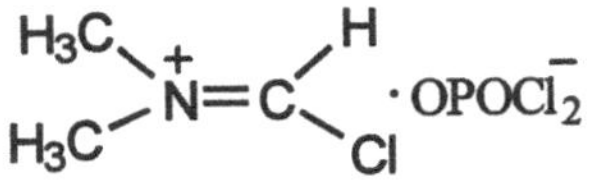

Fig. 2

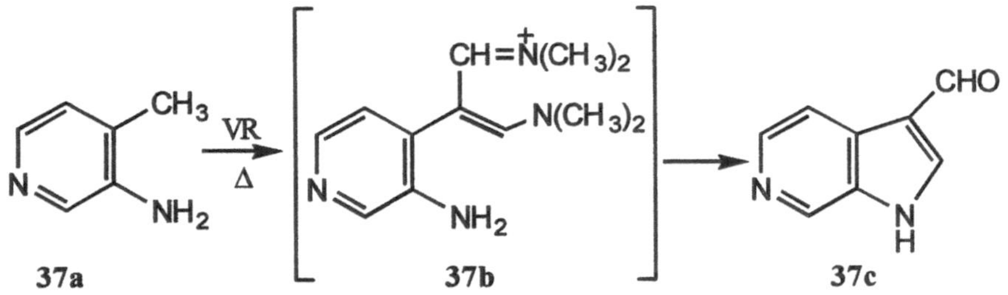

Scheme-43

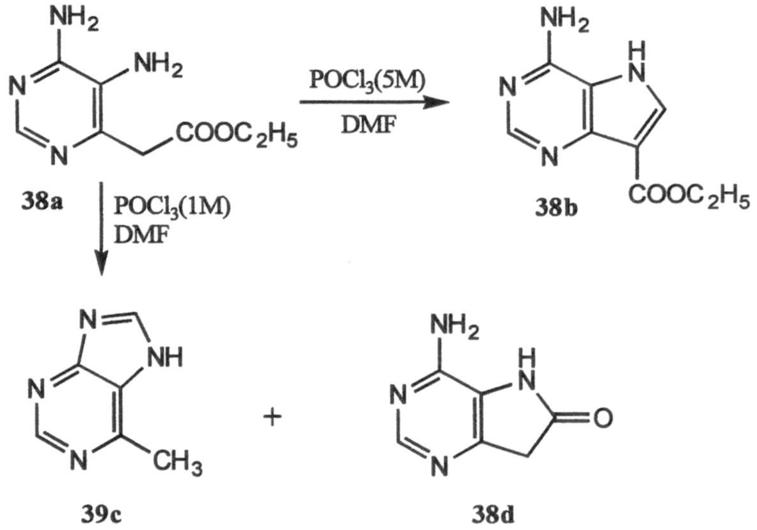

Scheme-44

Benzimidazole **42c** has been prepared by the reaction of Vilsmeier reagent with N, N'-diacetyl-*o*-phenylenediamine **42a** (scheme-47)[36]. 3-Aminopyrimidine-2-thiones **43a** with Vilsmeier reagent forms amidines **43b** which cyclize to give thiadiazolopyrimidines **43c** (scheme-48)[37].

6-Arylaminouracils on cyclization with 'aza-Vilsmeier reagent' (obtained by N-nitrosodimethylamine and phosphoryl chloride) forms alloxazines **44b** and isoalloxazines **45b** (scheme-49)[38].

Nucleophilic carbon–sulfur bond formation forms an important method for the synthesis of five- and six-membered sulfur containing heterocycles[39]. Bromine induced cyclization of a γ,δ-unsaturated thioamide **46a** yields 5-bromomethyl-thiolan-2-one **46c** (scheme-50).

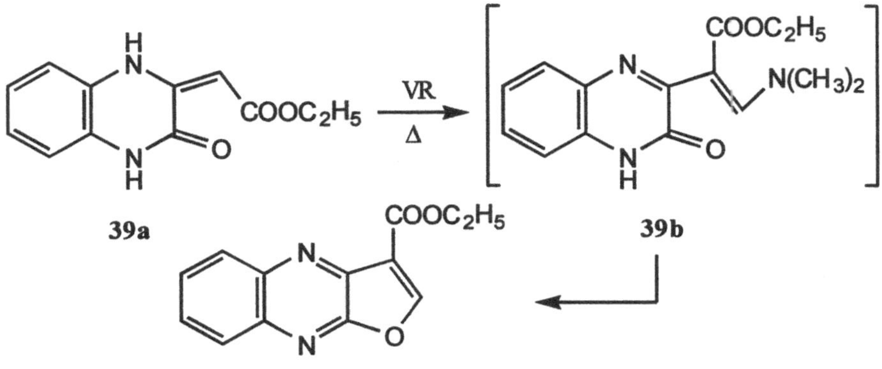

Scheme-45

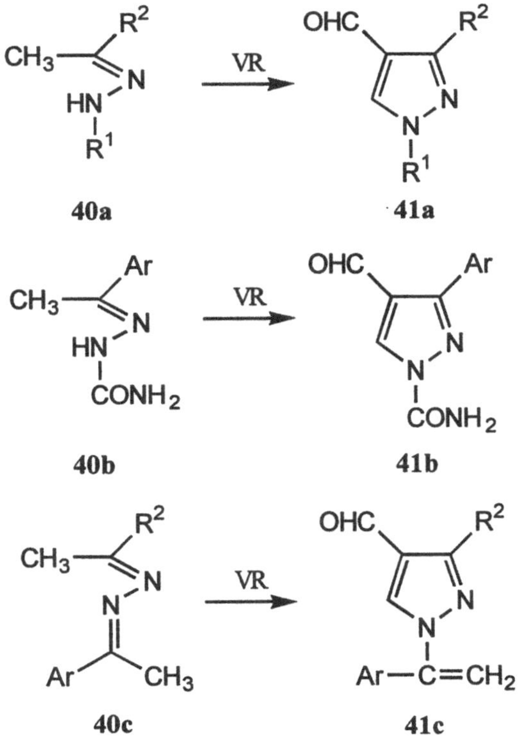

R^1 and R^2 = alkyl / aryl groups

Scheme-46

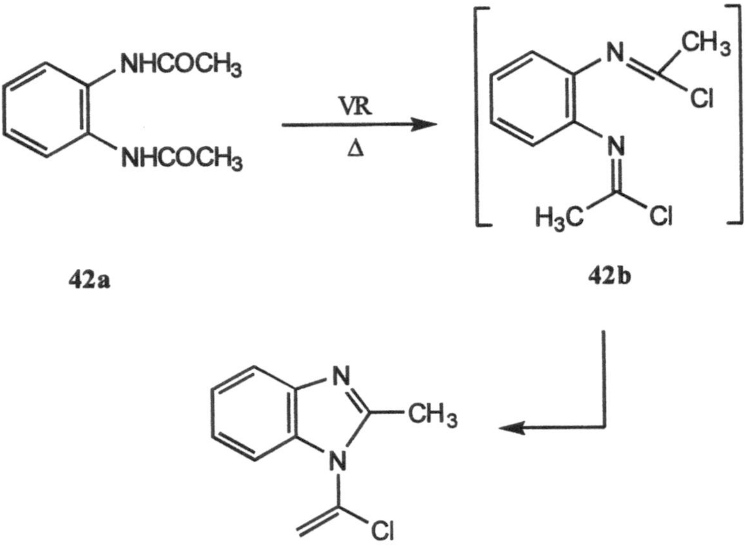

42a

42b

42c

Scheme-47

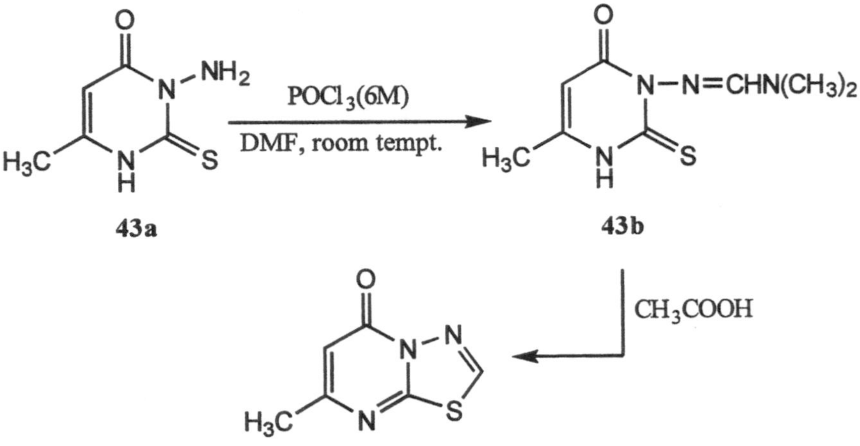

43a

43b

43c

Scheme-48

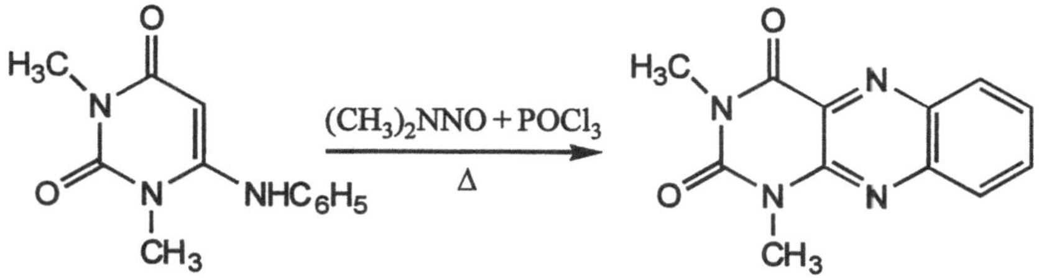

44a **44b** (42%)

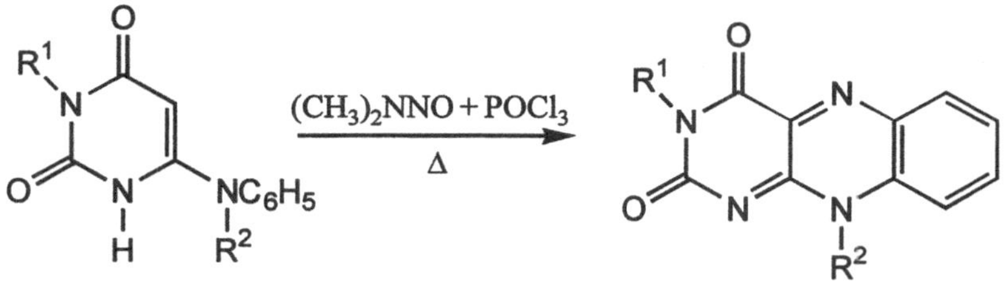

45a **45b** (60%)

Scheme-49

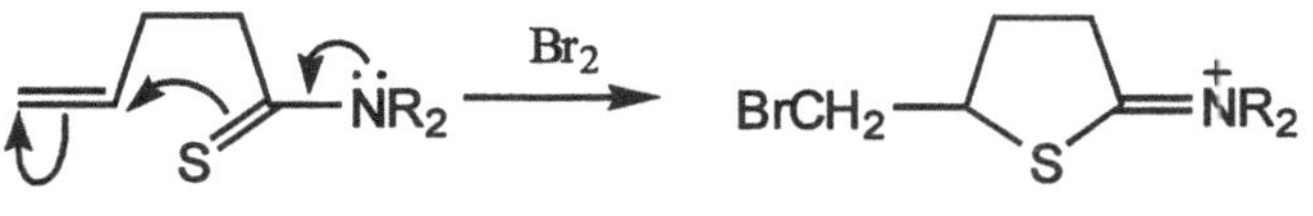

46a **46b**

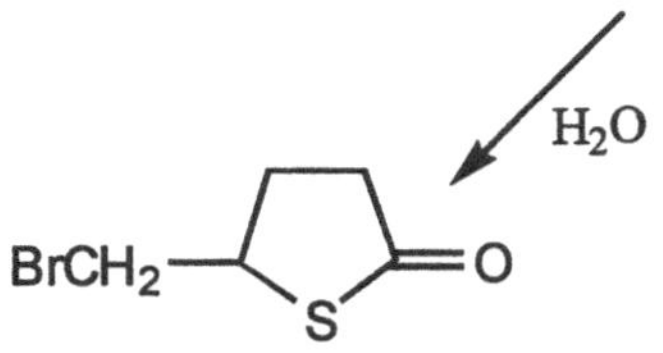

46c

Scheme-50

Methyl sulfide group ($-SCH_3$) also serves as a nucleophile in the cyclization step and the equilibrium with an intermediate sulfonium salt is driven to the right by the removal of methyl bromide (scheme-51). Cyclization via free thiolate is highly stereospecific and C–S bond formation takes place anti to the departing group (scheme-52).

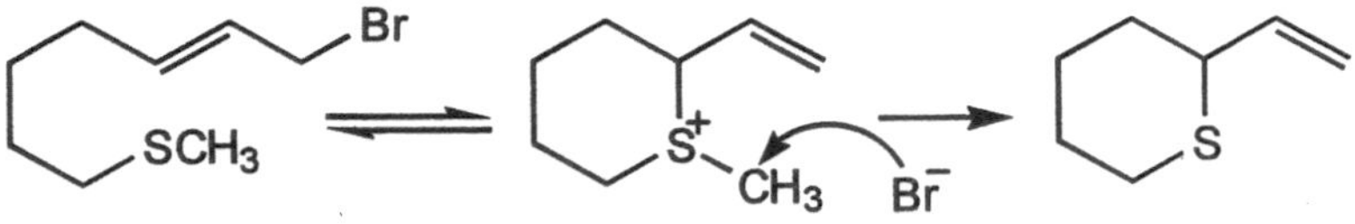

47a **47b** **47c**

Scheme-51

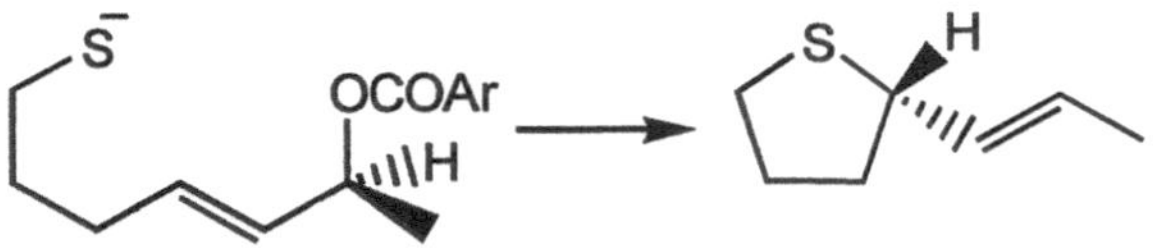

48a **48b**

Scheme-52

2.1.1.3 Intramolecular Nucleophilic Addition to Triple Bonds

The intramolecular cyclization of acetylenic compounds containing functional groups leads to the formation of five and six-membered heterocycles. Acetylenic compound having an alcoholic function at γ-position **49a** yields dihydrofuran **49c** on distillation over sodamide (scheme-53)[40]. If a heteroatom is present between the acetylenic and alcoholic functions, a convenient synthesis of various heterocycles can be achieved. N-Propargylethanolamines **50a** cyclize to give 4-alkyl-2-methylenemorpholine **50b** under the basic conditions (scheme-54)[41].

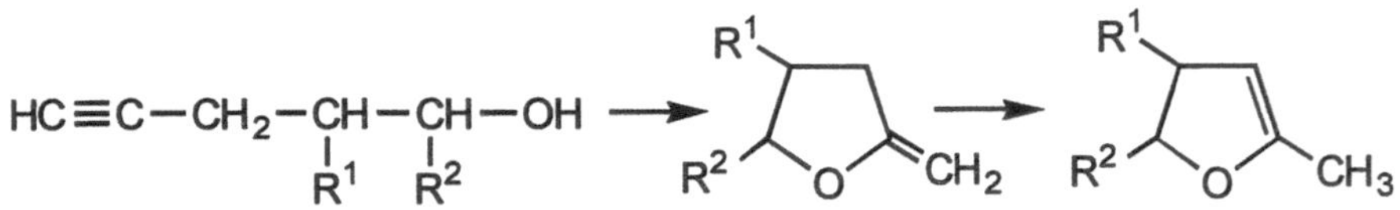

49a **49b** **49c**

Scheme-53

The intramolecular cyclization of acetylenic acid **51a** yields lactone **51b** (scheme-55)[42]. Phenylpropynamidoximes **52a** on heating undergo cyclization to yield isoxazoles **52b** (scheme-56)[43].

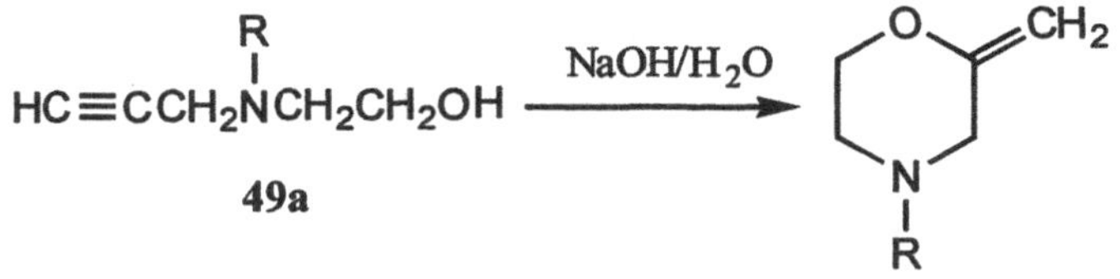

49a

50b

Scheme-54

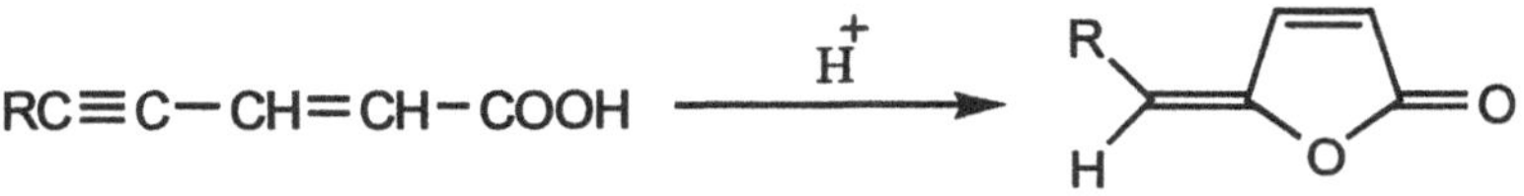

51a

51b

Scheme-55

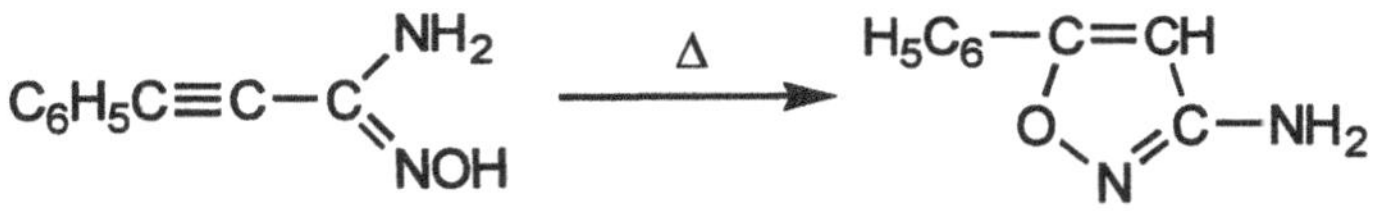

52a

52b

Scheme-56

Acetylenic amines **53a** under the action of heat cyclize to give pyrroles **53b** (scheme-57)[44]. Similarly, *o*-aminophenylacetylene **54a** on heating yields indole **54b** (scheme-58)[45]. The base catalyzed cyclization of thioureas containing acetylenic functions **55a** yield imidazolidinethiones **55b** (scheme-59)[46].

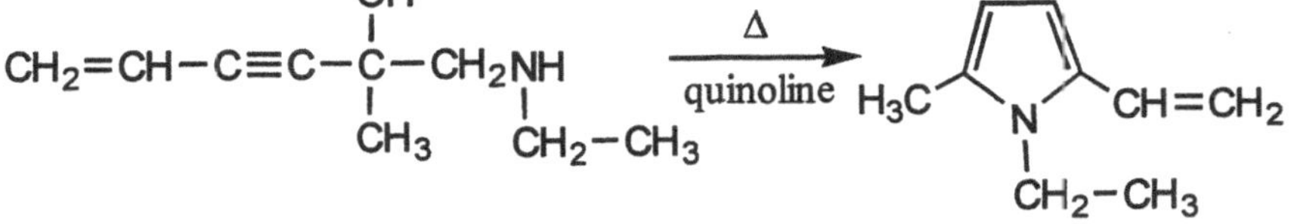

53a

53b

Scheme-57

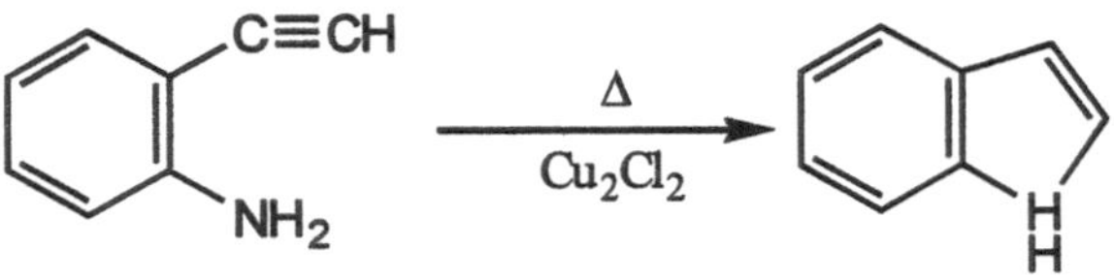

54a **54b**

Scheme-58

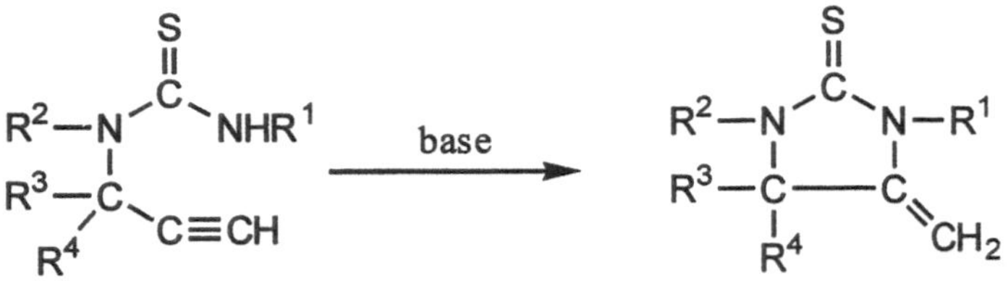

55a **55b**

Scheme-59

The nucleophilic addition to arynes is useful for the synthesis of some benzo-fused heterocycles e.g. the synthesis of phenanthridine **56d** by intramolecular nucleophilic addition to benzyne **56c** by ortho carbon atom of another benzene ring (scheme-60)[47].

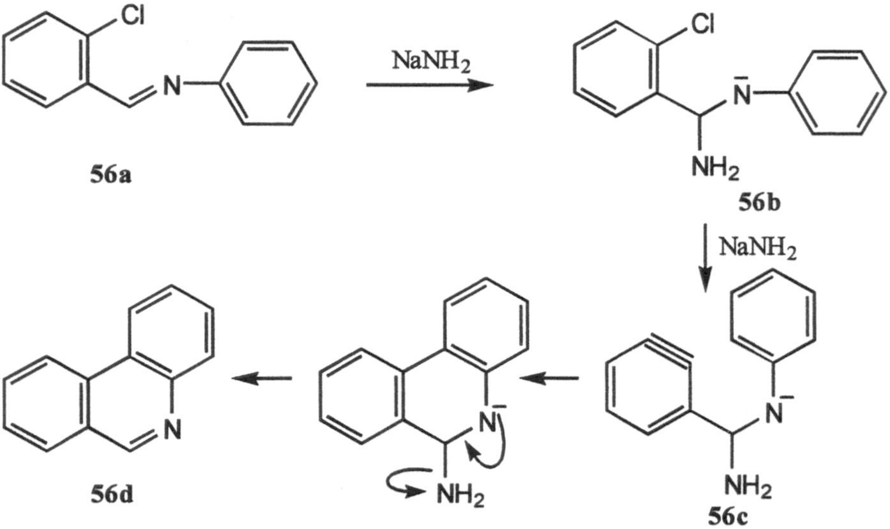

56a **56b** **56c** **56d**

Scheme-60

The nucleophilic addition to nitriles also comprises an important method for the synthesis of amino substituted heterocyclic compounds. The reaction of ethyl carboethoxyformimitate **57a** with α-aminonitriles **57b** yields amidines **57c** (scheme-61)[48].

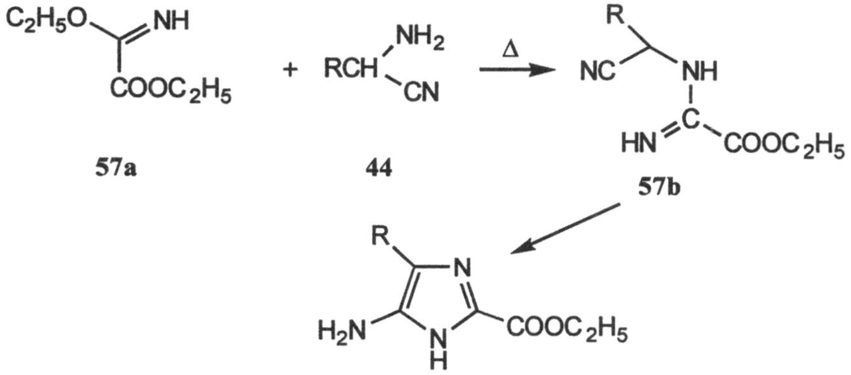

Scheme-61

A versatile synthesis of heterocycles involves the endo cyclization of isonitriles. A variety of heterocycles such as oxazolines, thiazolines, oxazoles, oligooxazoles, thiazoles, triazoles, imidazolines, imidazolinones, pyrroles, oxazines, thiazines *etc.*, can be obtained from α-metalated isonitriles[49]. The importance of α-metalated isonitriles lies in their ambivalent nature. They have a nucleophilic centre (the metalated anionized carbon atom) which can add to the polar multiple bonds and an electrophilic centre (the isonitrile carbon atom) which permits cyclization of the adducts **58b** on protonation to form heterocycles (scheme-62). With carbonyl compounds, the reaction of α-metalated isonitriles **59a** yields 2-oxazolines **59d**.

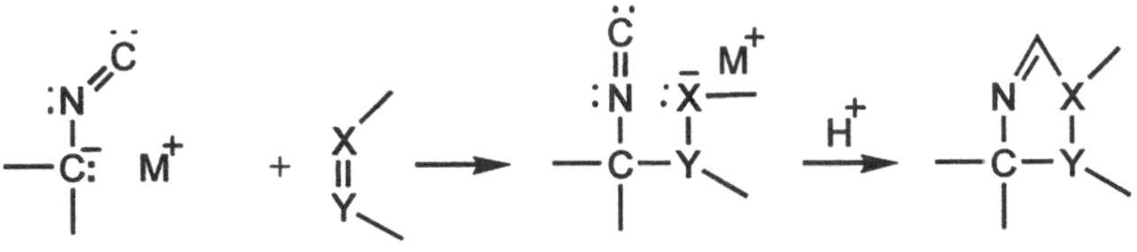

Scheme-62

The intermediate carbonyl adducts **59b** and the tautomeric 2-metalated 2-oxazolines
59c (the tautomeric equilibrium **59b** ⇌ **59c** lies almost entirely on the left hand
side) are transformed into 2-oxazolines **59d** (scheme-63).

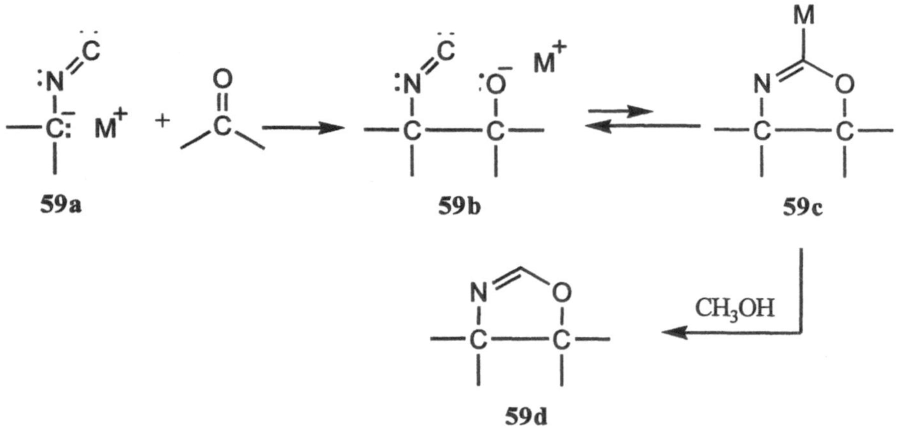

Scheme-63

The cyclization of 3-ethyl-2-isocyano-3-mercaptopentanoic ester **60c** (obtained
by the reaction of diethyl thioketone and lithio isocyanoacetic ester) provides
5,5-diethyl-2-thiazoline-4-carboxylic ester **60d** (scheme-64).

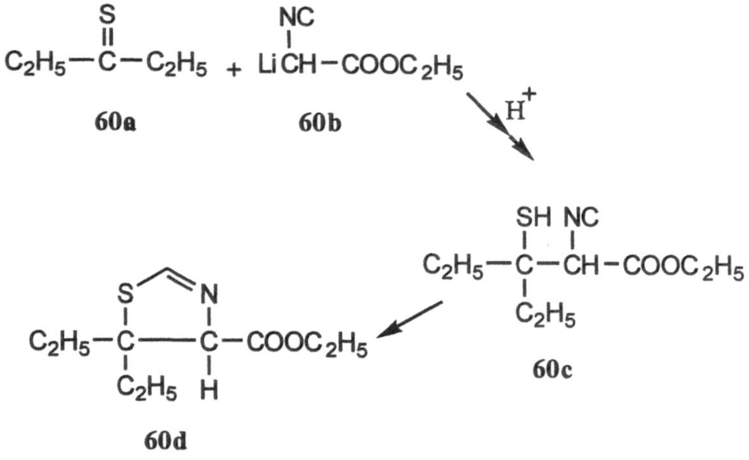

Scheme-64

α-Metalated isocyanides can also- add to azomethine group. The reaction of lithiomethyl or lithiobenzyl isocyanides **61a** with schiff bases yields 2-imidazolines **61d** (scheme-65).

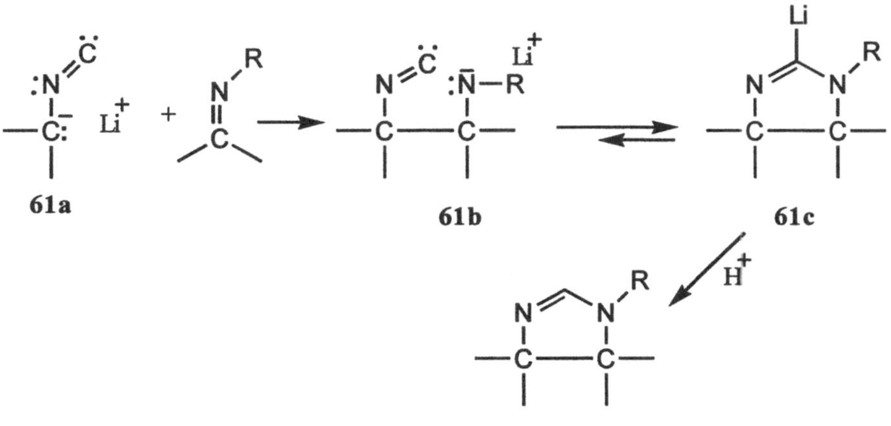

Scheme-65

Oxazoles **62d** are obtained from α-metalated isocyanides by their reaction with acylating agents (scheme-66). The intermediate α-isocyano ketones **62b** cyclize via

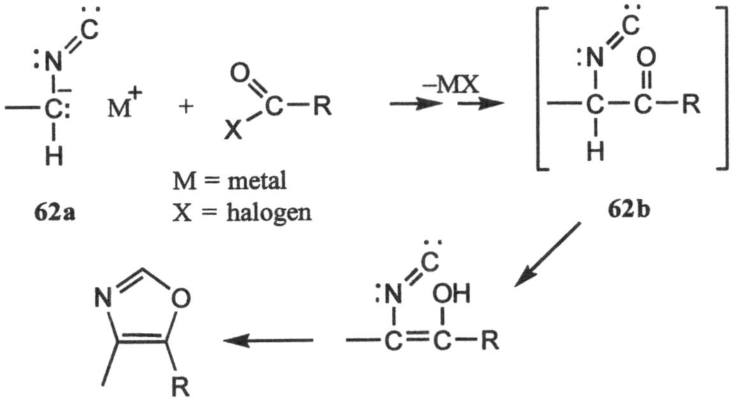

Scheme-66

enols **62c** to yield oxazoles **62d**. An important method of synthesis of imidazoles
63c includes the reaction of p-tolylthiomethyl isocyanide **63a** to alkyl or aryl
cyanides (scheme-67)[50].

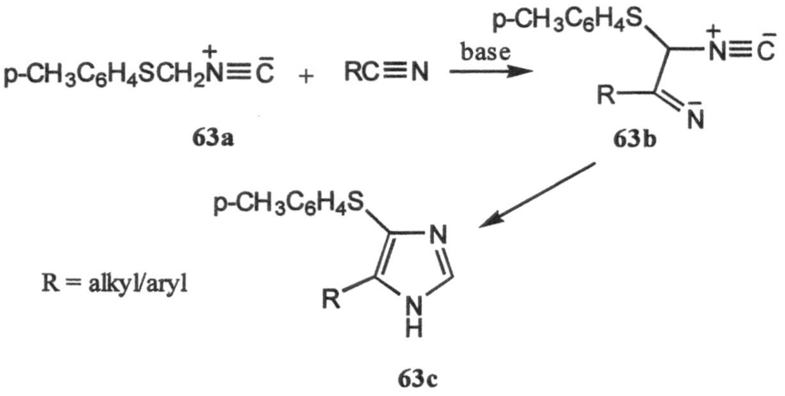

Scheme-67

A new synthesis of indole derivatives involves the ortholithiation of an alkyl
group in *o*-alkylphenyl isocyanides **64a** and subsequent intramolecular ring
closure (scheme-68)[51].

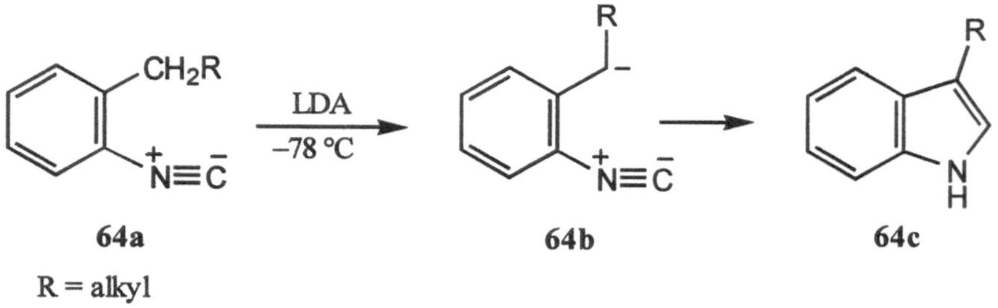

Scheme-68

2.1.2 Radical Cyclizations

The synthesis of heterocycles is also achieved by the intramolecular addition of
a radical to an olefinic bond or to an aromatic ring under kinetic control
preferentially in exo mode. However, the substituents at the double bond disfavour
the addition at the substituted position. The homolytic cleavage of
N-chloroalkylamines with double bond(s) yields chlorine and aminyl radicals

which undergo intramolecular cyclization reactions. The nature of the aminyl radical (neutral, protonated or complexed to metal ion) plays an important role in cyclization. A variety of β-functionalized, fused or bridged, azaheterocycles have been synthesized regio- and stereoselectively in good yield by the radical cyclization reactions[52]. The neutral aminyl radicals can be generated by the photolysis of N-chlorodialkylamines in neutral media. The sluggishness of the neutral dialkylaminyl radicals to add to olefins, as well as their strong tendency towards non-selective abstraction of hydrogen seems to rule out reactions that could be of preparative value. However, the photolysis of N-chloro-N-propyl-4-pentenyl amines **65a** in aqueous acetic acid provides only 2-chloromethyl-1-propylpyrrolidine **65d** (scheme-69).

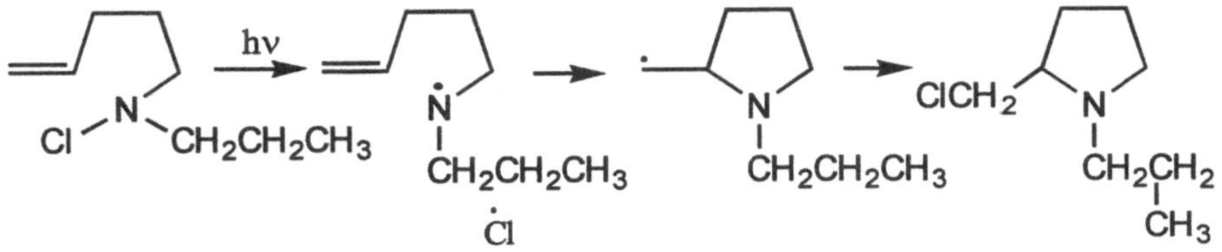

Scheme-69

The protonated aminyl radicals behave very differently from the neutral species. The main difference is the pronounced tendency of the protonated radicals to add efficiently to many types of the unsaturated hydrocarbons and arenes in preference to abstract allylic or benzylic hydrogen atoms. Protonated aminyl radicals can be obtained by the homolytic decomposition of N-chlorodialkylamines. Due to the electrophilic character of protonated aminyl radicals, they cyclize more readily than the neutral species. The intramolecular cyclization of protonated N-chloroamine **66a** is depicted in scheme-70.

The decomposition of N-chloroamines **67a** catalyzed by the reducing metal salts in the neutral solution produces aminyl radicals complexed to metal ions **67b**. The complexation is presumed to take place via the lone pair of electrons of the aminyl radicals. The reactivity is intermediate between neutral and protonated aminyl radicals. The cyclization of the aminyl radicals complexed to metal ions is depicted in scheme-71.

According to the redox mechanism, the cyclization of N-chloroalkenylamines in non-protonating media, where metal in a polydentate complex permits a relatively rigid transition state, an efficient stereospecific radical reaction should occur. Thus, *trans*- and *cis*-N-chloro-N-methyl-4-hexenylamines **69a** react with reducing metal salts to form *erythro*- and *threo*-2-(1-chloroethyl)-1-methylpyrrolidines **69b**, **69c** (scheme-72).

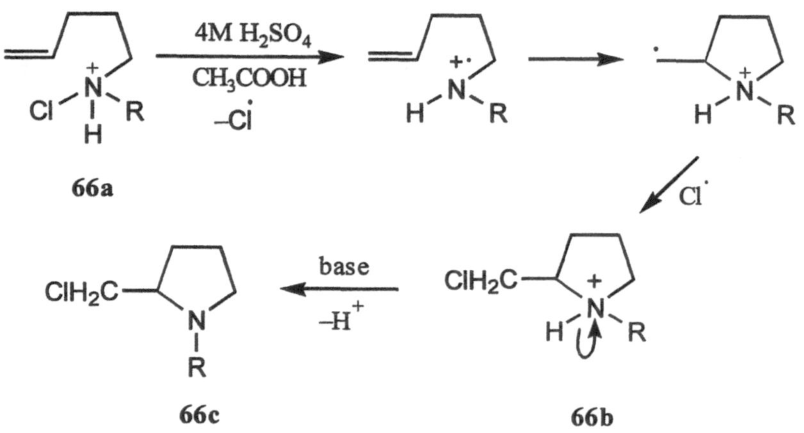

Scheme-70

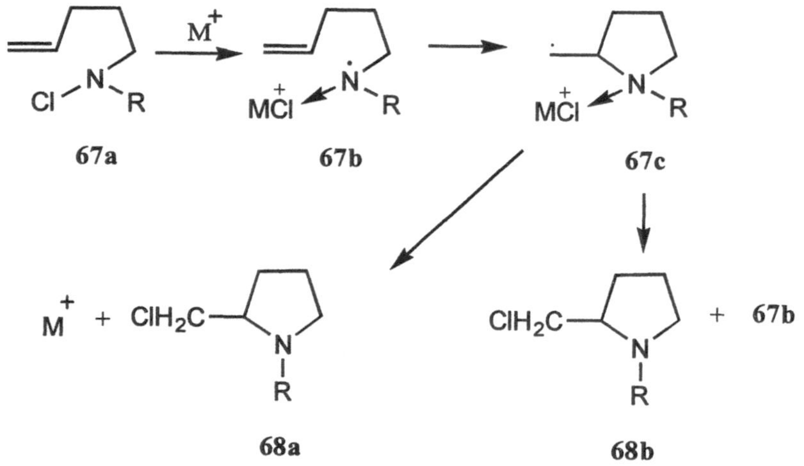

Scheme-71

The cyclization of acyclic N-chloroamines with two double bonds e.g. N-allyl-N-chloro-4-pentenylamine **70a** with titanium trichloride in aqueous acetic acid leads to 2-chloromethylpyrrolizidine **70c**. The cyclization of **70a** to the five-membered ring affords carbon centered radical **70b** which further undergoes homolytic cyclization and this appears to be faster than the transfer of chlorine. The reaction proceeds via a typical radical chain mechanism involving transfer of a chlorine atom from N-chloroamine **70a**. The compound **70a** in the same solvent with copper (I) chloride in the presence of copper (II) chloride does not form

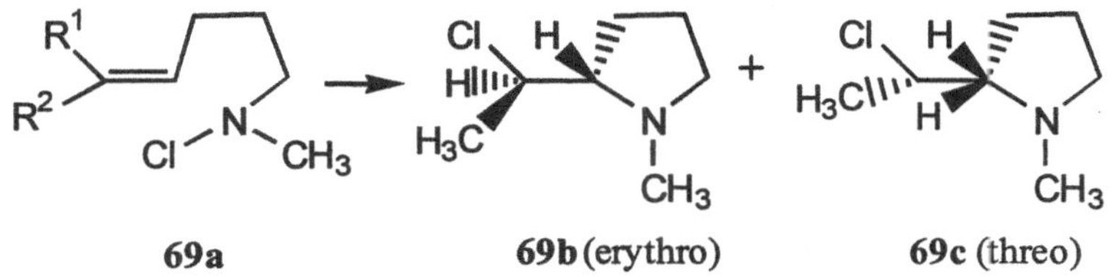

69a **69b** (erythro) **69c** (threo)

R^1=H ; R^2=CH_3 (trans)
R^1=CH_3 ; R^2=H (cis)

Scheme-72

bicyclic product, but N-allyl-2-chloromethylpyrrolidine **70e** is obtained exclusively, which indicates that the transfer of chlorine from copper (II) chloride is faster than the cyclization of 5-hexenyl radical **70d** (scheme-73).

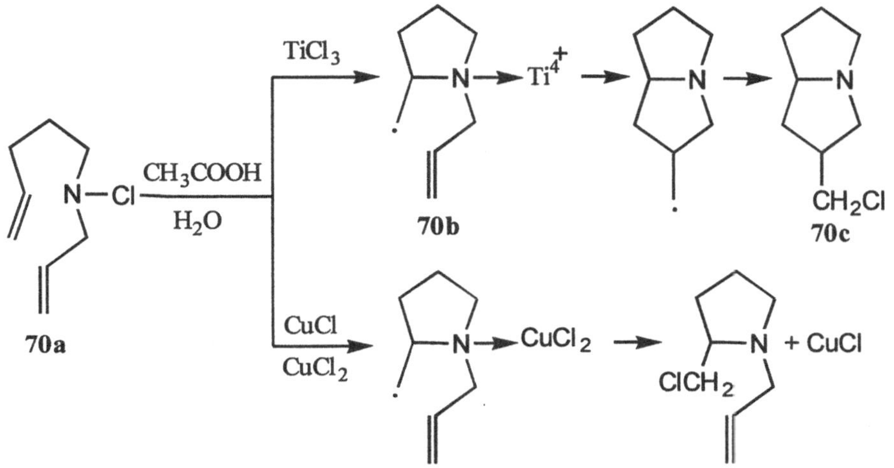

Scheme-73

The intramolecular radical addition reactions are not only limited to the olefinic bonds, but additions to the aromatic rings also occur. Indolines and tetrahydroquinolines can be prepared by such reactions. The decomposition of N-chloro-2-phenylethylamine **71** in conc. sulfuric acid catalyzed by Fe^{+2} yields N-methylindoline **72a** and benzyl chloride (scheme-74). However, in case of N-chloro-3-phenylpropylamine

73 β-scission does not occur and only N-methyltetrahydroquinoline **74** is produced (scheme-75). Similarly, photolysis of N-chloroamine in conc. sulfuric acid gives N-methylhexahydrocyclopentaquionline **76** (scheme-76).

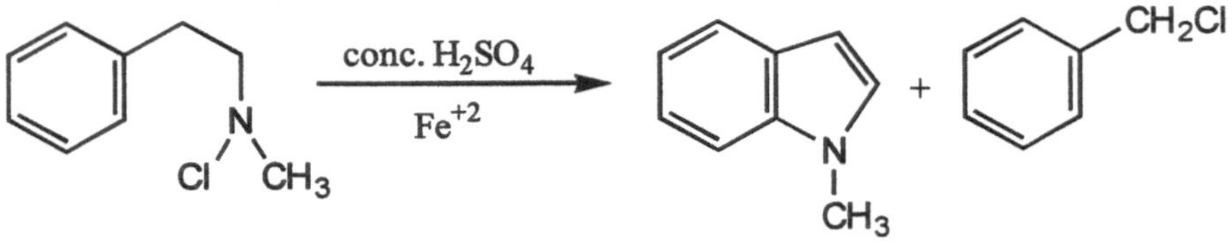

71   **72a** (27%)   **72b** (44%)

Scheme-74

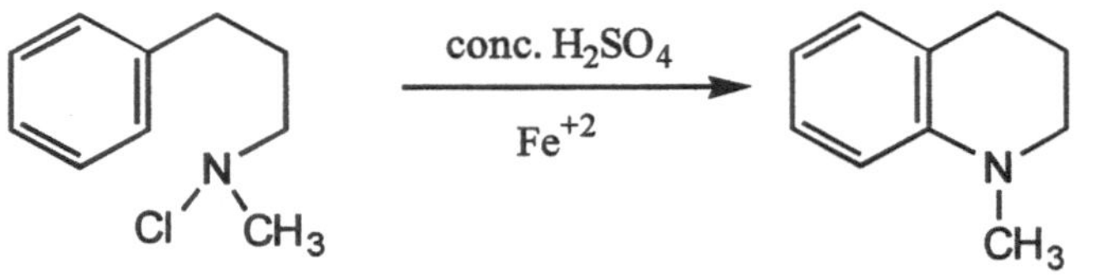

73   **74** (81%)

Scheme-75

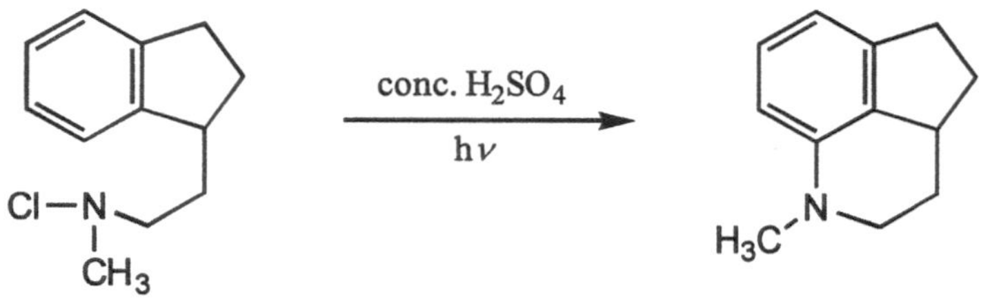

75   **76** (90%)

Scheme-76

The cyclization of N-chloro-N-methyl-2-(1,1-dimethyl-2-methylenecyclopropane)-ethylamine **77** in aqueous acetic acid catalyzed by titanium trichloride provides 3-azabicyclo[4.1.0] heptane derivative **78**. The overall endo-regioselectivity follows the Baldwin's rules and the general behaviour of the irreversible radical cyclization (scheme-77).

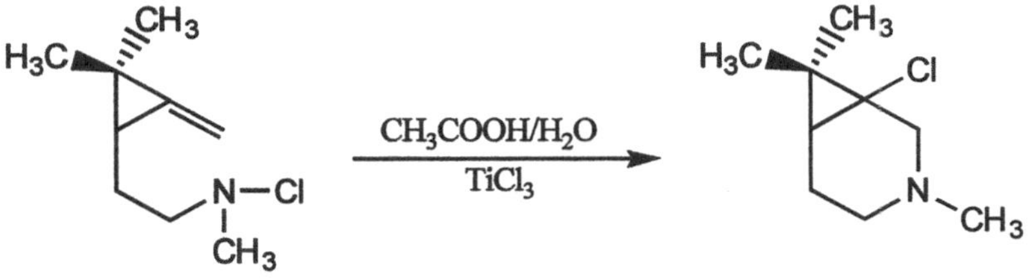

77 78

Scheme-77

Radical cyclization of N-chlorocycloalkenylamines **79** under solvolytic conditions forms a simple route to provide azapolycycles **80** (scheme-78).

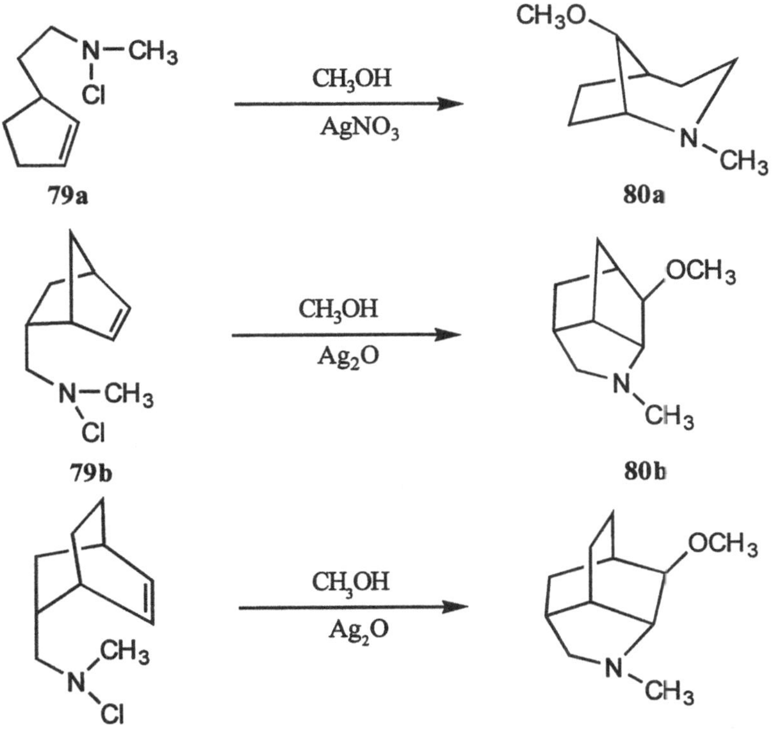

79c 80c

Scheme-78

Hofmann–Löffler reaction[53] which involves the transfomation of N-haloamines into the cyclic amines is also a radical cyclization reaction, in which a carbon–nitrogen bond is formed instead of addition to a multiple bond. The salt of N-chloroamine **81a** is homolytically cleaved under the action of heat, light, *etc.* to yield aminium radical **81c**, which, in turn, intramolecularly abstracts a sterically favoured hydrogen to give an alkyl radical **81d**. The alkyl radical in a chain reaction abstracts chlorine intramolecularly. The transformation of alkyl chloride **81e** into cyclic amine **81f** takes place in the presence of an alkali (scheme-79).

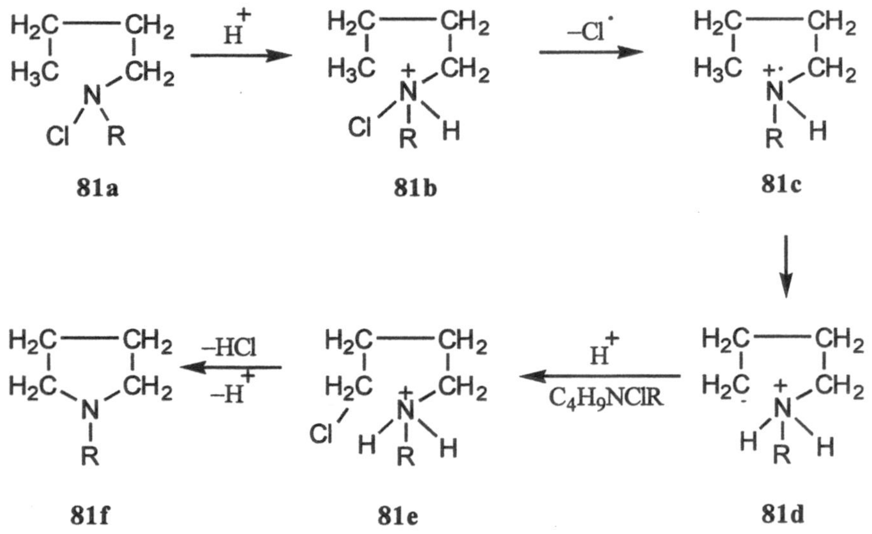

Scheme-79

2.1.3 Carbene and Nitrene Cyclizations

Carbene and nitrene are highly reactive species and are useful in synthetic organic chemistry. The tendency of the singlet carbene and nitrene to undergo insertion into unactivated C–H bond and the addition to multiple bonds are useful for the synthesis of heterocycles. The photolysis of azides yields nitrenes which undergo intramolecular addition to double bond(s) to yield azirines. The cyclization of the conjugated dienyl azides **82** gives 2*H*-azirines **83** which on isomerization at high temperature yields 2,2-dicyano-2*H*-pyrrole **84**. Over all, the reaction is a two-stage thermolysis of the azide (scheme-80)[54].

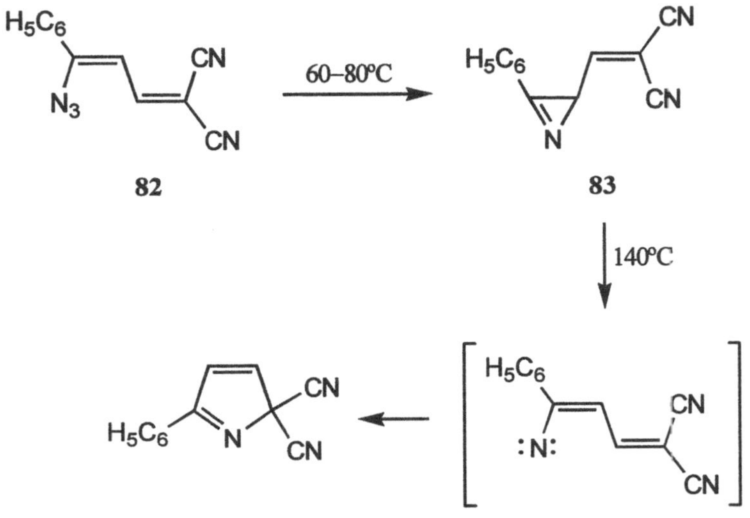

Scheme-80

2-(2-Benzofuryl)-2*H*-azirine **86** obtained by the photolysis of azide **85** undergoes thermal ring opening to yield nitrene. Vinyl nitrene **87** thus produced undergoes intramolecular addition reactions to yield a condensed pyrrole **88**, pyridine **89** or azepine **90** depending on the substituents (scheme-81)[55].

Annelated pyridines **92**, **94**, and **96** are obtained by the pyrolysis of aryl vinyl azides **91**, **93**, **95** involving nitrene insertion into methyl group (scheme-82)[56].

Lactam **99** is synthesized from acylnitrenes **98** which is obtained by the photolysis of sulfilimine **97**[57]. Isocyanate **100** accompany the photolysis and is considered to be formed by a non-nitrene process (scheme-83).

Photolysis of nitrile oxides **101**[58] also provides acylnitrene **102** which cyclizes to lactam **103** (scheme-84).

Variously substituted sulfilimines **104** on flash vacuum pyrolysis yields interesting cyclopenta[d]pyrimidines **106** along with carbodimides **107** (scheme-85)[59].

Azepines **113** are formed during the photolysis of aryl azide **108** in diethylamine. It is the singlet phenylnitrene which undergoes rearrangement (scheme-86)[60].

Thermolysis of O,N-bis(trimethylsilyl)-N-phenylhydroxylamine **114** at 100°C yields 2-diethylamino-3*H*-azepine **115** (scheme-87)[61].

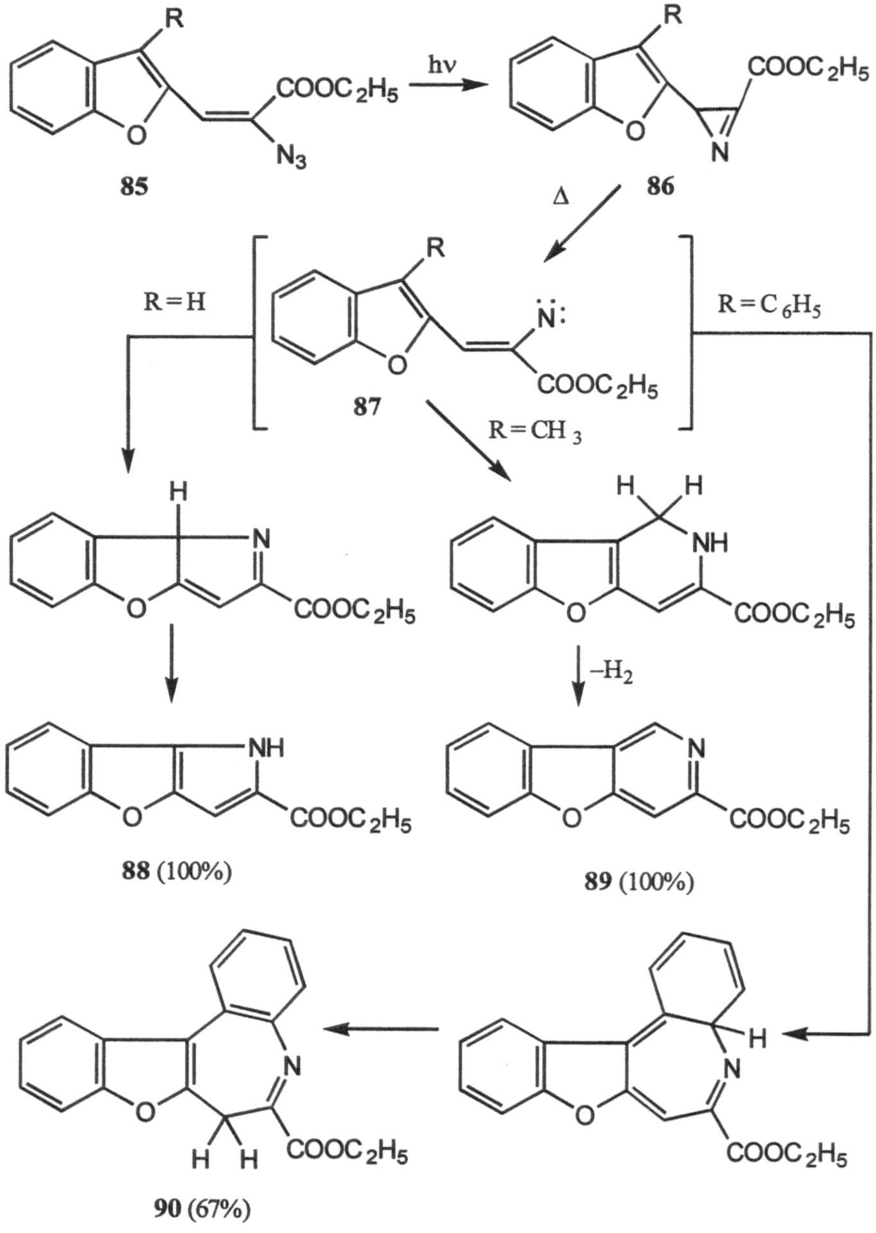

Scheme-81

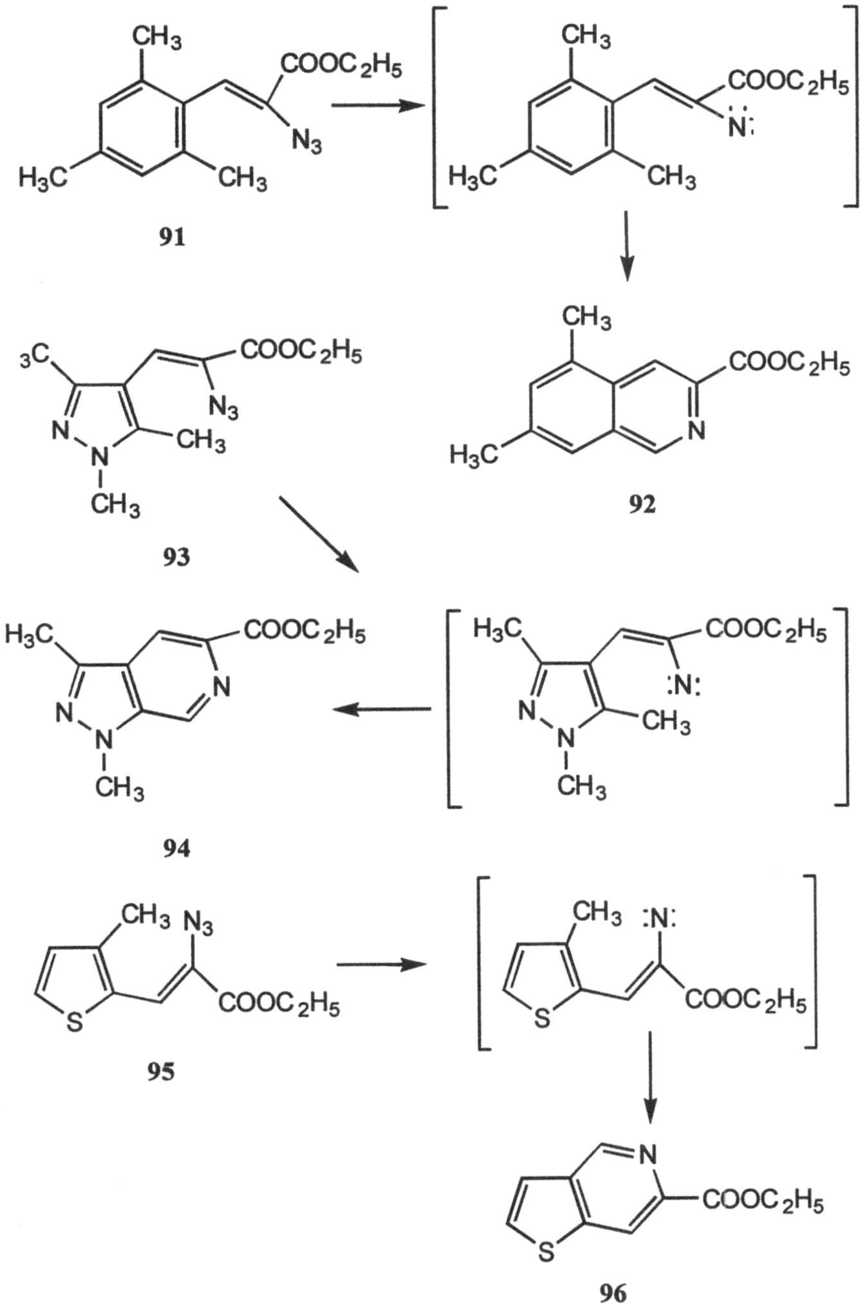

Scheme-82

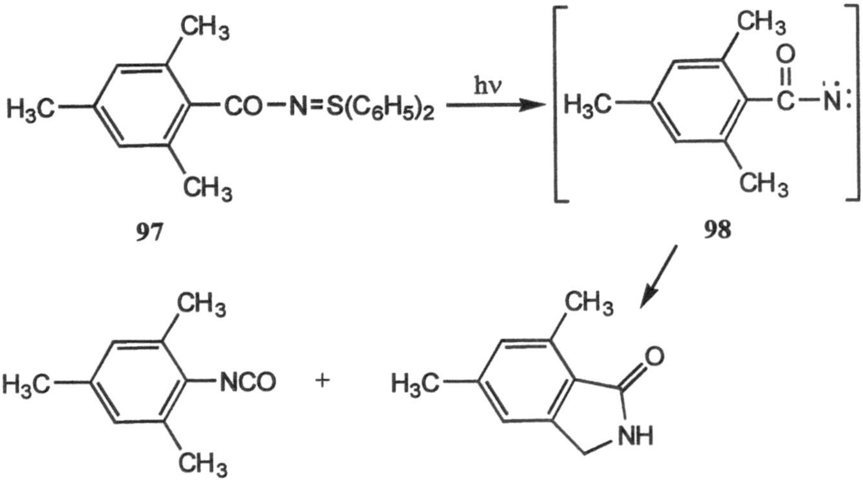

Scheme-83

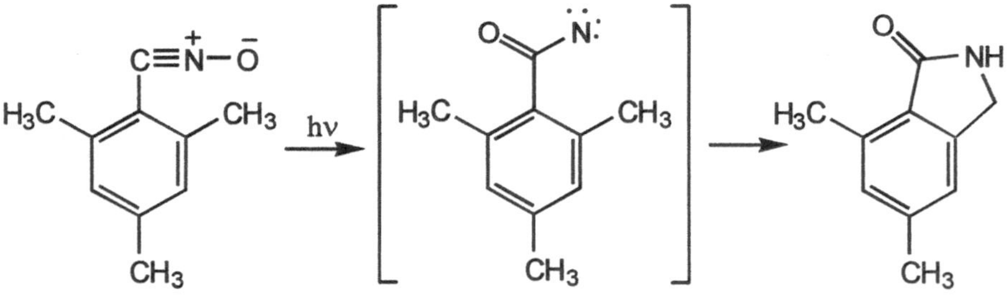

Scheme-84

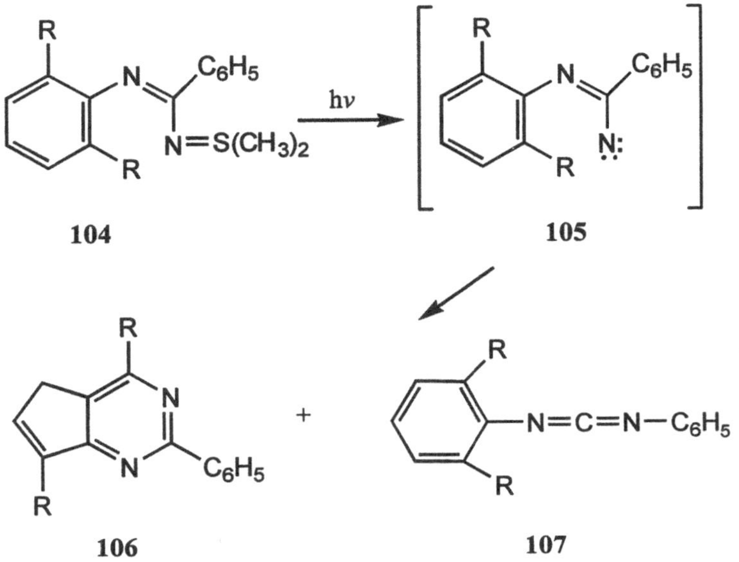

Scheme-85

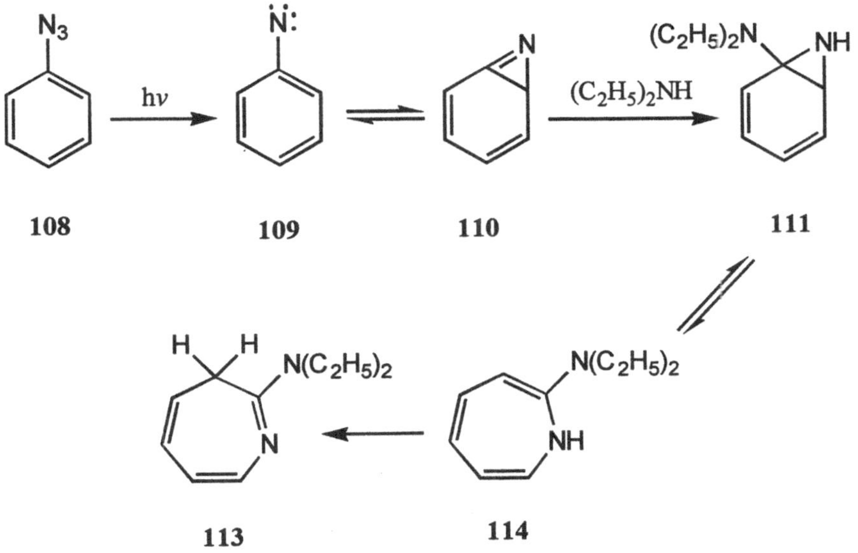

Scheme-86

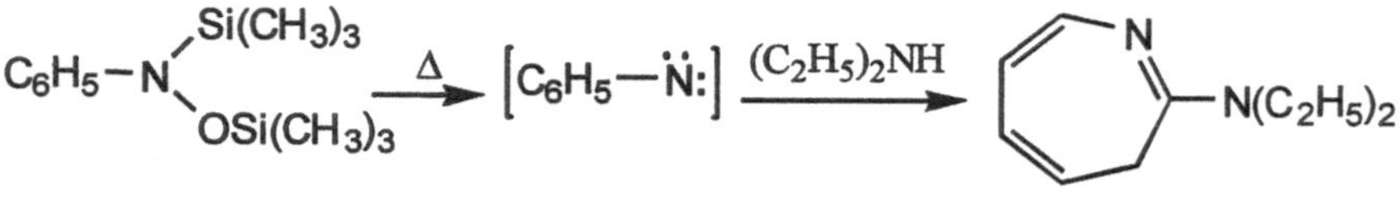

114 115

Scheme-87

Azepines **117** and **118** are also obtained from nitrobenzenes (deoxygenated by triethylphosphite) via arylnitrenes. A marked influence of the nucleophile upon the direction of the apparent migration of nitrene away or towards an ortho-subsititution has been noted (scheme-88)[62].

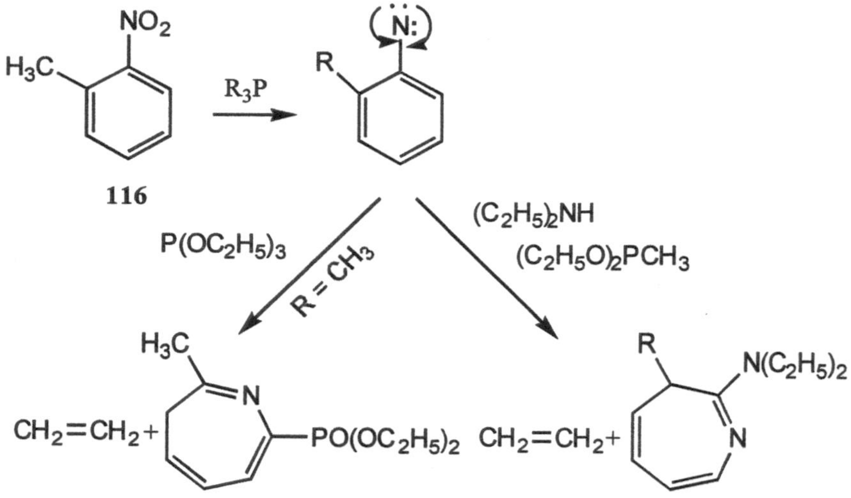

116 117 118

$R_3P = P(OC_2H_5)_3$ or $(C_2H_5O)_2PCH_3$

Scheme-88

Carbazoles **119c** are synthesized in high yields by thermolysis or photolysis of 2-azidobiphenyls, photolysis of 2-isocyanatobiphenyls and deoxygenation of nitro and nitroso aromatics. Singlet nitrene causes cyclization to carbazole (scheme-89)[63]. Sulfonylnitrene insertion into aromatic C–H bonds also yields heterocycles (scheme-90)[64]. Intramolecular addition reactions of carbenes (as α-

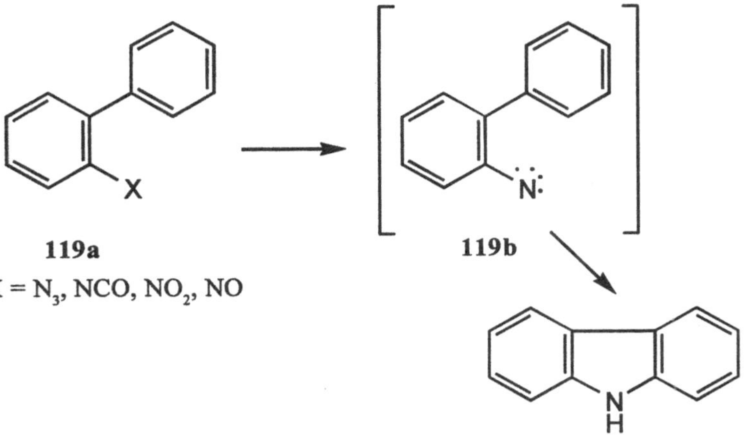

119a

X = N₃, NCO, NO₂, NO

119b

119c

Scheme-89

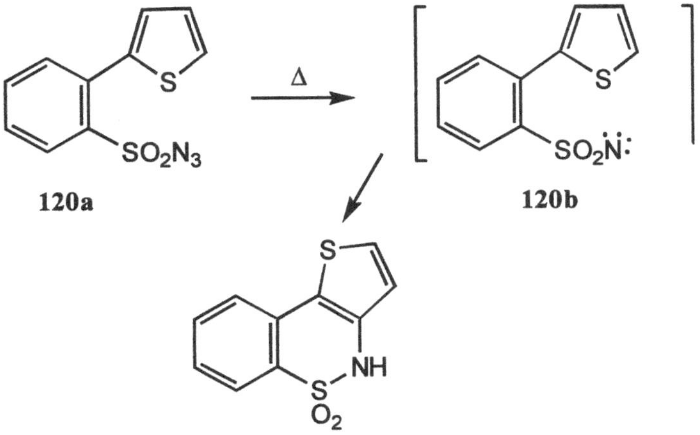

120a

120b

120c (61%)

Scheme-90

oxocarbenes) **121a,b** provide heterocycles **122a,b** (scheme-91)[65]. Thermolysis of sodium *o*-bromobenzenethiolate **123a** has been found to yield a 1 : 1 mixture of thianthrenes **124a** and **124b** via a benzothiirene intermediate **123b** (scheme-92)[66].

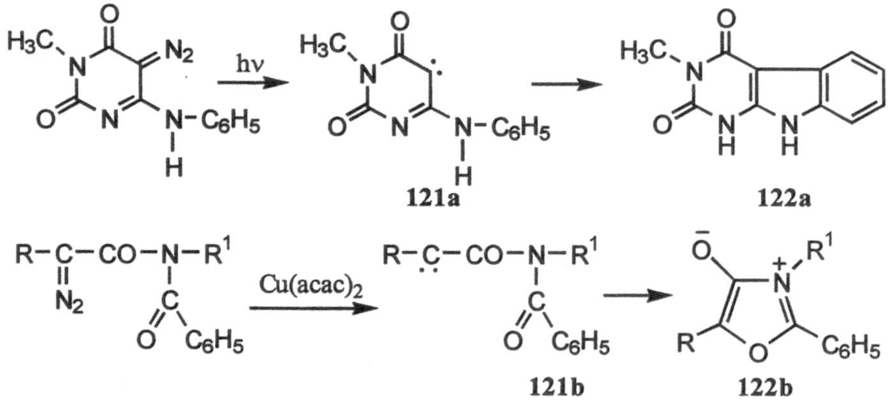

121a **122a**

121b **122b**

Scheme-91

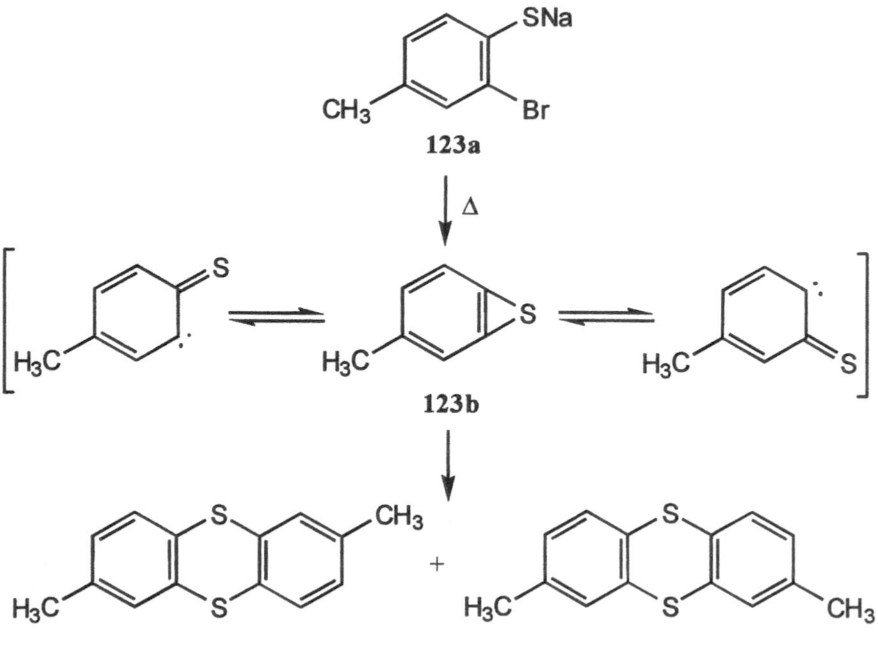

123a

123b

124a **124b**

Scheme-92

Synthesis of 1-sila, 1-germa and 1-stannaindanes **127** have been achieved in good yields by the multiple carbene-carbene rearrangements (scheme-93)[67]. The initial carbene **126** is obtained by flash pyrolysis of phenyldiazomethanes **125**.

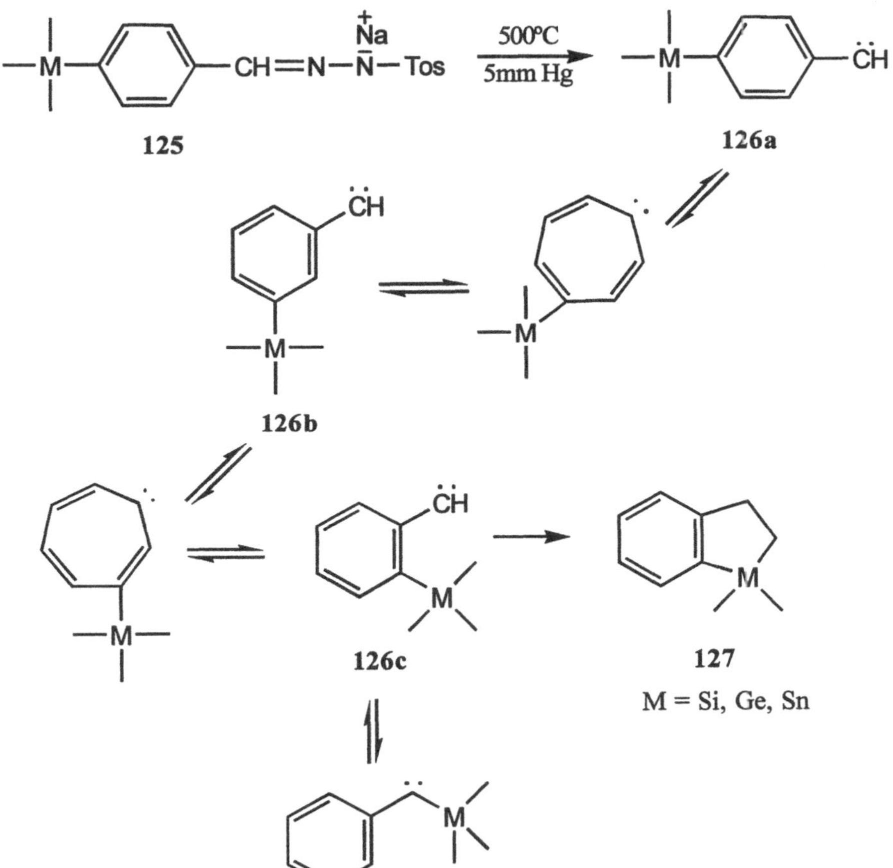

Scheme-93

The insertion of arylcarbenes **129** into ortho side chains yields dihydro-benzofurans and related compounds **130** (scheme-94)[68]. Thermolysis of diazo compound **132** results in aromatic C–H insertion of the resulting arylsulfonyl-carbene **133** to yield sulfoxide **134**. Addition to a benzenoid double bond also takes place and the ring expansion yields cycloheptatriene **135** (scheme-95)[69].

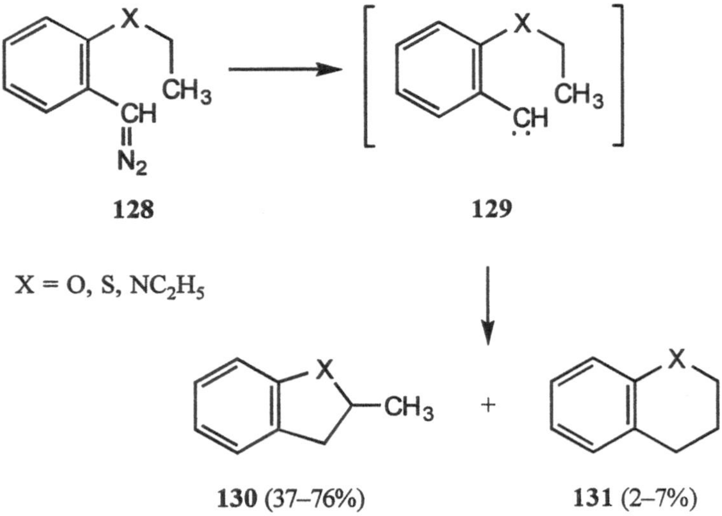

128   **129**

X = O, S, NC₂H₅

130 (37–76%)   +   **131** (2–7%)

Scheme-94

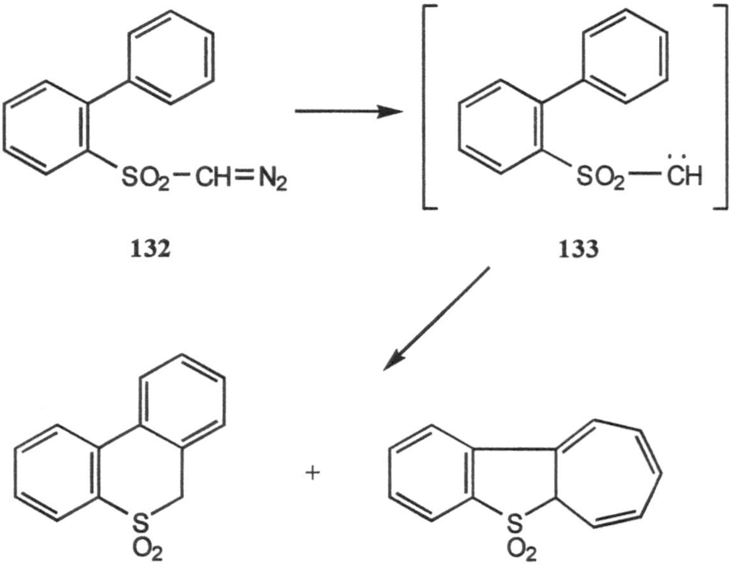

132   **133**

134   +   **135**

Scheme-95

2.1.4 Electrocyclic Reactions

An electrocyclic reaction involves the interconversion of two π-bonds into a σ-bond and a π-bond. An open chain reagent having kπ-electrons will give a cyclic product with (k–2)π-electrons and two electrons in a new σ-bond (Fig. 3). According to the principles of orbital symmetry conservation developed by Woodward and Hoffmann, the kπ-electrons must constitute a conjugated system and the formation of a new single bond and the rebonding in the resultant π-system of (k–2)π-electrons must occur in a concerted fashion.

6π-electrons   (6–2) = 4π-electrons + a σ-bond

Fig. 3

The reaction can be performed thermally or photochemically without any additional reagent and either condition is completely stereospecific. An equilibrium is set up between acyclic and cyclic counterparts. In cases where an acyclic counterpart predominates, the electrocyclic reaction may be a ring opening rather than a ring forming reaction. The electrocyclic ring closures are of two types. The first, in which p-orbitals of the π-electron system rotate in the same direction to form a new σ-bond is known as conrotatory process and the other, in which they rotate in the opposite direction is known as disrotatory process. The Woodward–Hoffmann rules for electrocyclic ring closure are based on the number of π-electrons in the open chain reagent and whether the reaction takes place in the ground state (thermal reaction) or in the first excited state (photochemical reaction). The selection rules for the electrocyclic reactions are summarized in Table 1.

Table 1. Selection rules for electrocyclic reactions

Number of π-electrons	Thermal	Photochemical
4n	Conrotatory	Disrotatory
4n + 2	Disrotatory	Conrotatory

However, when one of the terminal atoms of the open chain reagent is a heteroatom, stereochemical distinction between conrotatory and disrotatory process is lost. The commonly encountered electrocyclic reactions for the synthesis of heterocycles can be divided into four types :

(i) Reactions in which open chain reagent contains 4π-electrons in a 1,3-dipolar species. In general, such type of reaction can be represented as in Fig. 4. Here the

Fig. 4

ring opening is a more common process. Only a very few examples of the ring closure are available and are usually photochemical process. The electrocyclic ring closure of thiocarbonyl ylides **137** yields episulfides **138**. The best method of preparing thiocarbonyl ylides **137** is the thermal decomposition of thiadiazoline **136** which may or may not be isolated (scheme-96)[70].

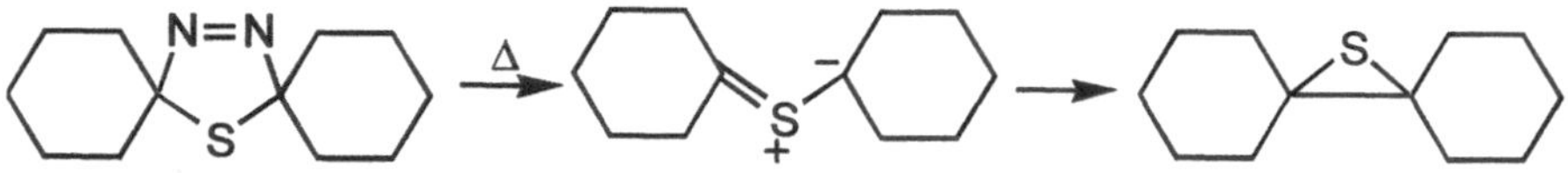

136a   **137a**   **138a**

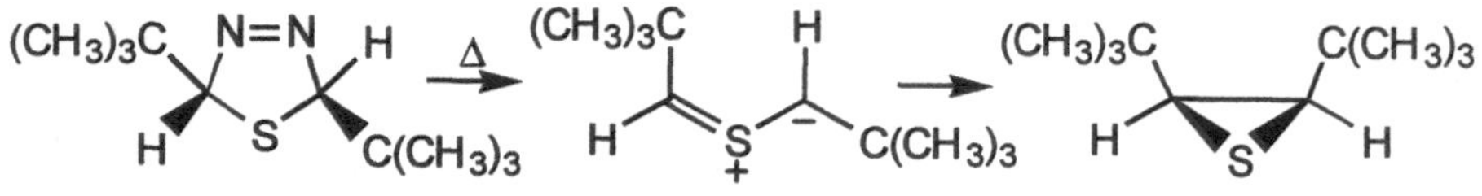

136b   **137b**   **138b**

Scheme-96

The ring closure of compound **139** yields 3-ethylthioxirane **140** which decomposes to give propionaldehyde (scheme-97)[71].

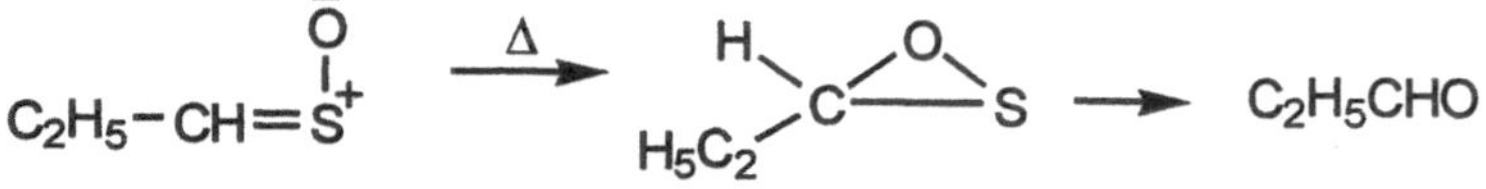

139   **140**

Scheme-97

An azoxy compound **141** is converted photochemically to an oxadiaziridine **142** which reverts thermally to the azoxy reactant (scheme-98)[72].

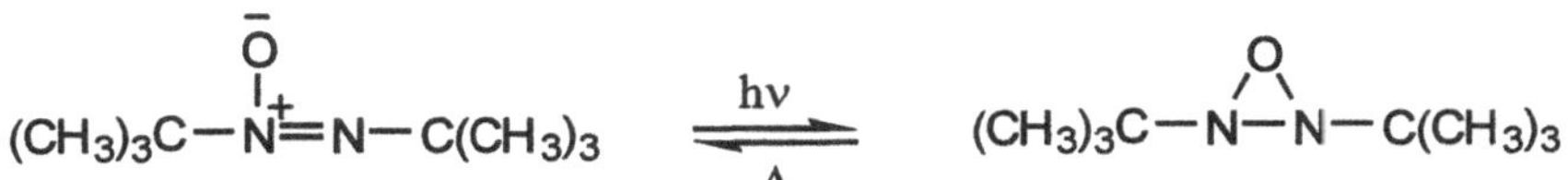

141 **141**

Scheme-98

(ii) Reactions in which the open chain reagent contains 4π-electrons in a heterodiene (Fig. 5). The ring opening is a more common process. A few example

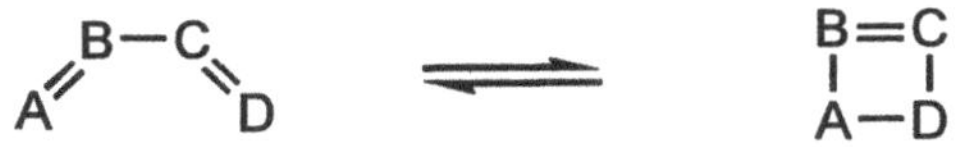

Fig. 5

of the electrocyclic ring closures are described; 2, 3, 4, 4-tetramethyloxetene **144** is obtained by the irradiation of 3,4-dimethyl-3-penten-2-one **143**. Oxetene **144** is reverted thermally to the reactant (scheme-99)[73]. In some cases, *o*-quinomethides **145** cyclize to benzoxetenes **146** (scheme-100)[74].

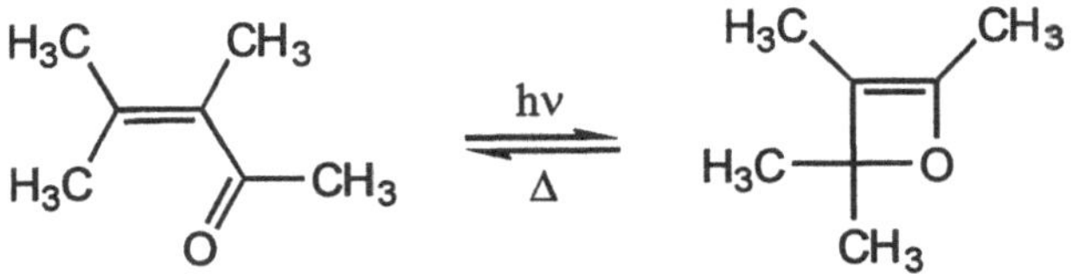

143 **144**

Scheme-99

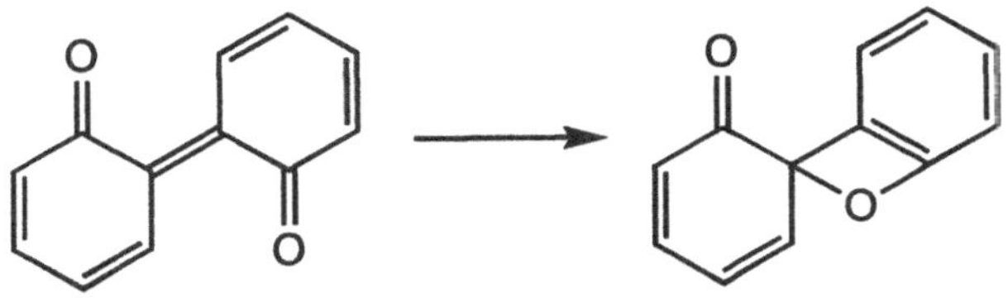

145 **146**

Scheme-100

A cyclobutene **148** having three heteroatoms is formed only as an intermediate in the thermal decomposition of compound **147** (scheme-101)[75].

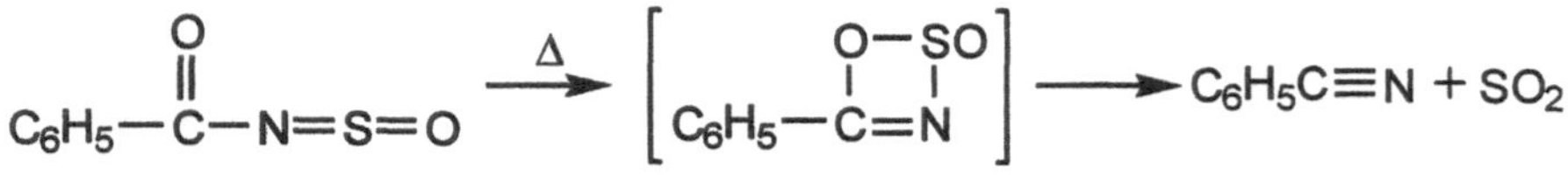

$$C_6H_5-\overset{\overset{O}{\|}}{C}-N=S=O \quad\xrightarrow{\Delta}\quad \left[C_6H_5-\overset{\overset{O-SO}{|\ \ \ |}}{C}=N \right] \quad\longrightarrow\quad C_6H_5C\equiv N + SO_2$$

      147   **148**

Scheme-101

(iii) Reactions in which an open chain reagent contains 6π-electrons in a 1,5-dipolar species (Fig. 6). Such types of reactions are more common and referred to as 1,5-dipolar cyclization reactions. Both ring opening and ring closure reactions are observed. Some examples of the ring closure reactions are described which are usually thermally induced process.

Fig. 6

The heterolysis of an oxirane **149** yields a dipolar intermediate **150** which undergoes ring closure with the formation of **151** (scheme-102)[76].

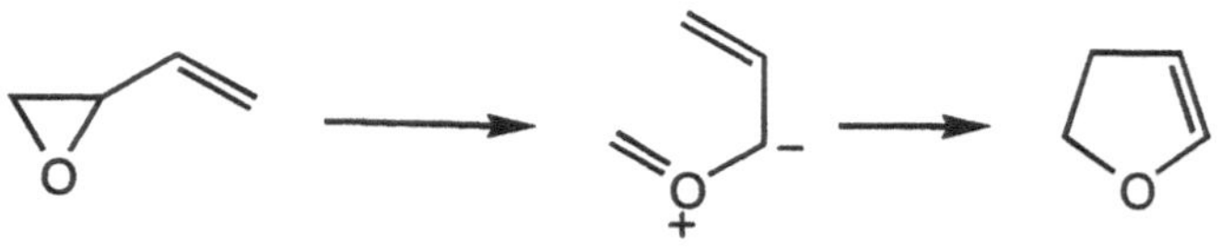

    149   **150**   **151**

Scheme-102

Indolizines **153** are obtained by the electrocyclization of pyridinium ylides **152** (scheme-103)[77].

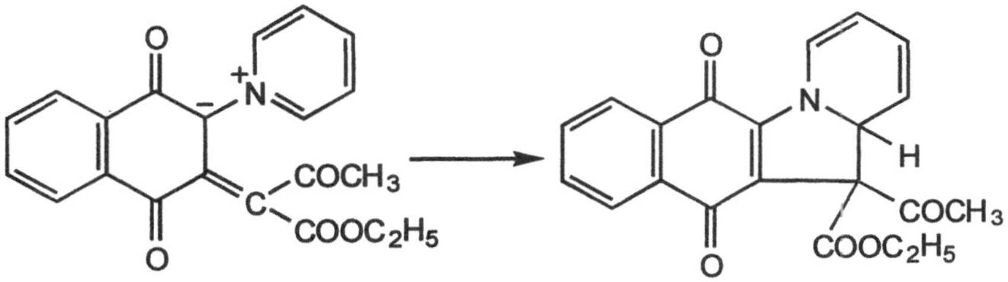

152   **153**

Scheme-103

1,5-Dipolar electrocyclization of thiocarbonyl ylides **156** yields dihydrothiophenes **157** and eventually thiophenes **158** (scheme-104)[78].

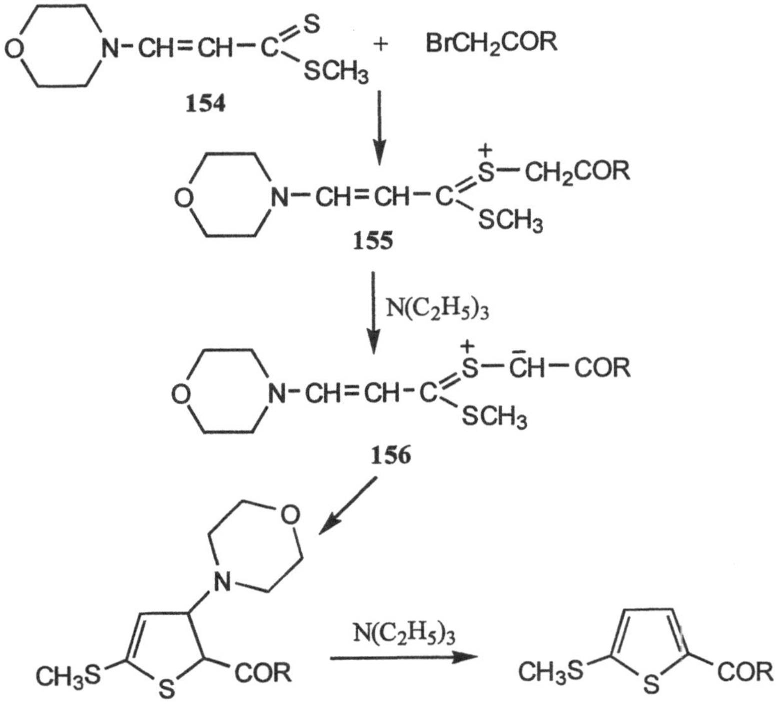

154

155

$N(C_2H_5)_3$

156

$N(C_2H_5)_3$

157   **158**

Scheme-104

Vinyl substituted nitrile ylides 160 obtained by the photolysis of arylazirines 159 cyclize to form arylpyrroles 162. The vinyl substituent causes 1,5-electrocyclization (scheme-105)[79].

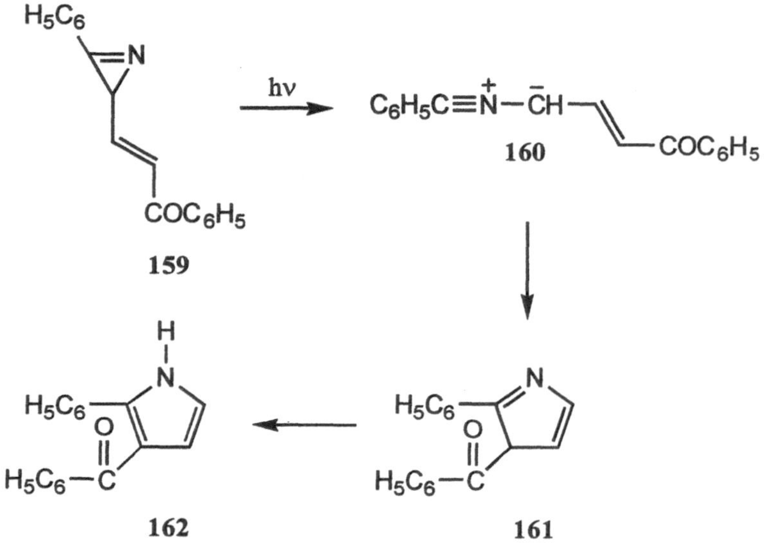

Scheme-105

Azomethine ylides 164 formed by the thermal opening of aziridines 163 yield oxazoles 165 (scheme-106)[80].

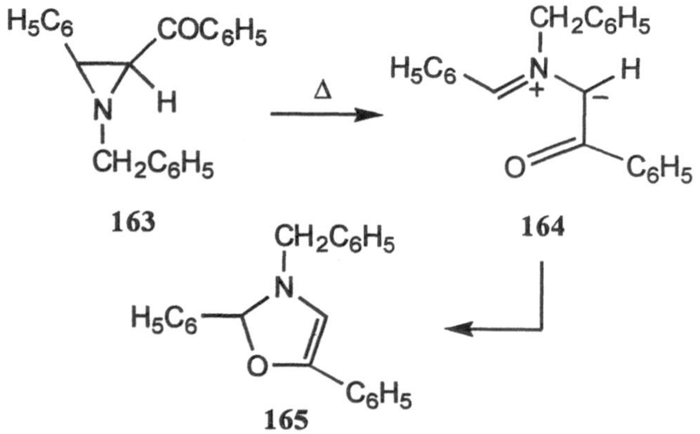

Scheme-106

Oxazoles **168** are obtained from acylazirines **166** via nitrile ylides **167** (scheme-107)[79]. Iminoazirine **169** on photochemical ring opening yields iminosubstituted

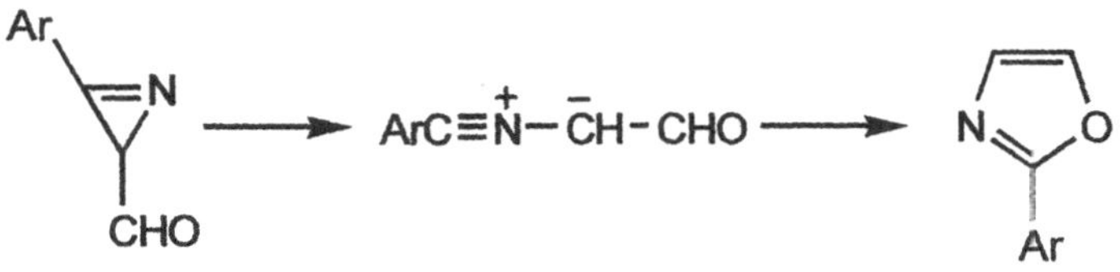

166 **168**

Scheme-107

nitrile ylide **170** which cyclizes to an imidazole **171** (scheme-108)[79]. Electrocyclization of thiocarbonyl imines (thione-S-imides) **172** affords heterocycles **173** (scheme-109)[81].

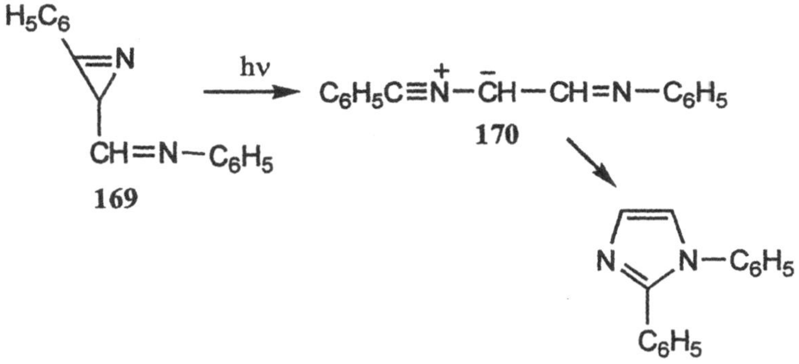

171

Scheme-108

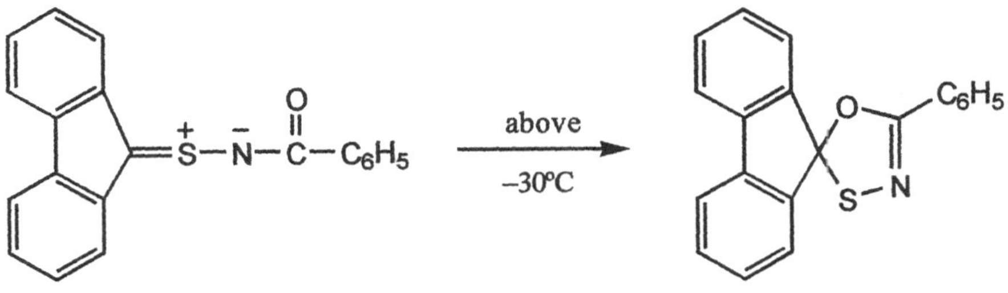

172 **173**

Scheme-109

(iv) Reactions in whicn open chain reagent contains 6π-electrons in a heterotriene (Fig. 7). Electrocyclic ring closure is a versatile method for the synthesis of six-membered heterocycles. The reverse reaction also occurs (scheme-110).

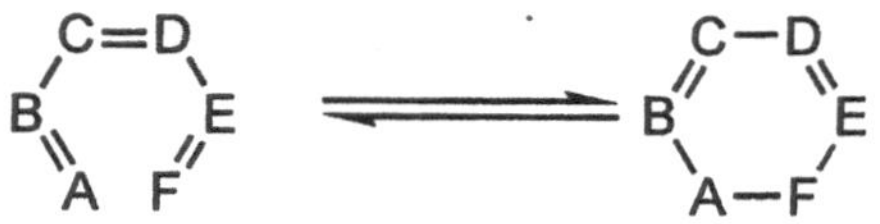

Fig. 7

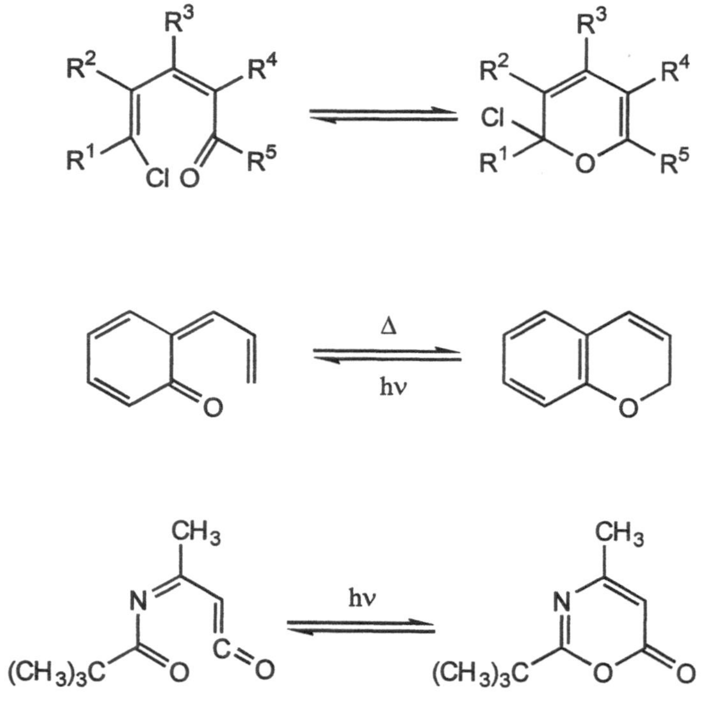

Scheme-110

Electrocyclization[82] of *o*-quinomethide **176** (obtained by Claisen rearrangement of **174** and 1,5-hydrogen shift) provides chromene **177** (scheme-111). 2-Pyrones **179** are synthesized from ketenes **178** having *cis*-enone substituent by the electrocyclization process (scheme-112)[83].

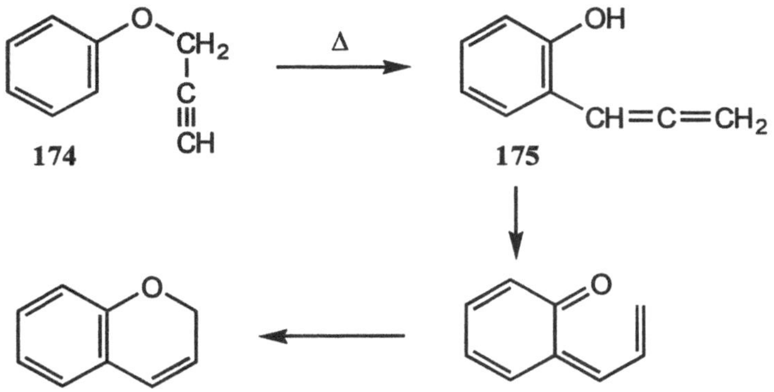

Scheme-111

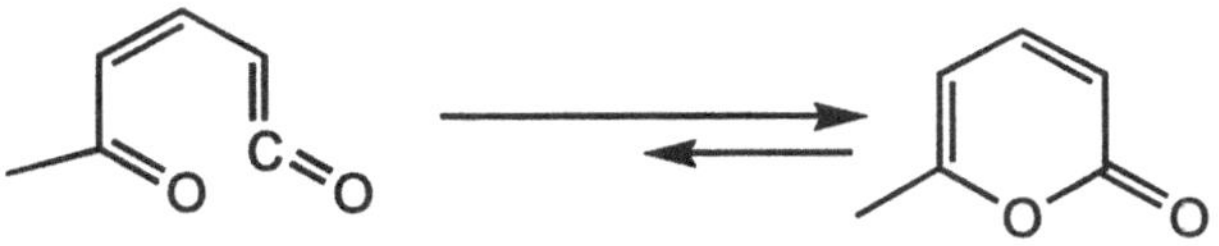

Scheme-112

Synthesis of coumarins **182** is achieved by the electrocyclic ring closure of isocyanates **181** (scheme-113)[84].

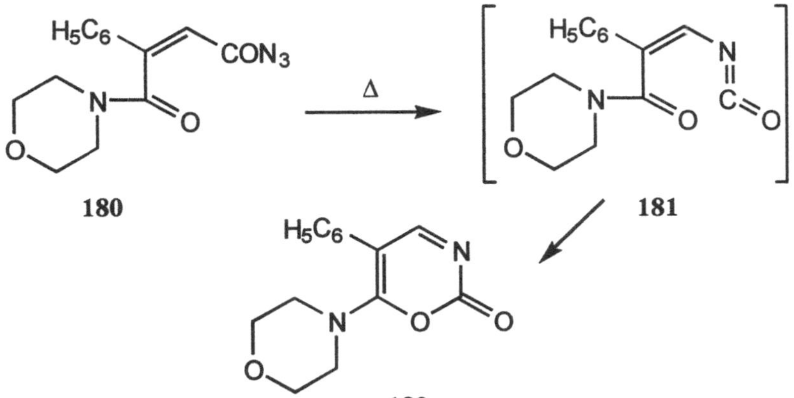

Scheme-113

Electrocyclization of dienthiones **184** yields thiapyrans **185** (scheme-114)[85].

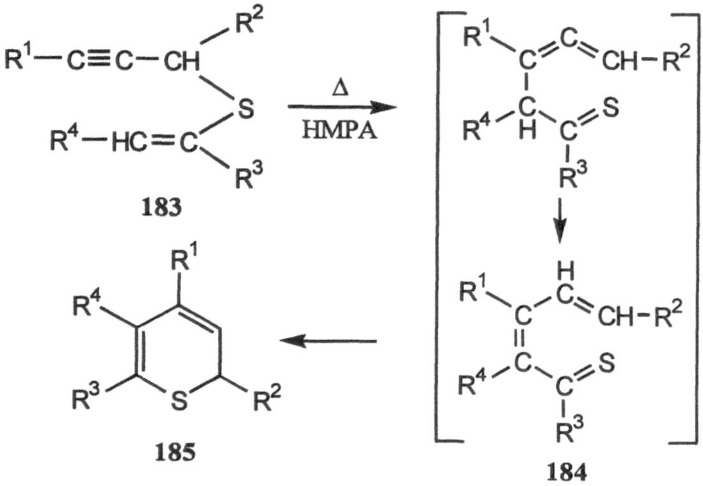

Scheme-114

Ring closure of konig's salts **186** by electrocyclic process yields N-phenyl-pyridinium ion **187** and aniline (scheme-115)[86].

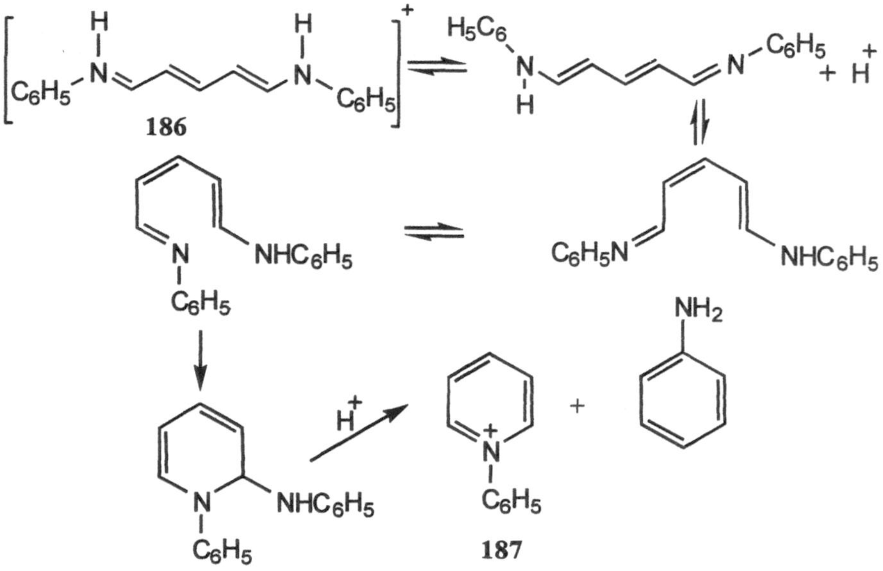

Scheme-115

Dienoximes **189** having a central cis double bond undergo reversible electrocyclic ring closure to N-hydroxydihydropyridines **190** which irreversibly yields pyridine-N-oxides **191** when a good leaving group is present at C-2 (scheme-116)[87].

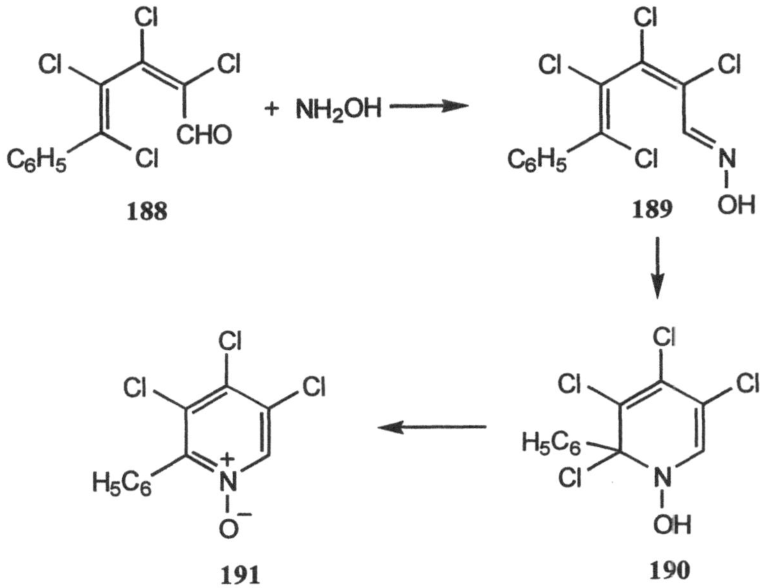

Scheme-116

2-Azatriene **193** undergoes electrocyclization to yield dihydropyridine derivative **194** (scheme-117)[88].

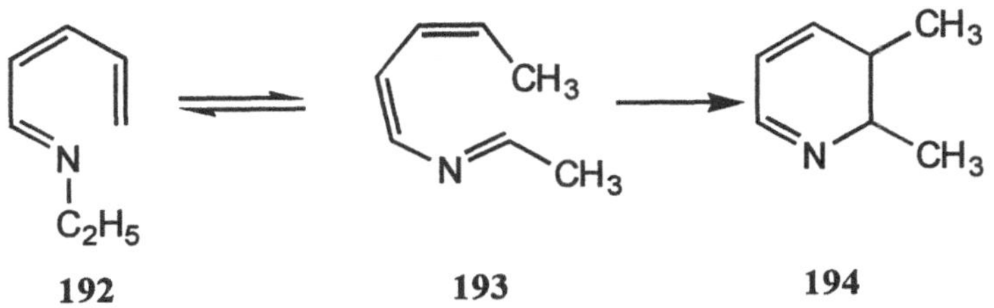

Scheme-117

The thermally induced electrocyclization of 1,4,8-triphenyl-2,5-diazooctatetraene **195** yields dihydropyrazine **196** (scheme-118)[89]. The extension of this reaction is used to synthesize condensed pyrazines **199** (scheme-119)[90].

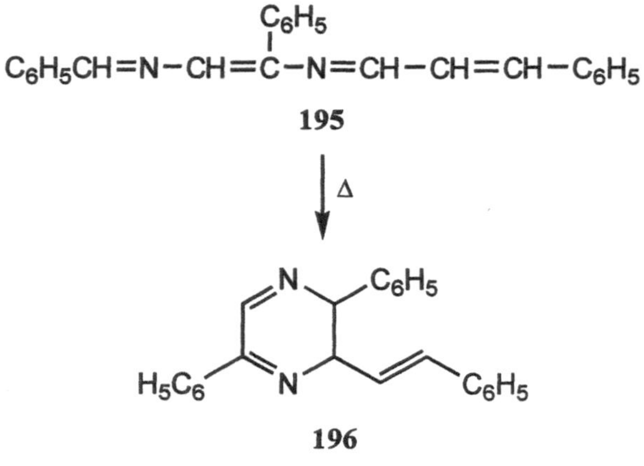

$$C_6H_5CH=N-CH=\underset{\underset{C_6H_5}{|}}{C}-N=CH-CH=CH-C_6H_5$$

195

$$\Delta$$

196

Scheme-118

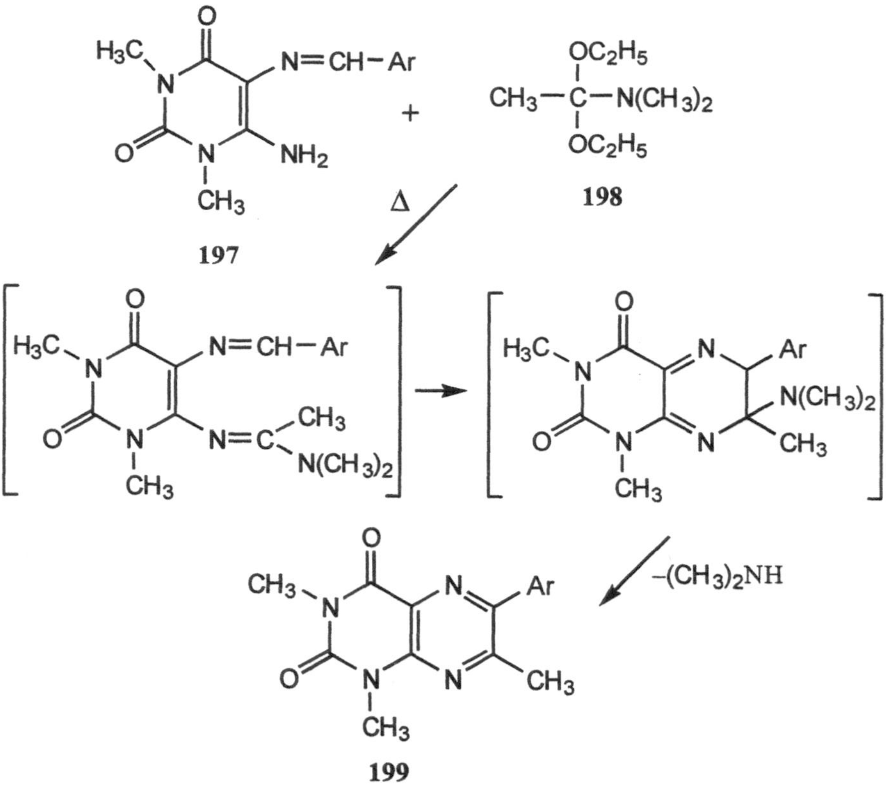

197 + **198**

$$\Delta$$

$$\rightarrow$$

$$-(CH_3)_2NH$$

199

Scheme-119

Isocyanates **201** and isothiocyanates 204 also undergo electrocyclization[91] to produce benzopyrimidine derivatives **203** and **205** (scheme-120).

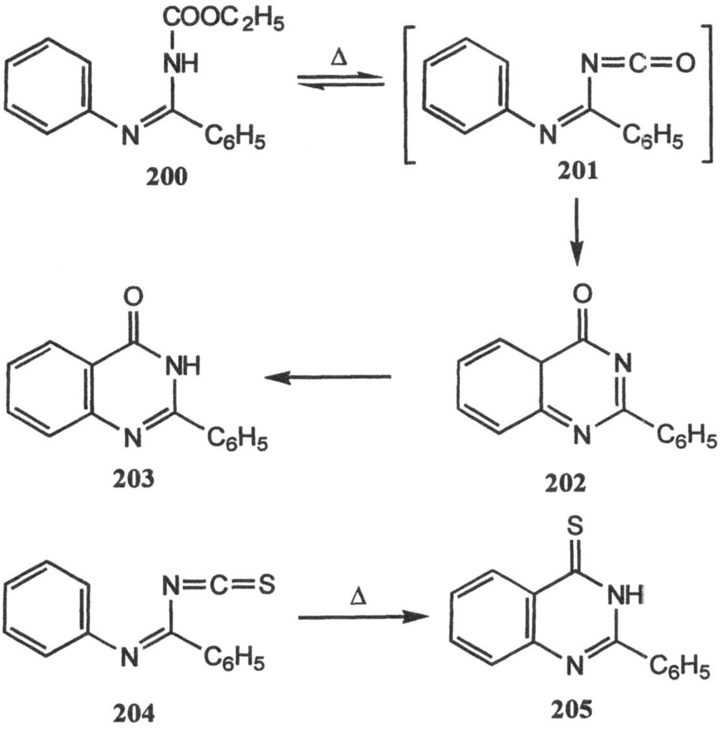

Scheme-120

Conversion of 2-azobenzaldoxime **206** into 1,2,3-benzotriazine type compound **208** can be considered as exocyclization which involves the addition of an unshared pair of electrons from oxime to diazonium ion (scheme-121)[92].

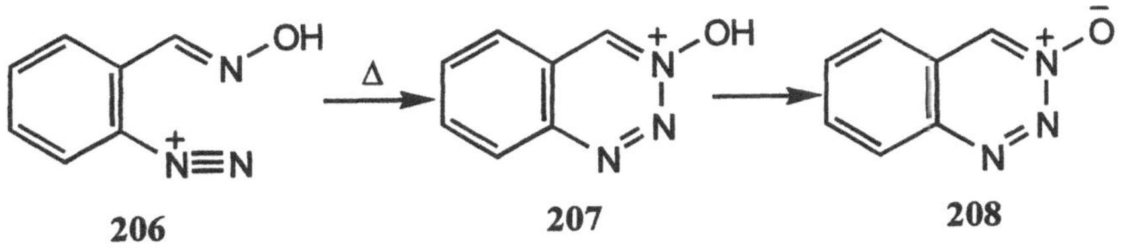

Scheme-121

The synthesis of pyrylium salts **212** from enamines **209** involves an electrocyclic-elimination reaction (scheme-122)[93]. The process of electocyclic-elimination has been extended for the synthesis of heterocycles e.g. quinolines **214**, pyrimidines **218**, *etc.* (scheme-123)[94].

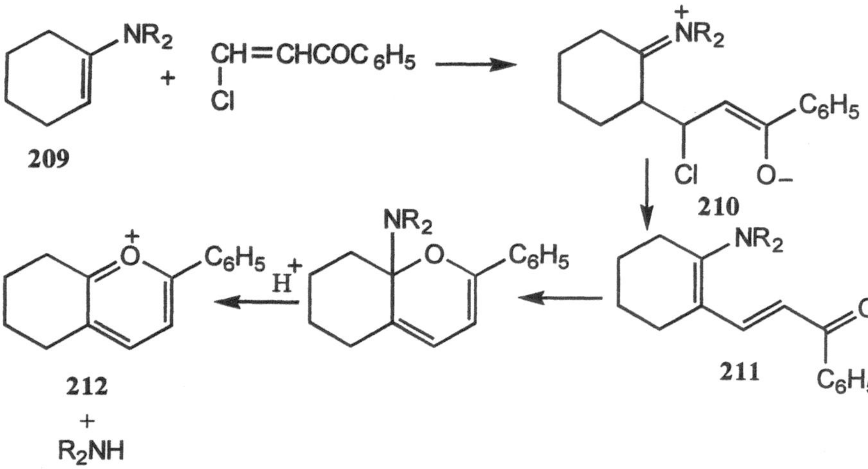

Scheme-122

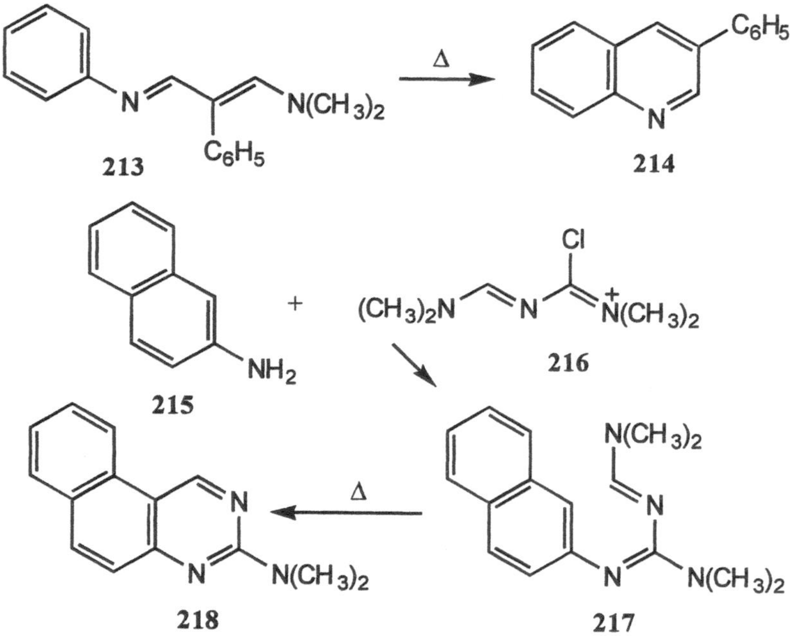

Scheme-123

2.2 Cycloaddition Reactions

The cycloaddition reactions represent a large group of reactions which involve the interaction of π-electron systems in a concerted manner and proceed via acyclic transition state. These are generally described as symmetry controlled reactions and the mechanisms are considered in terms of frontier orbital theory. The reactions occur by the interaction of the highest occupied molecular orbital (HOMO) of one component with the lowest unoccupied molecular orbital (LUMO) of the other. A wide variety of heterocycles especially containing four- to six-membered rings are prepared by the cycloaddition reactions. The important types of cycloaddition reactions for the construction of heterocyclic compounds includes 1,3-dipolar cycloaddition, hetero-Diels–Alder reactions, [2 + 2] cycloadditions, valence bond isomerization and cheletropic reactions.

2.2.1 1,3-Dipolar Reactions [3 + 2→5]

The cycloaddition reactions of type [3 + 2→5] constitute an extremely large number of examples, the major portion of which have been investigated by Huisgen and co-workers[95]. This group of transformations is more commonly referred to as 1,3-dipolar reactions. In such reactions, a neutral five-membered ring is formed by the combination of a dipolar species (generally a triatomic 1,3-dipole) and an unsaturated acceptor (dipolarophile). These reactions are analogous to Diels–Alder reaction in that they are concerted additions of [4 + 2]type[96]. These reactions are customarily represented as in Fig. 8 in which a–b–c is referred to as 1,3-dipolar molecule and d–e as dipolarophile, where 'a' has only six-electrons in the outer shell and 'c' has at least one pair of unshared electrons.

Fig. 8

1,3-Dipole can be defined as a three atom π-electron system with four π-electrons delocalized over three atoms. These 1,3-dipole species have more than one resonance structures. Each molecule has at least one resonance structure which indicates separation of opposite charges in 1,3-relationship (scheme-124). 1,3-Dipolar species contain a heteroatom as the central atom. In some cases the dipolar component is an isolable species, such as a diazoalkane or an azide, but usually it has to be generated as a reactive intermediate (scheme-124). Some important 1,3-dipolar compounds which undergo 1,3-dipolar cycloaddition are presented in Table 2.

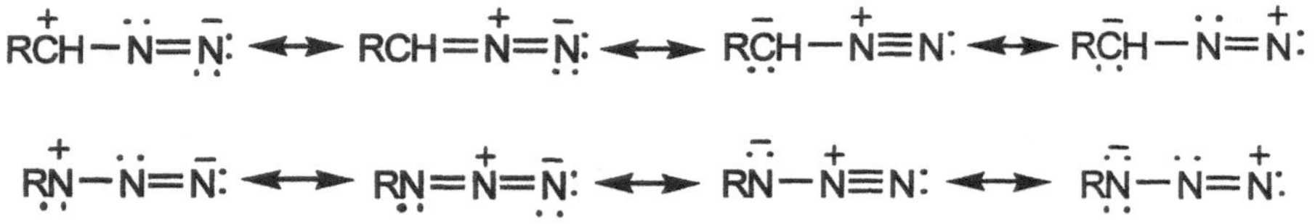

$$RCH-\overset{+}{N}=\overset{-}{N}: \longleftrightarrow RCH=\overset{+}{N}=\overset{-}{N}: \longleftrightarrow R\overset{..}{C}H-\overset{+}{N}\equiv N: \longleftrightarrow R\overset{..}{C}H-\overset{+}{N}=\overset{+}{N}:$$

$$R\overset{+}{N}-\overset{..}{N}=\overset{-}{N}: \longleftrightarrow R\overset{..}{N}=\overset{+}{N}=\overset{-}{N}: \longleftrightarrow R\overset{-}{N}-\overset{+}{N}\equiv N: \longleftrightarrow R\overset{-}{N}-\overset{..}{N}=\overset{+}{N}:$$

Scheme-124

Table 2. 1,3-Dipolar compounds

Name	Structure	
Diazoalkane	$\overset{-}{N}=N-\overset{+}{C}R_2 \longleftrightarrow \overset{-}{N}=\overset{+}{N}=CR_2$	
Azide	$\overset{-}{N}=N-\overset{+}{N}R \longleftrightarrow \overset{-}{N}=\overset{+}{N}=NR$	
Azomethine ylide	$R_2\overset{+}{C}-N-\overset{-}{C}R_2 \longleftrightarrow R_2C=\overset{+}{N}-\overset{-}{C}R_2$ $\quad\quad\; \mid \quad\quad\quad\quad\quad\quad\; \mid$ $\quad\quad\; R \quad\quad\quad\quad\quad\quad\; R$	
Azomethine imine	$R_2\overset{+}{C}-N-\overset{-}{N}R \longleftrightarrow R_2C=\overset{+}{N}-\overset{-}{N}R$ $\quad\quad\; \mid \quad\quad\quad\quad\quad\quad\; \mid$ $\quad\quad\; R \quad\quad\quad\quad\quad\quad\; R$	
Nitrone	$R_2\overset{-}{C}-\overset{+}{N}=O \longleftrightarrow R_2C=\overset{+}{N}-\overset{-}{O}$ $\quad\quad\; \mid \quad\quad\quad\quad\quad\quad\; \mid$ $\quad\quad\; R \quad\quad\quad\quad\quad\quad\; R$	
Azimine	$R\overset{+}{N}-N-\overset{-}{N}R \longleftrightarrow RN=\overset{+}{N}-\overset{-}{N}R$ $\quad\quad\; \mid \quad\quad\quad\quad\quad\quad\; \mid$ $\quad\quad\; R \quad\quad\quad\quad\quad\quad\; R$	

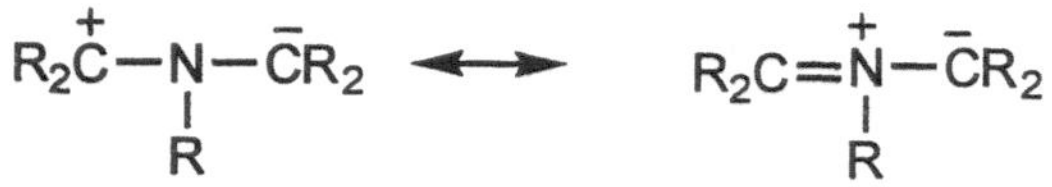

Name	Structure
Azoxy compound	$\overset{+}{R N}-\underset{\underset{R}{\mid}}{N}-\overset{-}{O}$ ⟷ $RN=\underset{\underset{R}{\mid}}{\overset{+}{N}}-\overset{-}{O}$
Nitro compound	$\overset{+}{O}-\underset{\underset{R}{\mid}}{N}-\overset{-}{O}$ ⟷ $O=\underset{\underset{R}{\mid}}{\overset{+}{N}}-\overset{-}{O}$
Nitroso imine	$\overset{+}{R N}-O-\overset{-}{N}R$ ⟷ $RN=\overset{+}{O}-\overset{-}{N}R$
Nitroso oxide	$\overset{+}{R N}-O-\overset{-}{O}$ ⟷ $RN=\overset{+}{O}-\overset{-}{O}$
Nitrile ylide	$R\overset{+}{C}=N-\overset{-}{C}R_2$ ⟷ $RC\equiv\overset{+}{N}-\overset{-}{C}R_2$
Nitrile imine	$R\overset{+}{C}=N-\overset{-}{N}R$ ⟷ $RC\equiv\overset{+}{N}-\overset{-}{N}R$
Nitrile oxide	$R\overset{+}{C}=N-\overset{-}{O}$ ⟷ $RC\equiv\overset{+}{N}-\overset{-}{O}$
Nitrile sulfide	$R\overset{+}{C}=N-\overset{-}{S}$ ⟷ $RC\equiv\overset{+}{N}-\overset{-}{S}$
Carbonyl ylide	$R_2\overset{+}{C}-O-\overset{-}{C}R_2$ ⟷ $R_2C=\overset{+}{O}-\overset{-}{C}R_2$
Carbonyl oxide	$R_2\overset{+}{C}-O-\overset{-}{O}$ ⟷ $R_2C=\overset{+}{O}-\overset{-}{O}$
Carbonyl imine	$R_2\overset{+}{C}-O-\overset{-}{N}R$ ⟷ $R_2C=\overset{+}{O}-\overset{-}{N}R$
Ozone	$\overset{+}{O}-O-\overset{-}{O}$ ⟷ $O=\overset{+}{O}-\overset{-}{O}$
Thiocarbonyl ylide	$R_2\overset{+}{C}-S-\overset{-}{C}R_2$ ⟷ $R_2C=\overset{+}{S}-\overset{-}{C}R_2$
Selenocarbonyl ylide	$R_2\overset{+}{C}-Se-\overset{-}{C}R_2$ ⟷ $R_2C=\overset{+}{Se}-\overset{-}{C}R_2$

R = alkyl / aryl

The other reactants utilized in dipolar cycloaddition are known as dipolarophiles and contain either double or triple bond between two carbon atoms, a carbon atom and a heteroatom, or two heteroatoms. Scheme-125 lists some commonly used dipolarophiles. Variations in the structures of both the 1,3-dipolar compounds

Ethylenic dipolarophiles

$CH_2=CH_2$, $CH_2=CHR$, $CH_2=CHOR$

$CH_2=CHNR_2$, $CH_2=CHCOR$, $CH_2=CHCO_2R$, $CH_2=CHCN$,

$R'CH=CHNO_2$, $R'C(Cl)=CHNO_2$, $R'CH=CH_2$, $R'CH=CHR'$,

$NCCH=CHCN$, $RCO_2CH=CHCO_2R$,

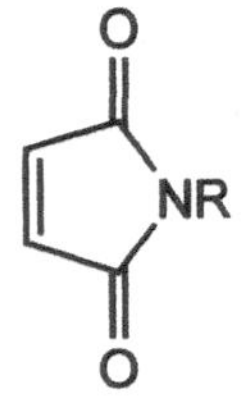

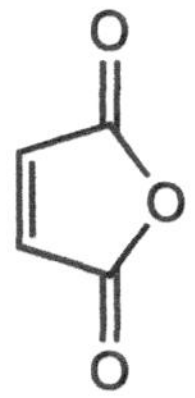

Heterocumulenes

$R'CON=C=O$, $R'CON=C=S$,

$RN=C=O$, $RN=C=S$

Acetylenic dipolarophiles

$R''C\equiv CH$, $R''C\equiv CR''$, $HC\equiv CCO_2R$,

$R''C\equiv CCO_2R$, $R'COC\equiv CCOR$,

$NCC\equiv CCN$, $CF_3C\equiv CCF_3$, $RO_2CC\equiv CCO_2R$, benzyne

R=alkyl, R'=alkyl/ aryl, R''=aryl / heteroaryl

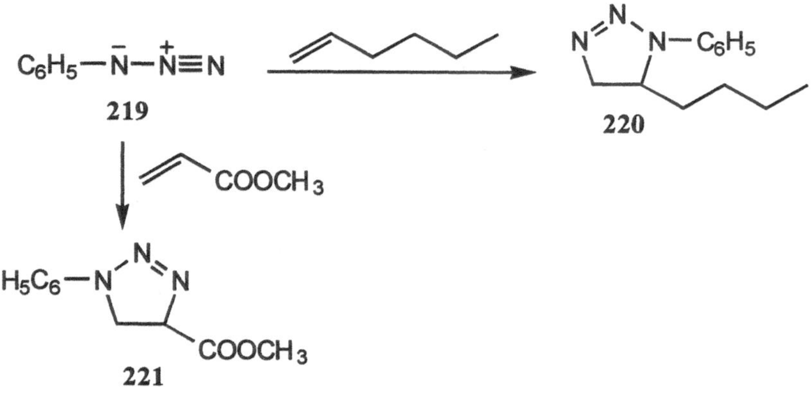

Scheme-125

and dipolarophiles make this a versatile and useful reaction in the synthesis of heterocyclic compounds. The most significant structural feature of 1,3-dipolar compounds is that they possess a π-system containing four electrons over three atoms and are isoelectronic with the alkyl anion. According to the Woodward–Hoffmann rules, 1,3-dipolar cycloaddition is a thermally allowed process[97]. The transition state of 1,3-dipolar cycloaddition can be represented as in Fig. 9, in which 4π-electron system of the dipole interacts with 2π-electron system of the dipolarophile. Frontier orbital theory provides a means of accounting for the variations in the reactivities of 1,3-dipoles and dipolarophiles[98]. The formation of transition state I (Fig. 9) will take place, if there is a favourable interaction between a filled

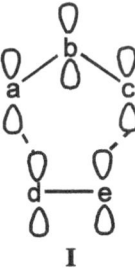

Fig. 9

π-orbital of one reactant and an empty π*-orbital of the other. For such favourable interactions, the conditions required are; the interacting orbitals are of the correct phase and energy and the interaction must be sterically feasible. These interactions are, therefore, dominated by the highest occupied π-orbitals (HOMO) and the lowest unoccupied π-orbitals (LUMO) of the two reactants. The relative energies of these so called frontier orbitals can vary as shown in Fig. 10.

Reactions are favoured if one component is strongly 'nucleophilic' and the other is strongly 'electrophilic'. The more electrophilic dipolarophiles have the lower energy LUMO values, whereas the more nucleophilic species have the higher energy HOMO values. Therefore, for a given 1,3-dipole molecule, the dominant interaction is of type :

(i) HOMO (dipole) and LUMO (dipolarophile) for electron-deficient dipolarophile, and

(ii) LUMO (dipole) and HOMO (dipolarophile) for electron-rich dipolarophiles.

Often the reactions are not performed with the simple unsubstituted dipoles and, therefore, the effect on the energies of the substituent(s) present have to be assessed. 1,3-Dipolar cycloadditions are highyl regioselective, for example, the reaction of phenyl azide with 1-hexene, which is a LUMO(dipole)–HOMO (dipolarophile) interaction, gives mainly 1-phenyl-5-butyltriazoline 220. But with methyl acrylate, 1-phenyl-4-carboxylic ester **221** is formed (scheme-125). The reaction is HOMO(dipole)–LUMO(dipolarophile) controlled because of the lowering of the energies of the frontier orbitals of dipolarophile by the ester group and change in the relative magnitudes of the atomic orbital coefficients in the interacting frontier orbitals of the two components[99]. In some reactions electronically preferred orientations may be disfavoured by the steric effects[100].

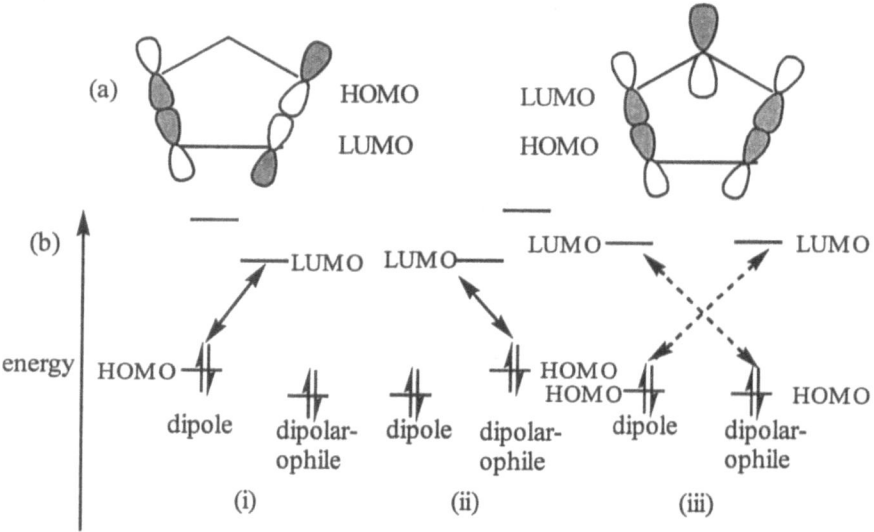

Fig. 10   (a) Frontier orbital combinations in 1,3-dipolar addition
(b) Types of interactions
   (i) HOMO (dipole)–LUMO (dipolarophile) dominant
   (ii) LUMO (dipole)–HOMO (dipolarophile) dominant
   (iii) Neither dominant.

1,3-Dipolar cycloaddition reactions are also highly stereoselective, however a few exceptions are due to the isomerization either before or after the cycloaddition[101]. For example, *cis* azomethine ylide **222** with acetylene dicarboxylic ester yields exclusively pyrrolidine tetracarboxylate **224** with cis ester groups at C-2 and C-5 (scheme-126). The isomeric *trans*-azomethine **225a** ylide provides only *trans*-pyrrolidine derivative **225b** (scheme-127)[102]. In cases, where two chiral centers are created in the cycloaddition interplay of two generally opposing forces in the transition state; attractive π-orbital overlap of unsaturated substituents favours an endo-transition state, and repulsive van der Waals steric interactions favour an exo-transition state. Thus, in the cycloaddition reactions of N-methyl-C-phenylnitrone **226** with a dipolarophile the stereochemical results are not consistent, sometimes favour *endo*-**227a** and sometimes *exo*-**228a** transition state (scheme-128)[103].

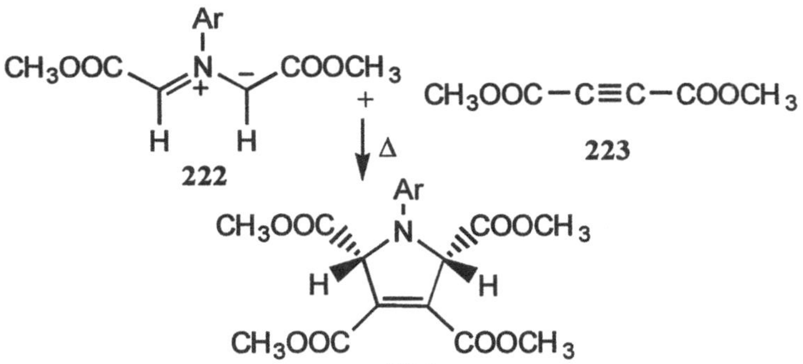

Scheme-126

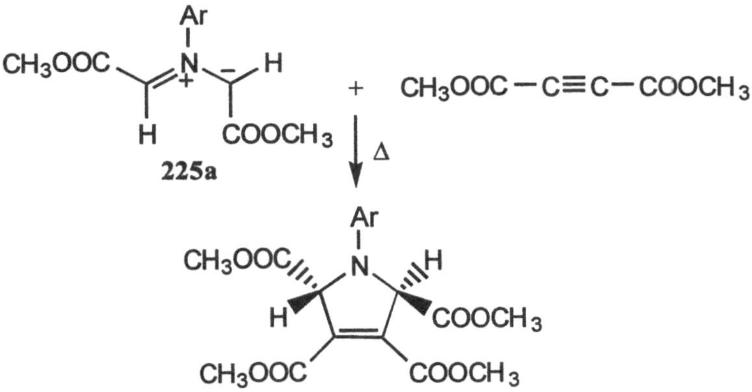

Scheme-127

The cycloaddition reactions of some widely used 1,3-dipolar compounds e.g., azomethine ylides, nitrile oxides, nitrones, diazoalkanes, azides, *etc.* are discussed. Azomethine ylides **229** on cycloaddition with olefinic and acetylenic dipolarophiles yield pyrrolidines **230** and Δ³-pyrrolines **231**, respectively. Δ³-Pyrrolines can be easily converted to pyrrole (scheme-129).

Azomethine ylides required for the cycloaddition are generated in situ. The synthesis of stabilised azomethine ylides with an electron-withdrawing substituent involves the thermolysis or photolysis of substituted aziridines. The generated azomethine ylides in the presence of dipolarophiles react in a stereospecific manner to yield pyrrolidines or pyrrolines directly in one step. The cycloaddition reactions of monosubstituted olefins are regioselective (scheme-130).

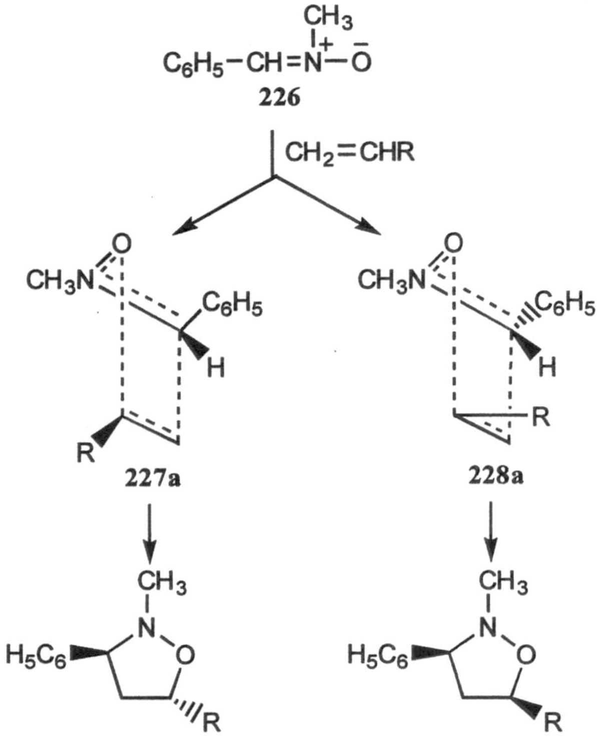

Scheme-128

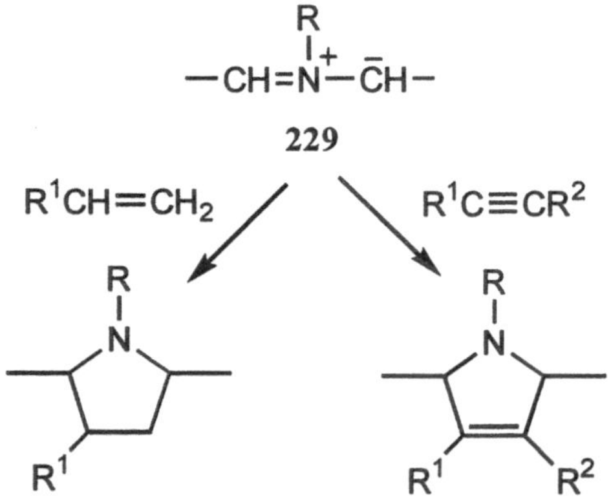

Scheme-129

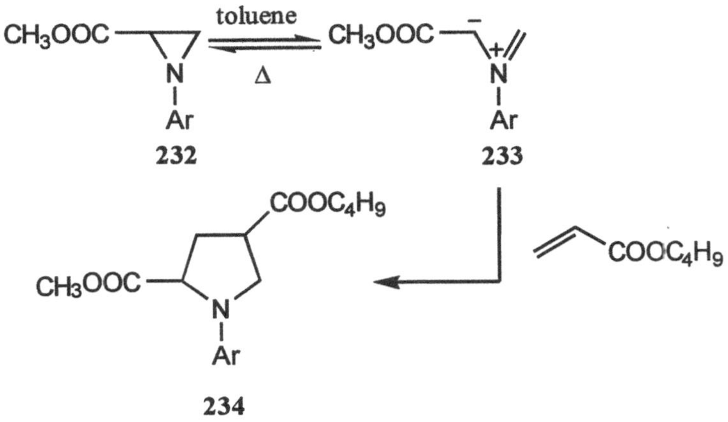

Scheme-130

Intramolecular cycloaddition reaction of azomethine ylide **236** (obtained from aziridine **235**) occurs readily yielding pyrrolidine **237** as a single stereoisomer (scheme-131)[104]. As compared to stabilized azomethine ylides non-stabilized ones

Scheme-130

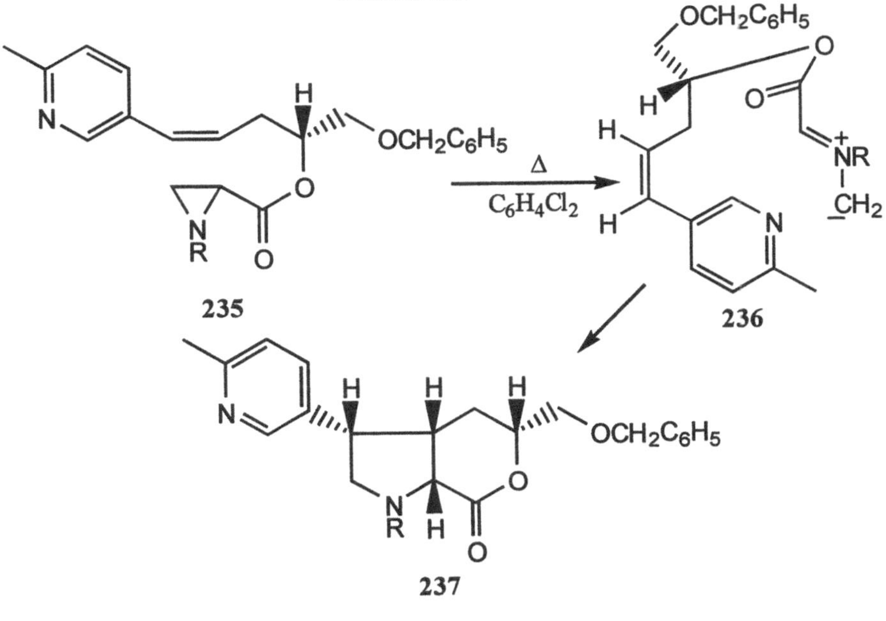

Scheme-131

are not as readily available, however one important method involves the desilylation of trimethylsilylmethyl iminium salts. Azomethine ylide generated by the reaction of imine **238** and trimethylsilylmethyl triflate in acetonitrile, and treated subsequently with cesium fluoride and with dimethyl acetylenedicarboxylate to give pyrroline derivative **241** (scheme-132)[105].

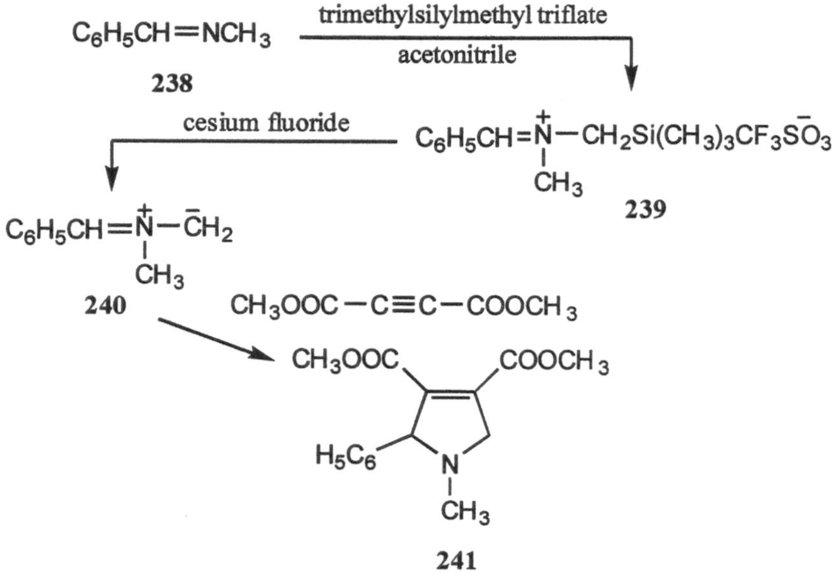

$C_6H_5CH=NCH_3$ $\xrightarrow[\text{acetonitrile}]{\text{trimethylsilylmethyl triflate}}$

238

$\xrightarrow[\text{cesium fluoride}]{}$ $C_6H_5CH=\overset{+}{N}-CH_2Si(CH_3)_3CF_3\overset{-}{S}O_3$
$\qquad\qquad\qquad\qquad\qquad\qquad\quad |$
$\qquad\qquad\qquad\qquad\qquad\qquad CH_3$

239

$C_6H_5CH=\overset{+}{N}-\overset{-}{C}H_2$
$\qquad\quad |$
$\qquad\quad CH_3$

240 $CH_3OOC-C\equiv C-COOCH_3$

241

Scheme-132

Nitrile oxides react readily with ethylenic and acetylenic dipolarophiles to form Δ^2-isoxazolines and isoxazoles (scheme-133)[106]. Usually, nitrile oxides are generated

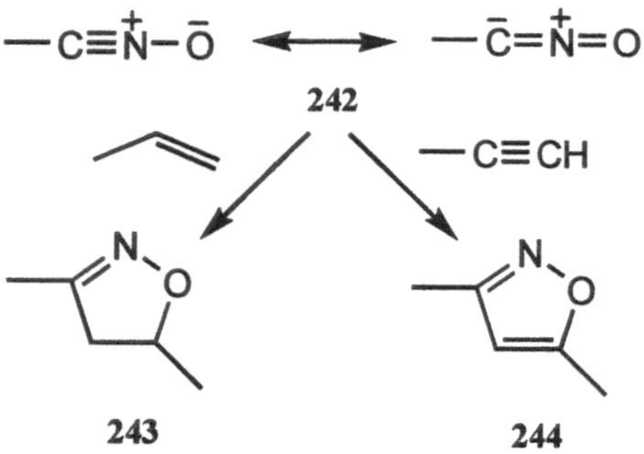

$-C\equiv\overset{+}{N}-\overset{-}{O}\quad\longleftrightarrow\quad-\overset{-}{C}=\overset{+}{N}=O$

242

$-C\equiv CH$

243   **244**

Scheme-133

in situ. The commonly used methods include, the dehydrochlorination of hydroxymoyl chlorides with triethylamine, dehydration of primary nitro compounds with an aryl isocyanate or by oxidation of aldoximes. The cycloaddition of nitrile oxide **246** obtained from tetrahydropyranyl derivative of 2-nitroethanol **245** provides heterocycle **247** (scheme-134)[107].

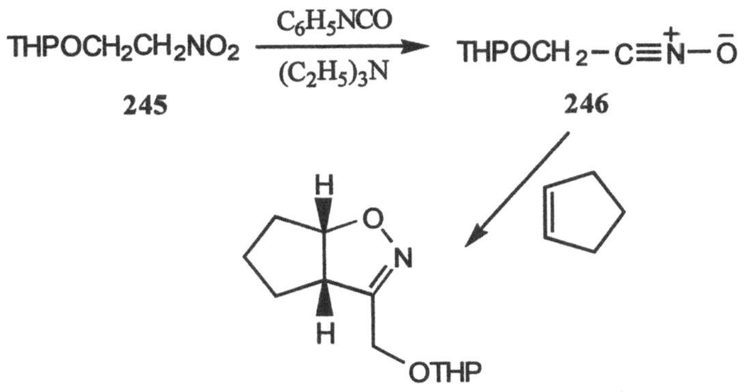

$$\text{THPOCH}_2\text{CH}_2\text{NO}_2 \xrightarrow[\text{(C}_2\text{H}_5)_3\text{N}]{\text{C}_6\text{H}_5\text{NCO}} \text{THPOCH}_2-\text{C}\equiv\overset{+}{\text{N}}-\bar{\text{O}}$$

245   **246**

247

Scheme-134

Isoxazolines **250** have been obtained in good yields by the reaction of nitrile oxide **249** with methyl vinyl ketone (scheme-135)[108]. The nitrile oxide **252** obtained

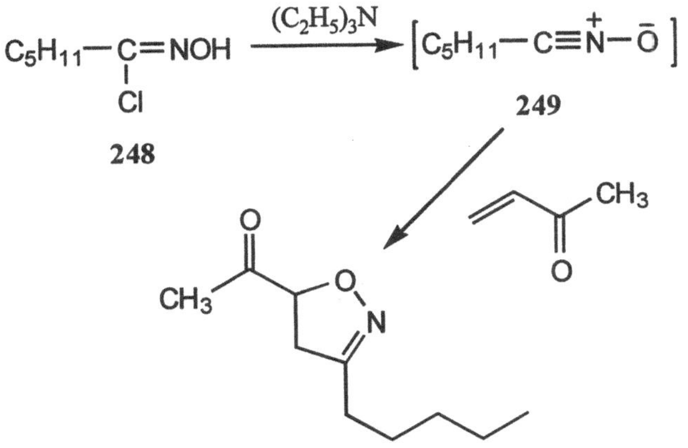

$$\text{C}_5\text{H}_{11}-\underset{\underset{\text{Cl}}{|}}{\text{C}}=\text{NOH} \xrightarrow{\text{(C}_2\text{H}_5)_3\text{N}} \left[\text{C}_5\text{H}_{11}-\text{C}\equiv\overset{+}{\text{N}}-\bar{\text{O}}\right]$$

248   **249**

250

Scheme-135

from ω-nitroalkene **251** undergoes intramolecular cycloaddition to provide heterocycle **253** (scheme-136)[109]. Reaction of (+)-(S)-isopropylidene-3-butene-1,2-

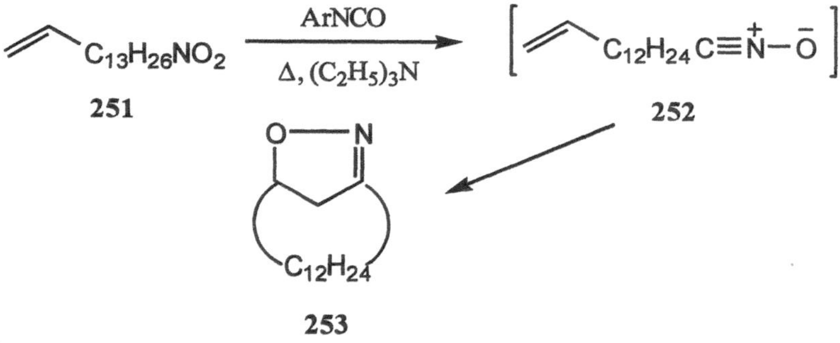

Scheme-136

diol **254** and carbethoxyformonitrile oxide provides a 4:1 mixture of *erythro-* and *threo*-isoxazolines **255** and **256** (scheme-137). As compared to intermolecular addition, intramolecular cycloaddition is often achieved with an excellent selectivity e.g. cycloaddition of nitroalkene **257** yields cycloadduct **258** in excellent yield (scheme-138).

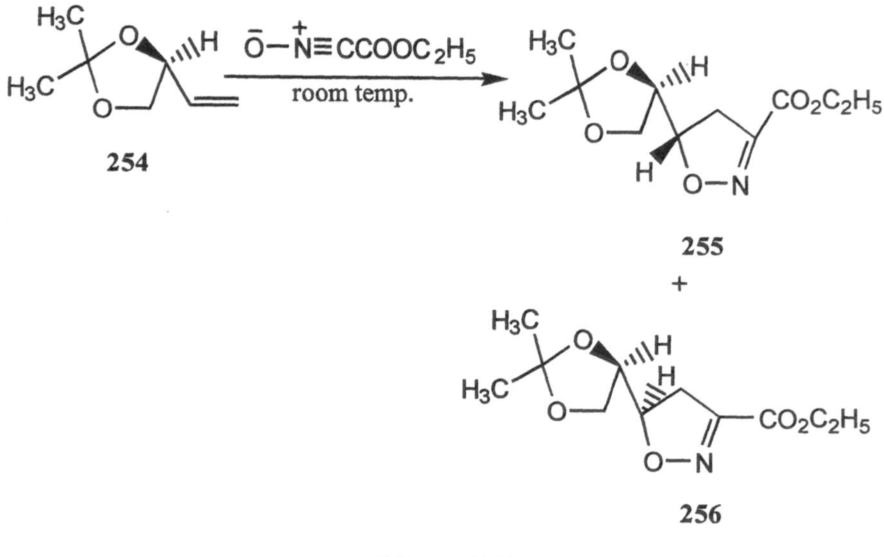

Scheme-137

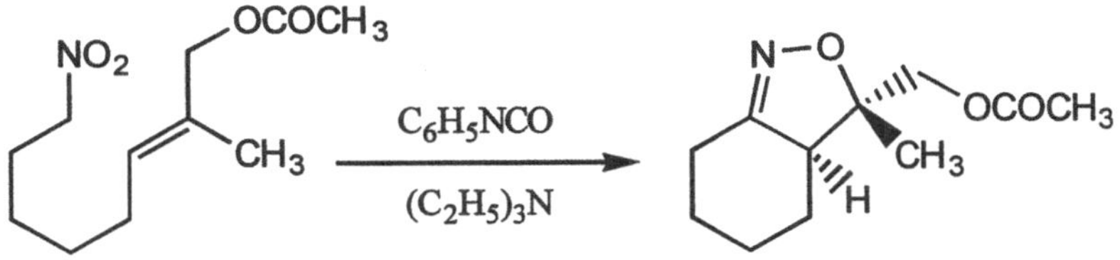

Scheme-138

Cycloaddition reactions of nitrones with ethylenic and acetylenic dipolarophiles yield isoxazolidines **260** and Δ^2-isoxazolines **261**, respectively (scheme-139)[110].

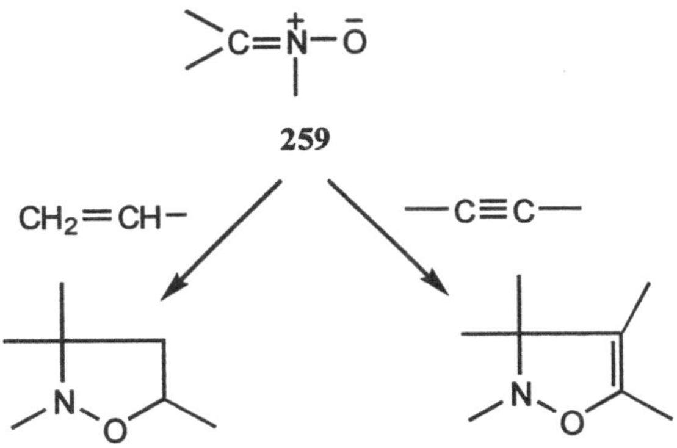

Scheme-139

Cycloaddition of N-methyl-C-phenylnitrone **262** with acrylonitrile yields predominantly *trans*-product **263**, whereas with nitroethylene yields *cis*-isomer **265** as the main product (scheme-140)[111]. Cis-trans isomerisation of nitrone is ruled out in case of cyclic nitrones and stereoselectivity is higher e.g. the cycloaddition of 4-phenylbut-1-ene **268** yields only one adduct **269** via exo transition state (scheme-141)[112].

Intramolecular cycloaddition of C-alk-4-enylnitrone **271** yields exclusively *cis*-fused cycloadduct **272** (scheme-142). However, C-alk-5-enylnitrone **274** yields a mixture of *cis*- and *trans*-ring-fused products **275** due to the greater flexibility allowed by the longer connecting chain. The bridged ring product **276** is also formed by the attack of oxygen atom of nitrone at the end of the double bond (scheme-143)[113].

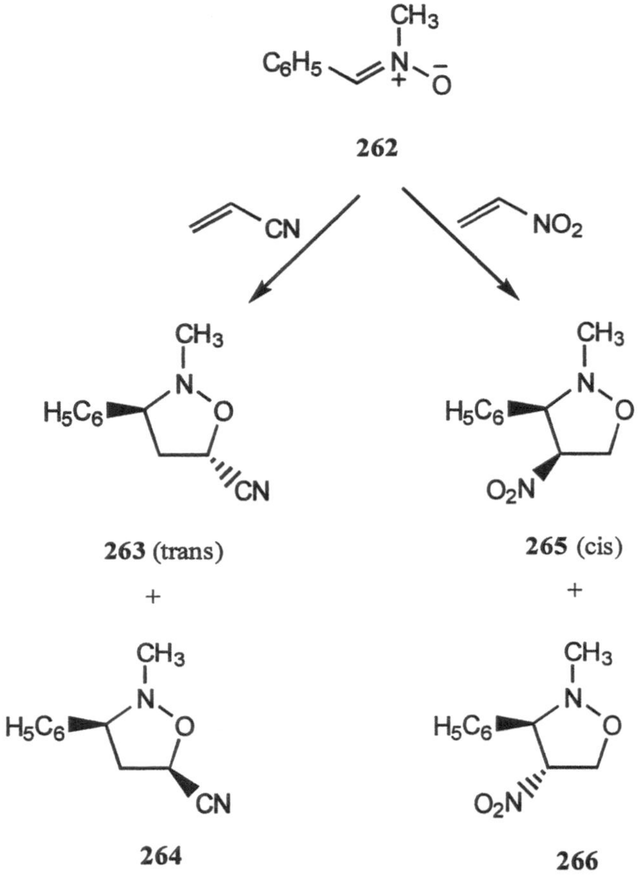

Scheme-140

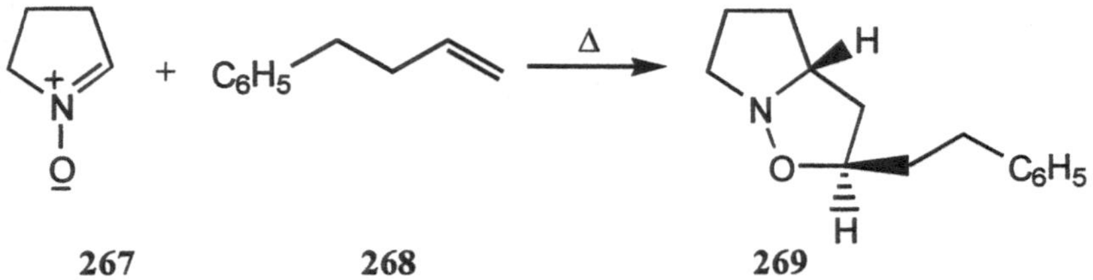

Scheme-141

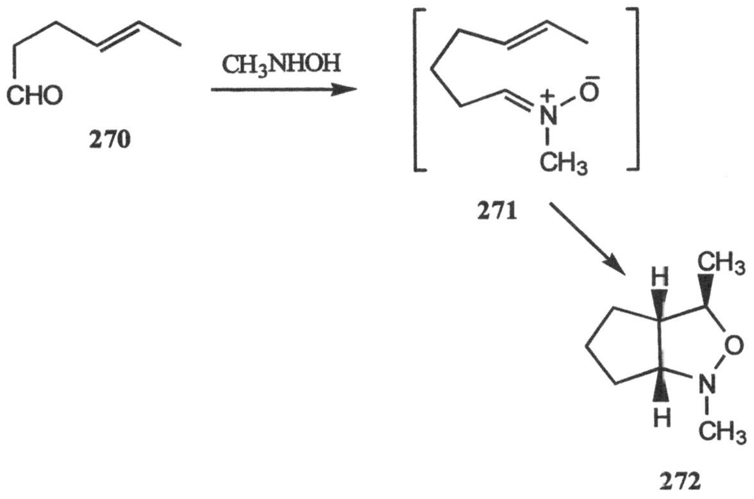

Scheme-142

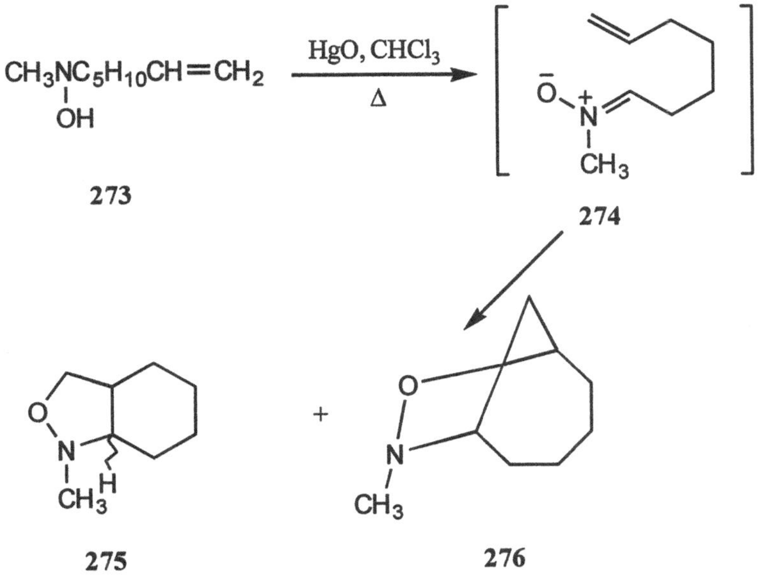

Scheme-143

Nitrones are generally prepared in situ by the oxidation of a disubstituted hydroxylamine with yellow mercuric oxide or by the reaction of an aldehyde or ketone with a monosubstituted hydroxylamine. The cycloaddition reaction of cyclic nitrone **278** (obtained by silver ion catalyzed cyclization of γ- and δ-oximinoallenes **277**) with styrene yields heterocycle **279** (scheme-144).

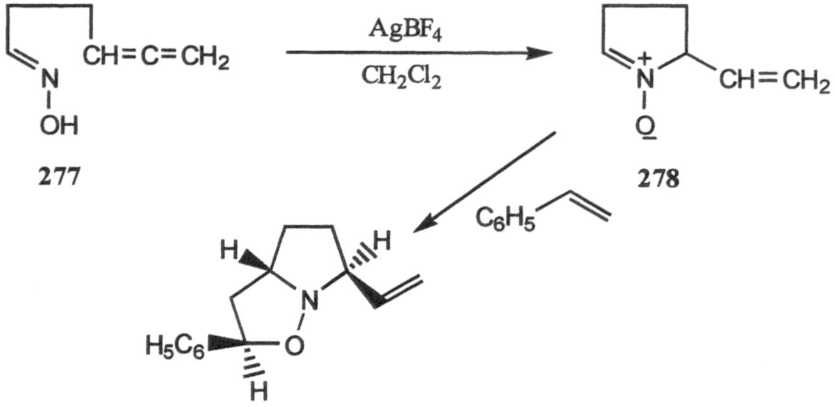

Scheme-144

[3 + 2] Cycloaddition of ethyl vinyl ether to optically active nitrone **281** yields heterocycle **282** with excellent facial and endoselectivity (scheme-145)[114].

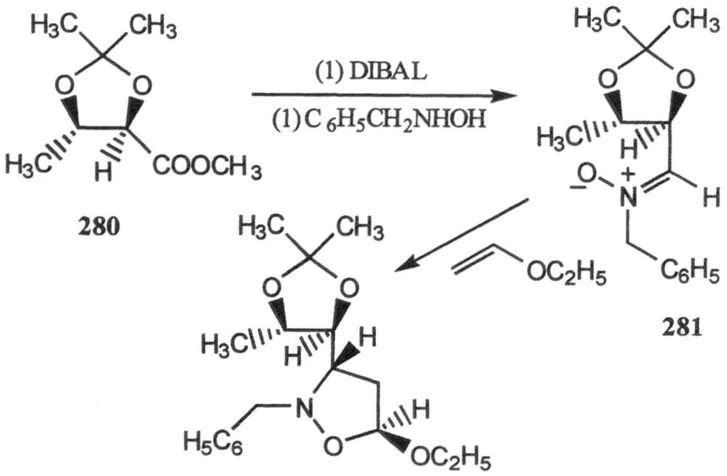

Scheme-145

An equimolecular mixture of heterocycles **284** and **285** is obtained from nitrone **283** having an opticaly active auxillary (–)-menthyl group with *trans*- benzyl crotonate (scheme-146)[115]. Intramolecular cyclization of nitrone **286** yields exclusively isoxazolidine **287** (scheme-147).

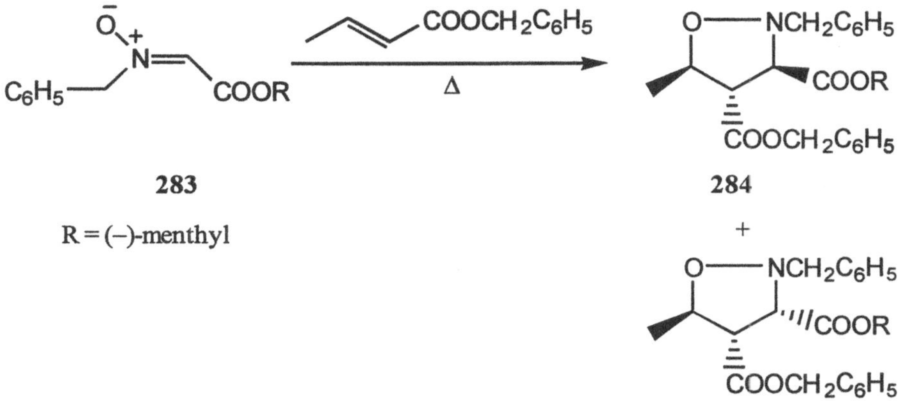

283 **284**

R = (–)-menthyl

285

Scheme-146

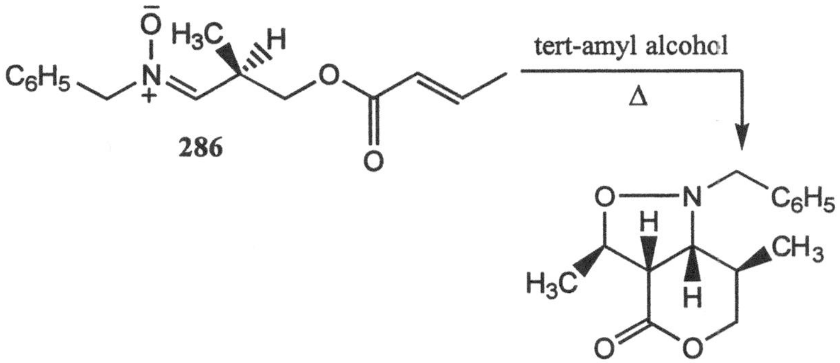

286

287

Scheme-147

Azides react with olefinic and acetylenic dipolarophiles to yield 4,5-dihydrotriazoles and triazoles, respectively. In an intramolecular cycloaddition, 6-azide **288** yields dihydrotriazole **289** (scheme-148). The olefinic azides **290** undergo thermal decomposition yielding cyclic imines **292** and 1-azabicyclo[3.1.0]hexane **293** via the intermediate formation of labile triazolines **291** (scheme-149)[116]. Azide **294** on heating in *o*-dichlorobenzene yields imine **295** with complete regiospecificity (scheme-150)[117].

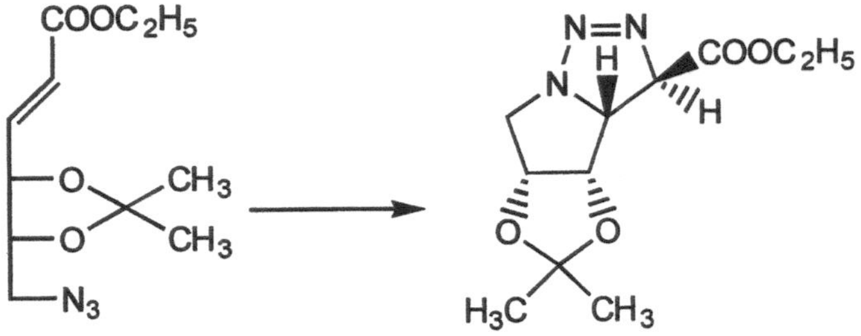

288   **289** (68%)

Scheme-148

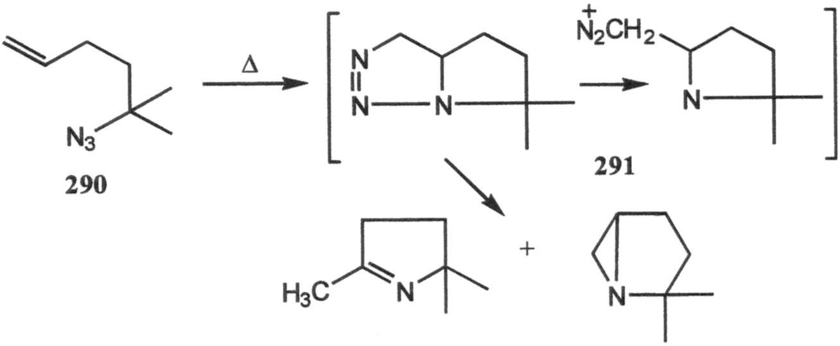

290   **291**

                    292   **293**

Scheme-149

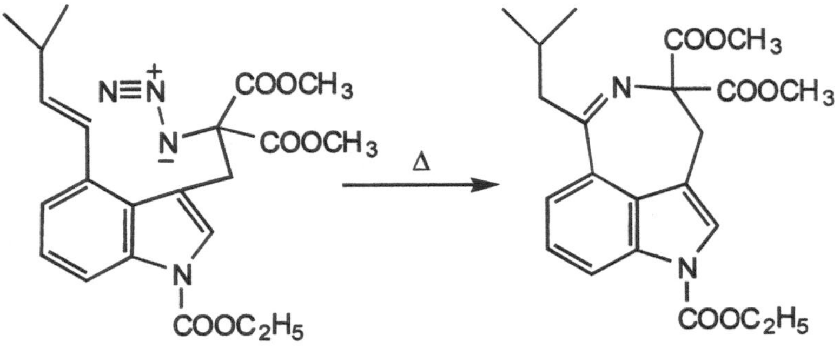

294   **295**

Scheme-150

Azomethine imine **297** (obtained in situ by the reaction of N-acyl-N'-alkylhydrazines **296** with an aldehyde) reacts with alkenes to provide pyrazolidines (scheme-151)[118].

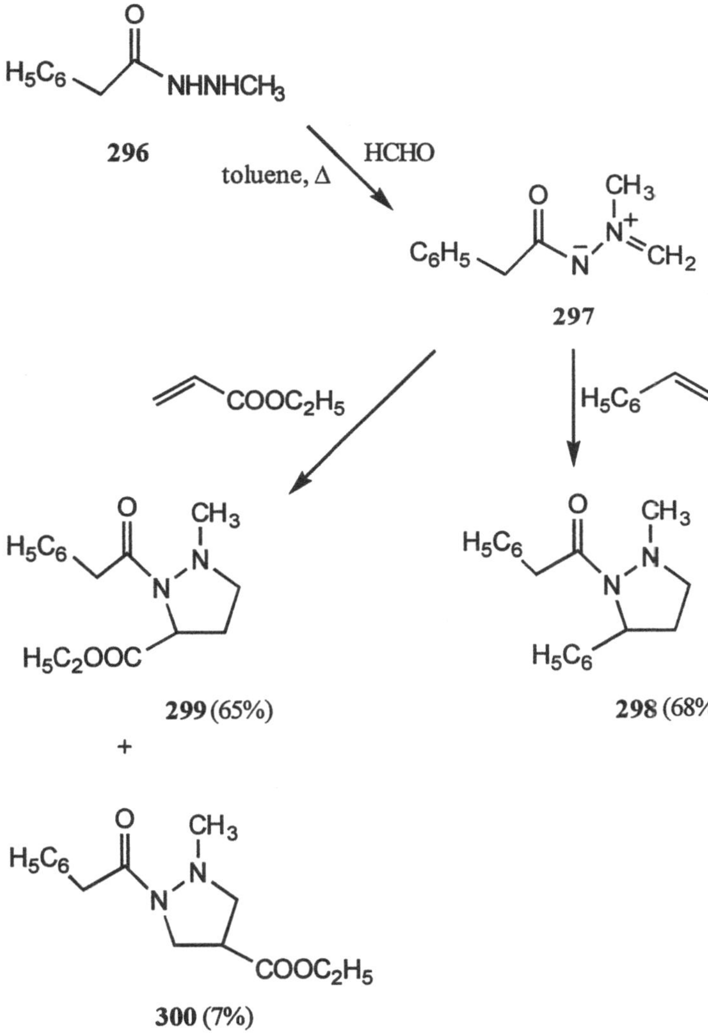

Scheme-151

Intramolecular cycloaddition reaction of N-alkenoyl-N'-alkylhydrazine **301** with benzaldehyde provides bicyclic pyrazolidine **303** (scheme-152)[119].

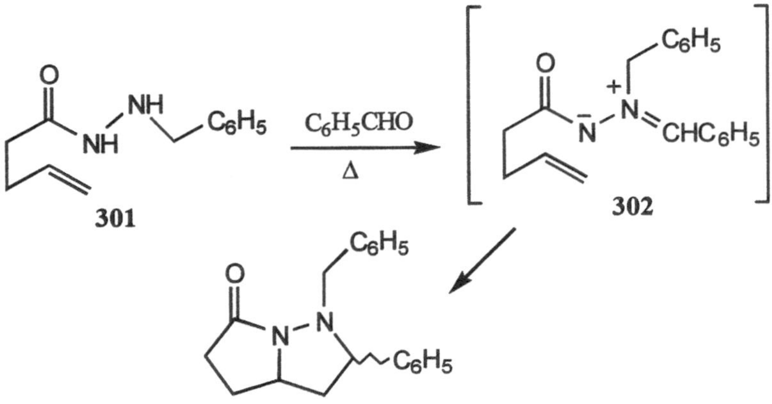

Scheme-152

N-Imide of pyridine generated in situ by deprotonation of N-amino salt reacts with acetylenic dipolarophiles to yield adduct **305** which subsequently on dehydrogenation provides heterocycle **306** (scheme-153)[120].

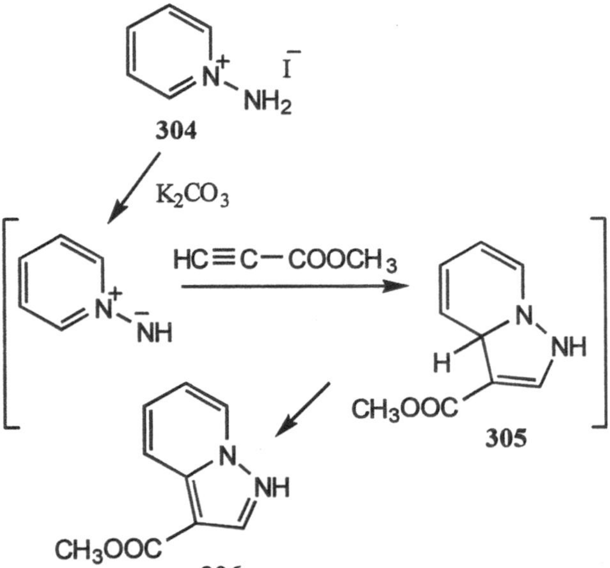

Scheme-153

Cycloaddition reaction of diazoalkanes with olefinic and acetylenic dipolarophiles yields pyrazolines and pyrazoles, respectively. The reaction of methyl acrylate with diazomethane yields heterocycle **308** (scheme-154).

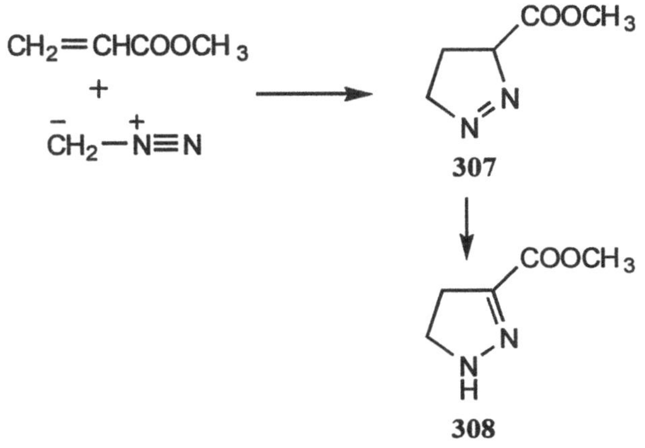

Scheme-154

Cycloaddition of methyl diazoacetate to N-(1-propenyl)pyrrolidine provides pyrrolidinopyrazoline **311** which on elimination of pyrrolidine gives pyrazole **312** (scheme-155)[121]. Intramolecular cycloaddition of diazo compound **313** yields pyrazoline **314** (scheme-156)[122].

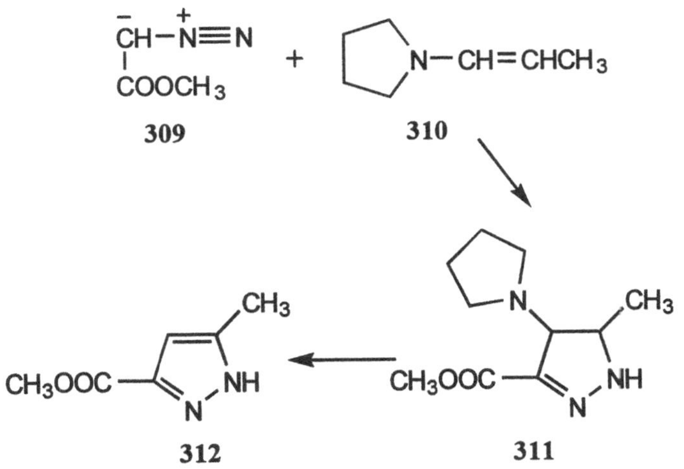

Scheme-155

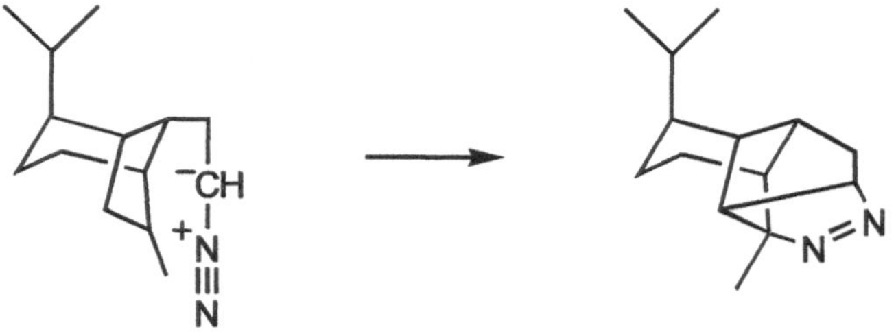

313   314

Scheme-156

Cyclization of tosylhydrazone **315** with sodium hydride yields pyrazoline **316** (scheme-157).

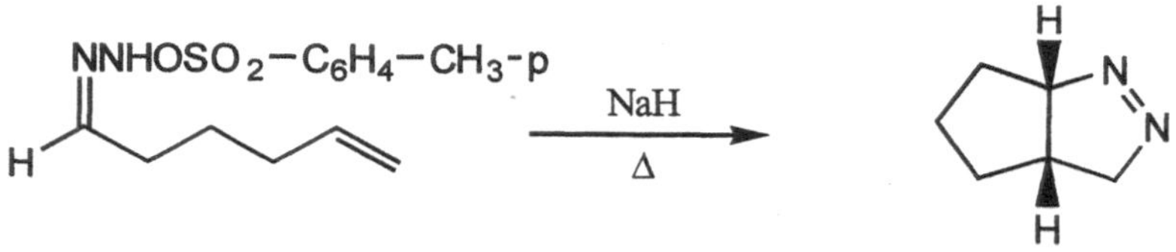

315   316

Scheme-157

2.2.2 [2 + 2→4] Cycloadditions

Four-membered heterocycles can be prepared by concerted or non-concerted [2 + 2] cycloadditions[123]. These reactions often compete with the side reactions. The important processes are cycloreversion either to starting materials (a) or cycloreversion with dismutation (b), and the expansion to yield six-membered rings (c) as shown in Fig. 11. The alkenes undergo thermal [2 + 2] cycloadditions mostly via a step-wise pathway involving diradical or zwitterionic intermediates[124]. The widely employed reactions involve ketenes and heterocumulenes such as isocyanates, isothiocyanates *etc.* The reaction of ketenes with imines has been used for the synthesis of β-lactams **319** (scheme-158)[125]. The reaction of oxazolidone **320** with benzyl imines **321** in the presence of triethylamine is almost completely diastereoselective yielding heterocycles **322a** and **322b** (scheme-159)[126]. Ketene **324** reacts with optically active imine **323** providing β-lactam **325** with high diastereoselectivity (scheme-160). The addition of alkenes to diformylmethyl

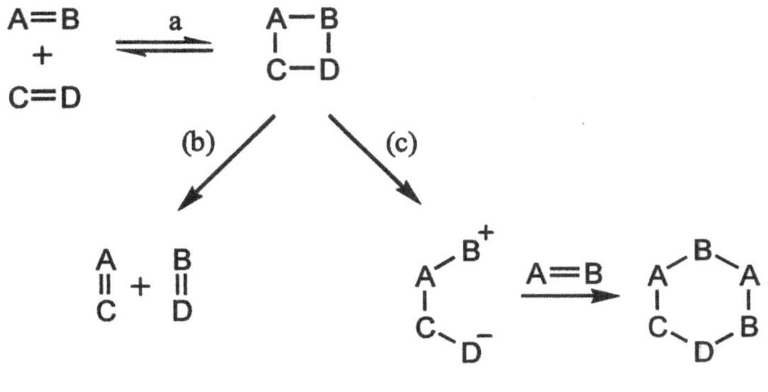

Fig. 11

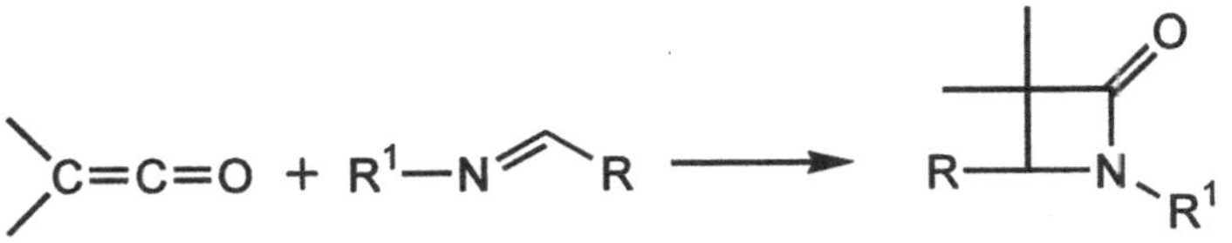

317 318 319

Scheme-158

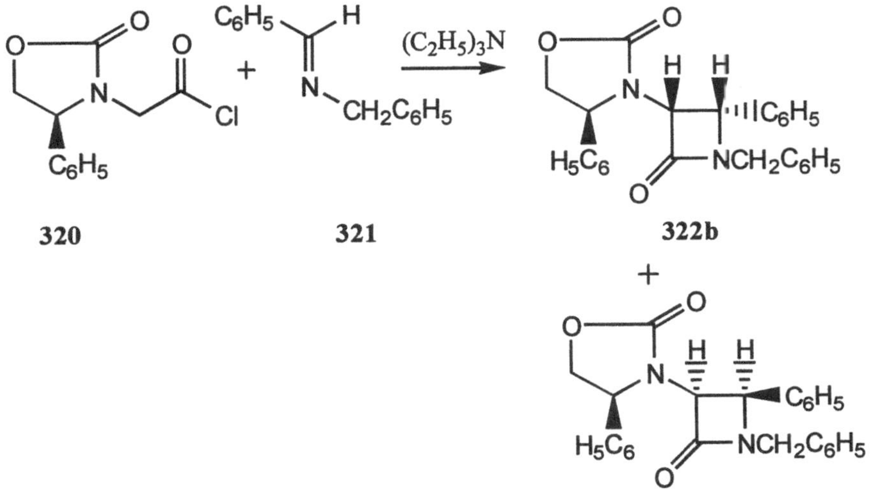

320 321 322b

322a

Scheme-159

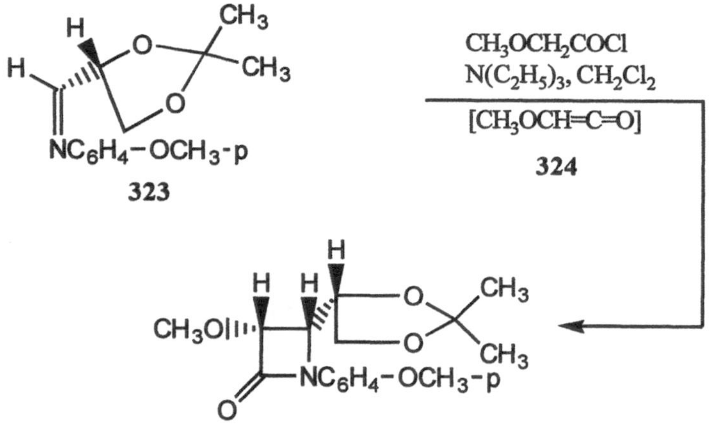

Scheme-160

acetate constitutes an important application for the synthesis of pyran ring system **328** (scheme-161)[127]. Paterno–Büchi photochemical cycloaddition of olefins to

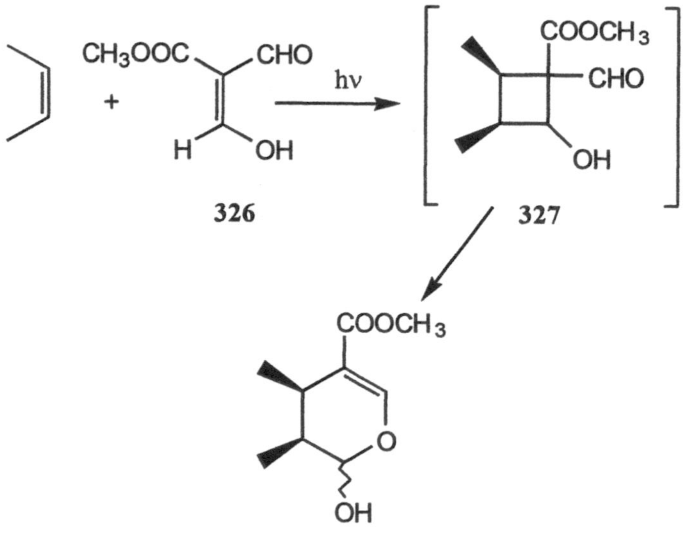

Scheme-161

aldehydes and ketones yields oxetanes. A stepwise radical mechanism is suggested (scheme-162)[128]. The photochemical cycloaddition of aldehydes to furans provides head to head *exo*-products **330** with high regio- and stereoselectivity (scheme-163)[129].

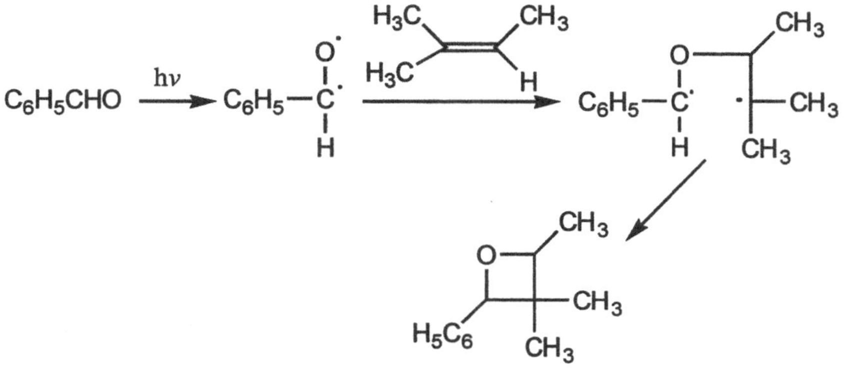

329

Scheme-162

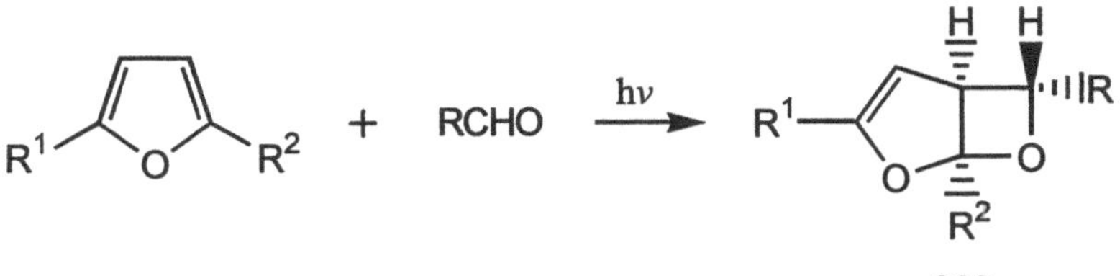

330

Scheme-163

2.2.3 Hetero–Diels–Alder Reactions: [4 + 2→6] Cycloadditions

The cycloaddition of a conjugated diene to a dienophile (a double or triple bond) to form an adduct with a six-membered carbocyclic or heterocyclic ring is known as Diels–Alder reaction. The formation of two new σ-bonds takes place in the cycloadduct at the expense of two π-bonds in the starting compound. The reaction is depicted in Fig. 12. It is known as [4 + 2] cycloaddition reaction, since it involves interaction of a four π-electrons system with a two π-electrons system. According to the Woodward–Hoffmann rules, it is thermally allowed process. Its synthetic usefulness arises from its versatility and high regio- and stereoselectivity. Both inter- and intramolecular reactions are widely used. The reaction conditions used are diversified; some reactions occur at ambient or slightly elevated temperatures,

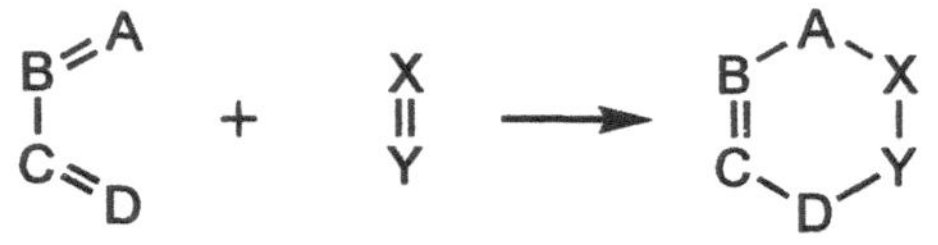

Fig. 12

some are accelerated by the catalysts and some occur under high pressure. In Hetero-Diels–Alder reaction, one of the components, either diene or dienophile contains one or more heteroatoms. However, so far heterodienes have not been extensively employed. The majority of the reactions utilize electron-rich diene and an electron deficient dienophiles, however some reactions are also known between an electron rich dienophile and an electron-deficient diene[130]. In a normal Diels–Alder reaction, i.e. interaction between electron-rich diene and electron-deficient dienophile, the main interaction is between HOMO and LUMO of diene and dienophile, respectively. Electron-withdrawing substituents on the double bond of dienophile accelerate the reaction by lowering the energy of the LUMO, whereas the electron donating substituents on diene facilitate the reaction by raising the energy of the HOMO. In case of inverse electron demand, the interaction takes place between the HOMO of dienophile and LUMO of diene. Many of these features also apply to Hetero-Diels–Alder reactions of dienes with heterodienophiles, although much less mechanistic work has been done on these reactions. In heterodienophiles, one or both the atoms of the multiple bond may be heteroatom(s)[131, 134]. Most widely used heterodienes are carbonyl, thiocarbonyl imine, nitroso compounds, *etc.* The cycloaddition reactions of certain aldehydes to conjugated dienes under thermal conditions to provide dihydropyrans **331** are well known (scheme-164). These reactions occur at elevated temperatures. Certain reactive 1,3-dienes, bearing alkoxy or silyloxy substituents **332** react under mild conditions with a wide range of aldehydes under the catalytic action of Lewis acid to afford substituted 2,3-dihydro-4-pyrones **333** (scheme-165).

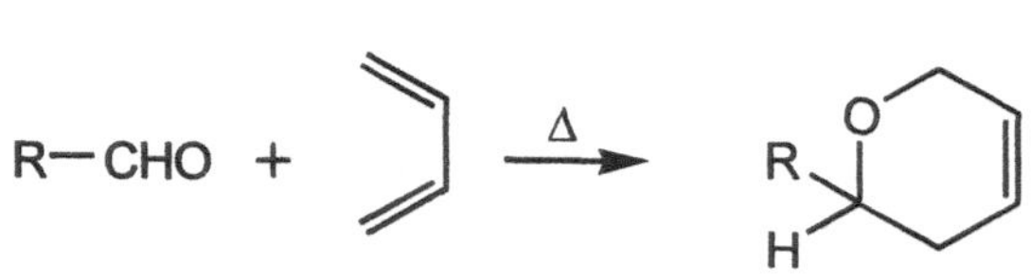

331

Scheme-164

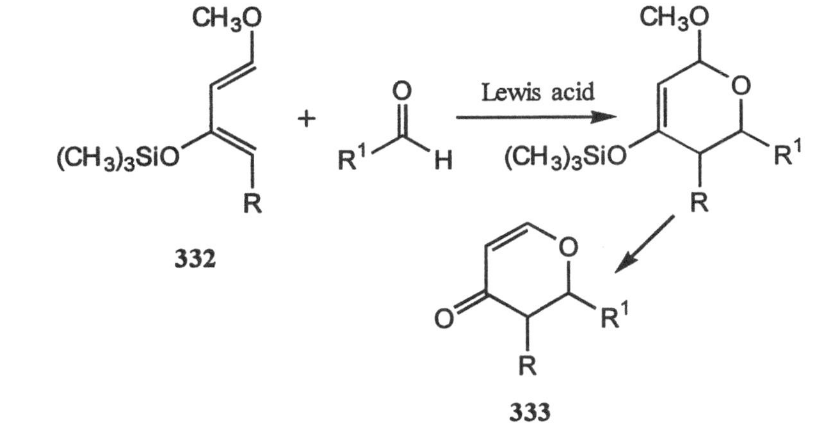

Scheme-165

The catalysts used in such cycloadditions include zinc chloride, magnesium chloride, boron trifluoride etherate, europium complexes Eu(fod)$_3$ and Eu(hfc)$_3$, titanium tetrachloride, *etc.* The course of the reactions and stereochemistry of dihydropyrones obtained depends on the reaction conditions, catalyst and geometry or the substitution pattern of diene. The cycloaddition reactions of *trans*, *trans*-1,4-disubstituted dienes **334** catalyzed by zinc chloride are highly stereoselective and yield mostly *cis*-2,3-disubstituted 4-pyrone **337** along with **338** as a minor product. The reaction proceeds through the formation of an isolable intermediate (1 : 1 adduct) **336** (scheme-166)[132].

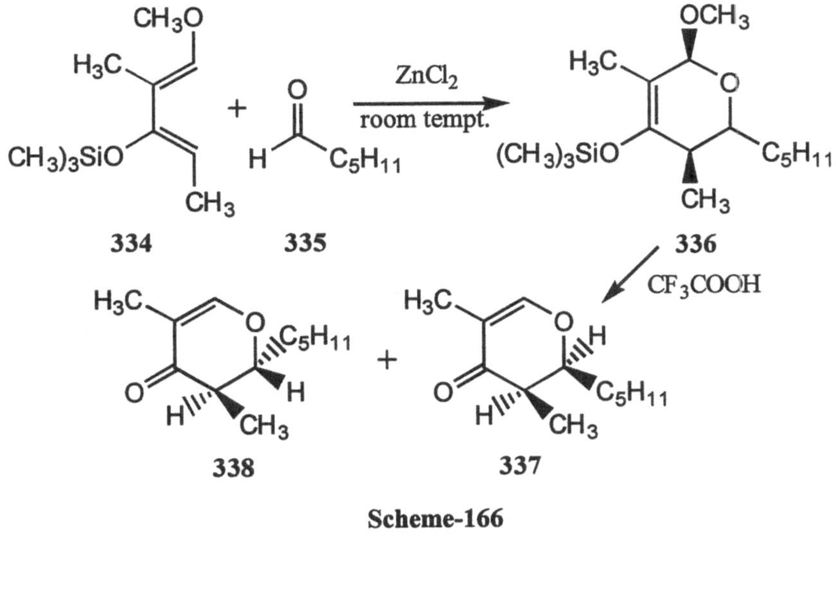

Scheme-166

The reaction of acetaldehyde with diene **334** catalyzed by Eu(fod)$_3$ occurs with complete endo-specificity to yield dihydropyran **339** involving stereocontrolled formation of three chiral centres. Dihydopyran **339** on further treatment with trifluoroacetic acid affords *cis*-2,3-disubstituted dihydropyrone **340**. Treatment of **340** with methanol–triethylamine causes axial protonation of silyl enol ether introducing a new chiral centre at C-5 affording tetrahydropyrone **341** (scheme-167)[133]. Cycloaddition reactions with α-alkoxy aldehydes catalyzed by magnesium bromide lead preferentially to *trans*-2,3-disubstituted 4-pyrones **343** (scheme-168)[134].

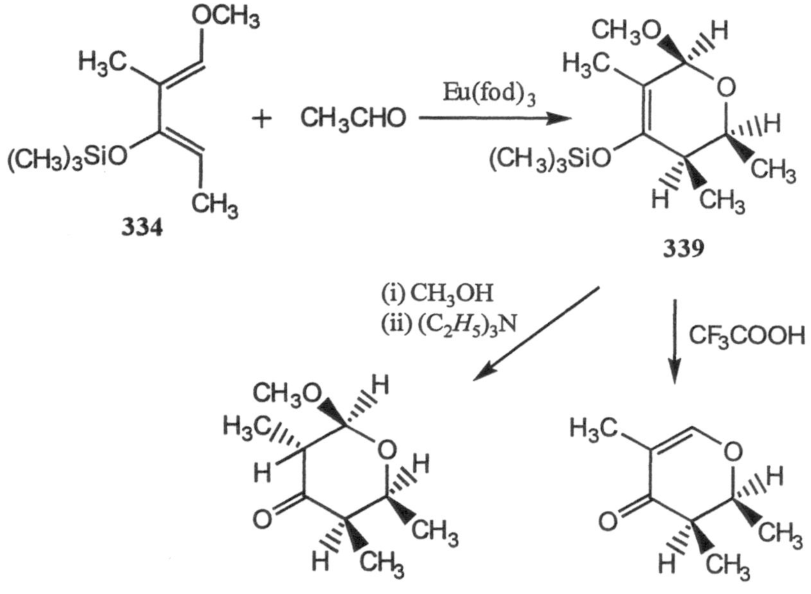

Scheme-167

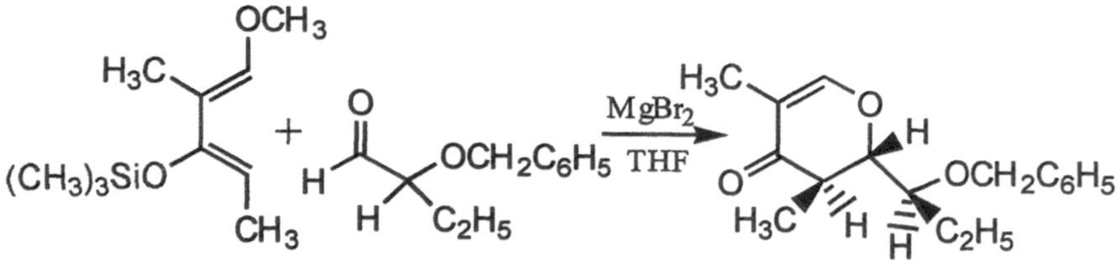

Scheme-168

Thiocarbonyl compounds are more reactive than the carbonyl compounds and provide a good method of synthesizing six-membered sulfur heterocycles by their cycloaddition with conjugated dienes (scheme-169)[134]. Thiobenzaldehyde **348**

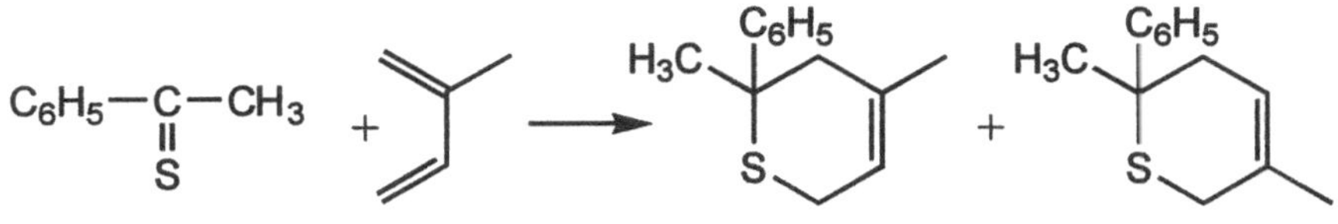

Scheme-169

generated in situ by thermolysis of thiosulfinate **347** on cycloaddition with anthracene and 2,3-dimethylbutadiene give heterocycles **349** and **350**, respectively (scheme-170). The cycloaddition reaction of thioaldehydes **352** and **355** with

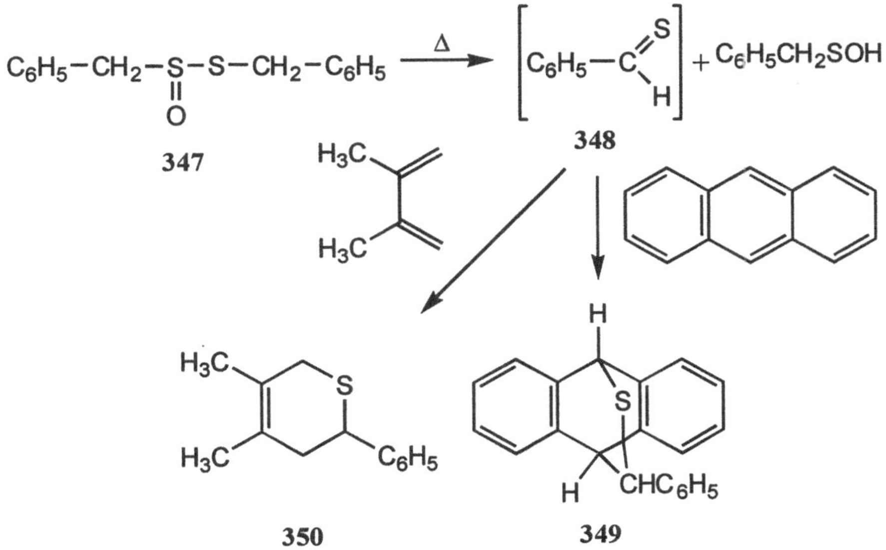

Scheme-170

butadiene **351** provides thiopyranes **354** and **356** as major products, respectively (scheme-171). N-Sulfinylsulfonamides **359** which can be obtained readily from aryl sulfonamides **358** and thionyl chloride on cycloaddition with conjugated dienes afford 3,6-dihydro-1,2-thiazine derivatives **360** (scheme-172).

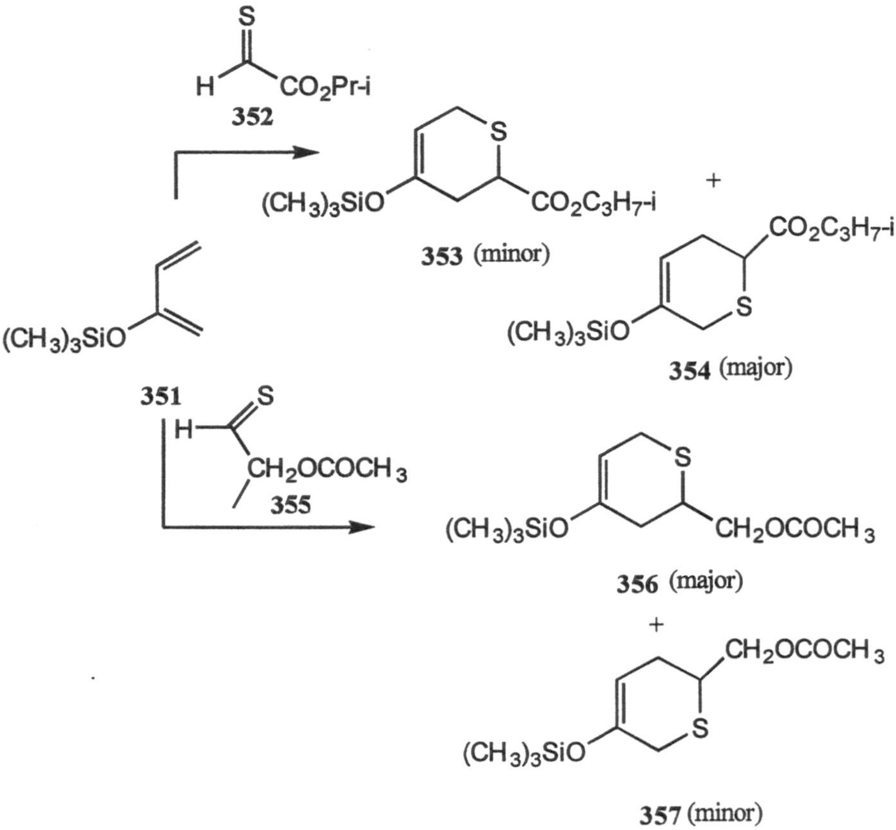

Scheme-171

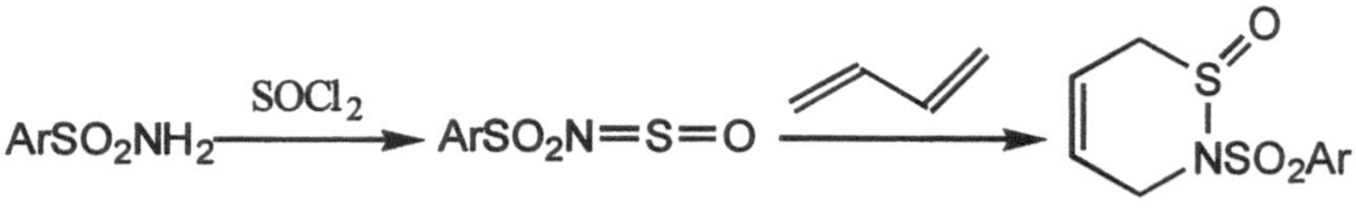

358 **359** **360**

Scheme-172

Imines constitute another useful group of heterodienophiles and react with dienes to form tetrahydropyridines **362** (scheme-173). Cycloaddition of activated imine **364** with 1-methoxy-3-trimethylsilyloxybutadiene **363** yields dihydropyridone **365** (scheme-174). The reactions are highly regio- and stereoselective. Ordinarily

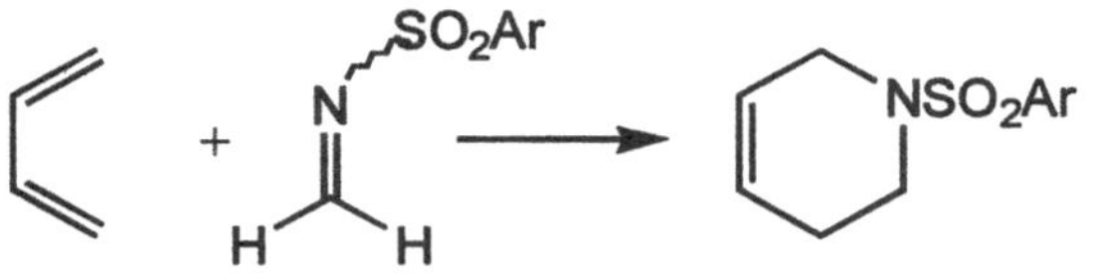

361   **362**

Scheme-173

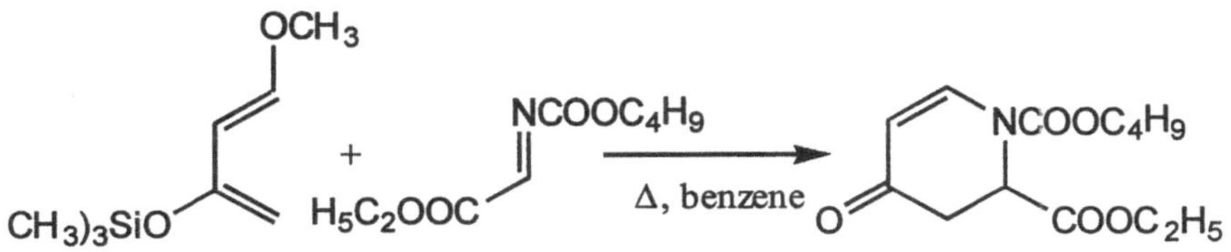

363   **364**   **365**

Scheme-174

unactivated Schiff bases **366** react smoothly with the reactive diene, 1-methoxy-3-trimethylsilyloxybutadiene **363** under the catalytic action of zinc chloride to form 2,3-dihydro-4-pyridones **367** (scheme-175). Intramolecular Diels–Alder reaction of iminoacetate **368** affords bicyclic lactam **370** (scheme-176).

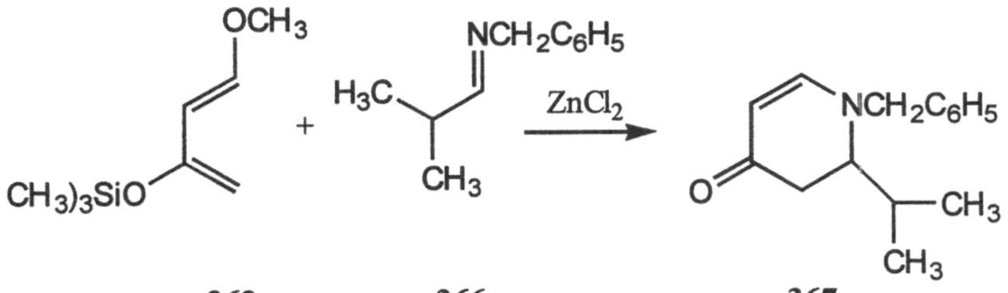

363   **366**   **367**

Scheme-175

Another useful group of heterodienophiles includes nitroso compounds **372** which react with 1,3-dienes to form 3,6-dihydro-1,2-oxazine derivatives **373** (scheme-177)[134, 135]. The reaction of benzoyl nitrosoformate **375** generated in situ from the corresponding carbamic ester **374** with 1-cyanocyclohexadiene provides a single adduct **376** (scheme-178).

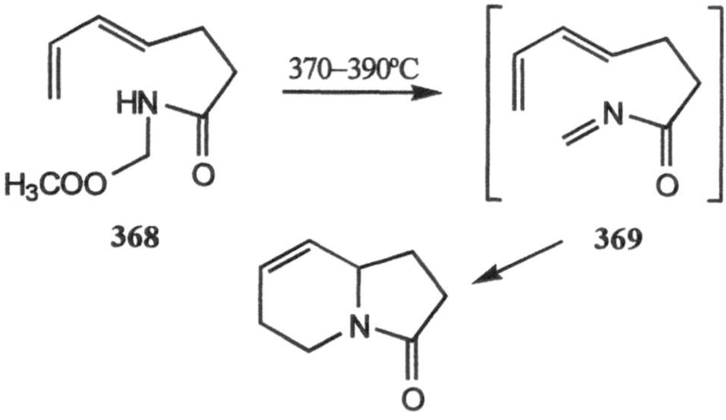

368   **369**

370

Scheme-176

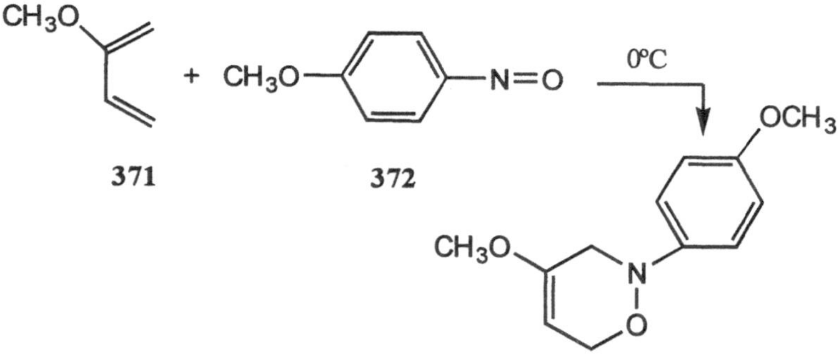

371   **372**   **373**

Scheme-177

Heterodienes with electron-rich dienophiles easily undergo hetero-Diels–Alder reactions, however they have not been extensively investigated so far. Most commonly used heterodienes are α,β-unsaturated carbonyl compounds. Diels–Alder reaction of α,β-unsaturated carbonyl compounds with enol ethers and enamines yields dihydropyran derivatives with inverse electron demand[136]. The reaction with enol ether requires elevated temperture. An electron-withdrawing group at C-2 of α,β-unsaturated carbonyl compound makes the cycloaddition easier. A facile cycloaddition is achieved with α-quinone methide generated in situ either by the decomposition of a Mannich base or from the analogous alcohol[136] (scheme-179). Cycloaddition of methyl vinyl ketone with enamine yields dihydropyran **382**. The reaction proceeds with the formation of intermediate zwitterions (scheme-180).

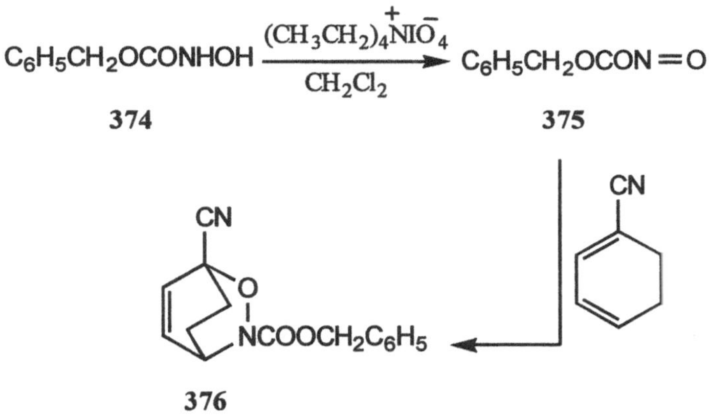

Scheme-178

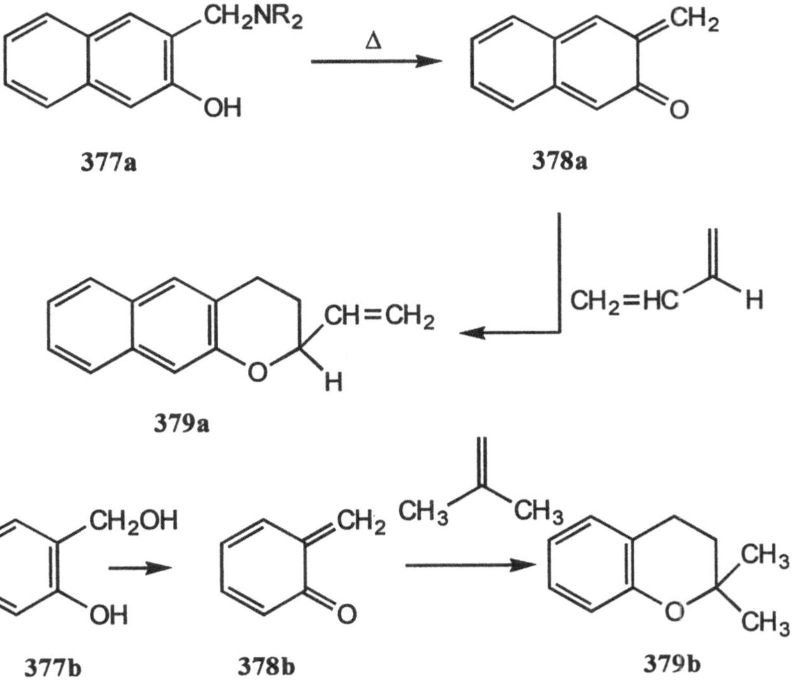

Scheme-179

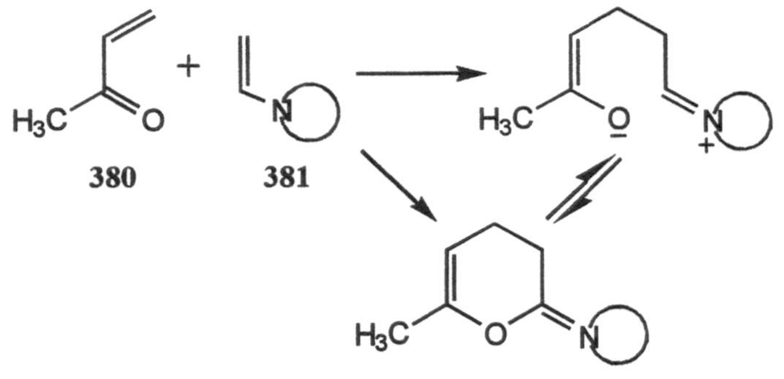

Scheme-180

Azabutadienes also undergo Diels–Alder reaction[137] either with inverse or normal electron demand depending on the presence of electron-withdrawing or electron donating substituents. α,β-Unsaturated imine (1-azabutadiene) **383a** reacts preferentially through its enamine tautomer **383b** (scheme-181).

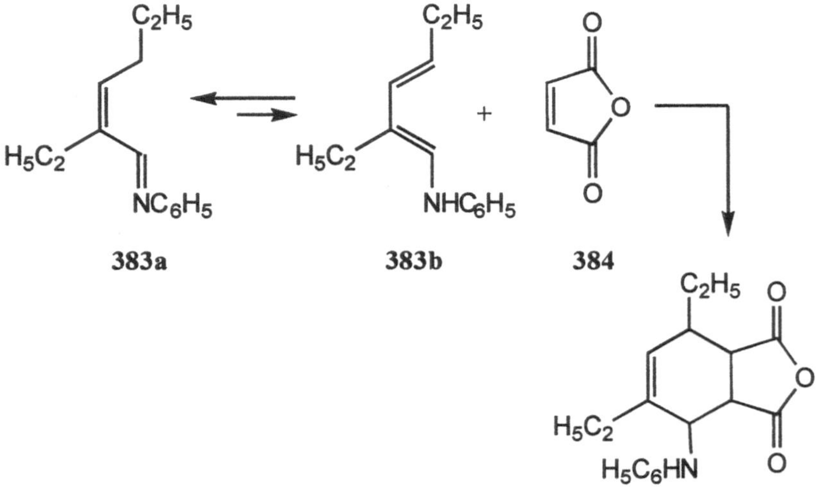

Scheme-181

α,β-Unsaturated hydrazones undergo cycloaddition with electron-deficient dienophiles e.g. with diphenyl ketene to afford adduct **388** (scheme-182).

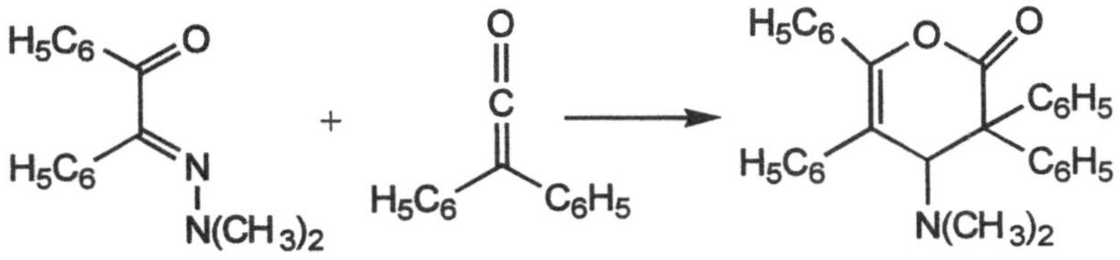

386 387 388

Scheme-182

2-Substituted-4-methyl-1,3-oxazin-6-ones (2-azabutadienes) **389** react exothermally with electron-rich dienophiles e.g. 1-(diethylamino)propyne **390** to provide substituted 4-aminopyridine (scheme-183).

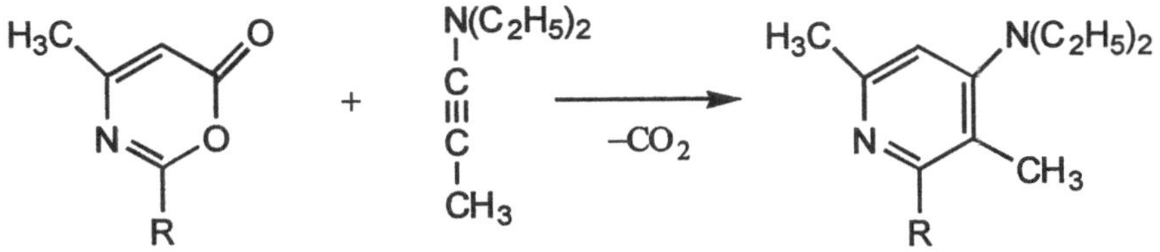

389 390 391

Scheme-183

Azadienes with additional heteroatoms undergo cycloaddition with electron-rich dienophiles. Vinyl nitroso compounds **392** and **395** constitute a class of electron-deficient hetro-2-azabutadienes which undergoes a number of Diels–Alder reactions (scheme-184).

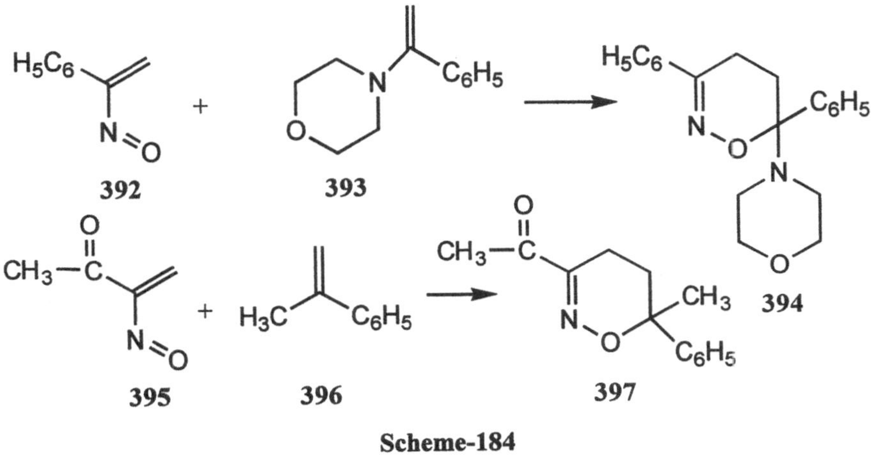

392 393 394

395 396 397

Scheme-184

2.2.4 Cheletropic Reactions

Cheletropic reactions can be defined as the reactions in which both new σ bonds are made to the same atom. However, they are of the limited value in the construction of heterocycles.

Cheletropic reactions can be classified into two groups: (2 + 1) and (4 + 1) cycloadditions.

2.2.4.1 [2 + 1] Cycloadditions

In general, the reaction can be represented as in Fig. 13 :

Fig. 13

The important examples include the addition of carbenes and nitrenes to the double bonds. Some aziridines and thiiranes have been prepared by the addition of carbene to the carbon–nitrogen and carbon–sulfur multiple bonds[138].

Aziridines have been prepared by the addition of nitrenes to olefins. An important example includes the oxidation of 2,4-dinitrobenzene sulfenamide with lead tetraacetate (LTA) in the presence of elecron-rich alkenes to give rise aziridine in good yield (scheme-185)[139].

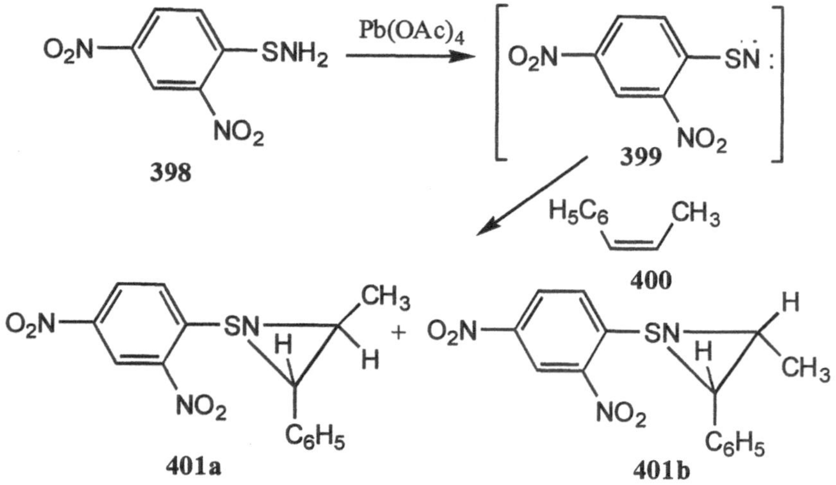

Scheme-185

2.2.4.2 [4 + 1] Cycloadditions

In general [4 + 1] electrophilic reactions can be represented as in Fig. 14 :

Fig. 14

The example includes cycloaddition of butadiene to sulfur dioxide yielding 2,5-dihydrothiophene-1,1-dioxide (scheme-186). Many other acyclic dienes react similarly retaining stereochemistry in the adduct[140].

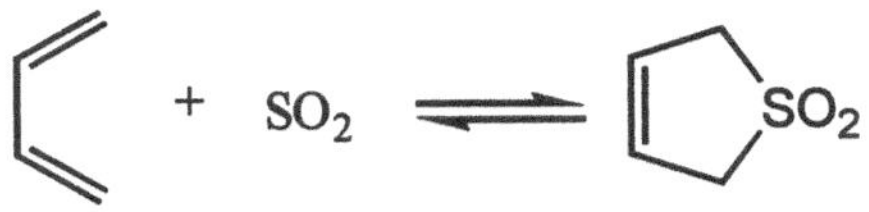

402

Scheme-186

2.2.5 Valence Bond Isomerization

The reaction consists in reorganization of σ and π-electrons within the frame work of a molecule without the migration of atoms or groups of atoms and hence involves the changes in atomic distances and bond angles and is known as valence bond isomerization. The application of such reactions provides a good synthetic tool for the synthesis of various heterocyclic compounds. Photolysis of ethyl azide formate **403** yields carbethoxynitrene **404** which reacts with benzene to yield N-carbethoxyazepine **406** (scheme-187)[141].

Lithium chloride catalyzed pyrolysis of **407** at 200°C yields one or both the isomers of 1,2-divinylethylene oxide. The less stable *cis*-isomer rearranges to its valence isomer 4,5-dihydrooxepin **408** (scheme-188)[142].

Cyclization of 3-ethylamino-4-hydroxy-1,5-hexadiene **409** gives rise to a mixture of *trans*-N-ethyl-2,3-divinylaziridine **410** and its valence isomer, N-ethyl-4,5-dihydroazepine **411** (scheme-189)[143].

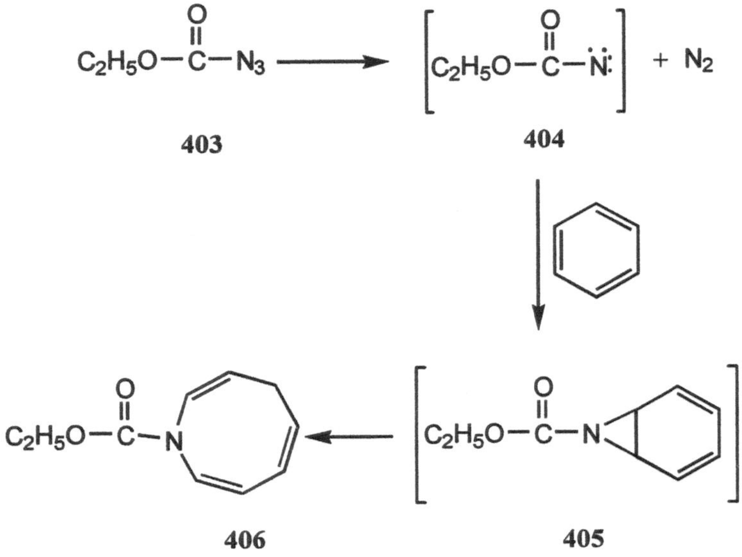

Scheme-187

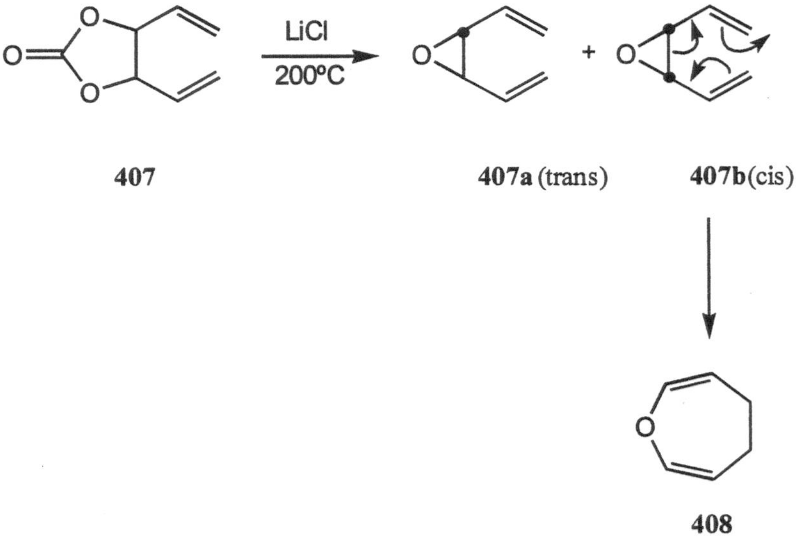

Scheme-188

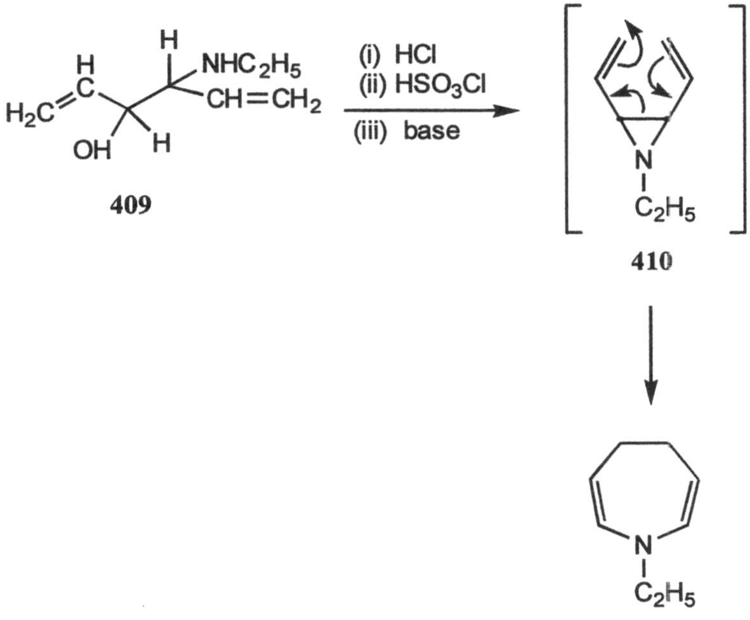

409 **410** **411**

Scheme-189

The reaction of sodium salts of 2,6-disubstituted phenols **412** with chloramine leads to the ring enlargement of phenoxide moiety yielding 1,3-dihydro-2*H*-azepin-2-one **413** (scheme-190)[144].

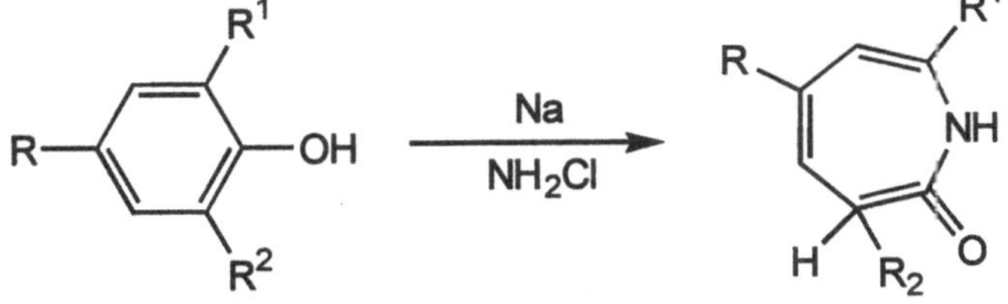

412 **413**

Scheme-190

2,7-Dimethyloxepin **417** is obtained via valence bond isomerization when dibromo compound is dehydrobrominated with potassium tert-butoxide in ether at 0°C (scheme-191)[145]. Compound **421** is also prepared by the valence bond isomerization (scheme-192)[146].

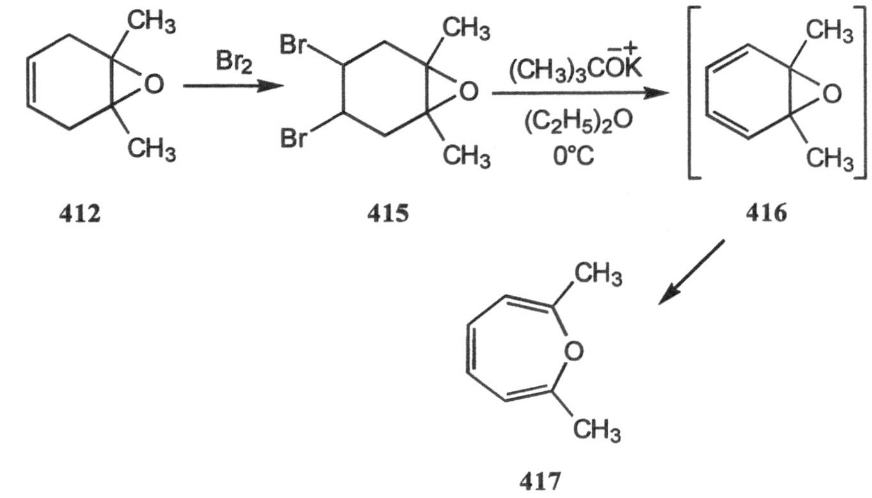

Scheme-191

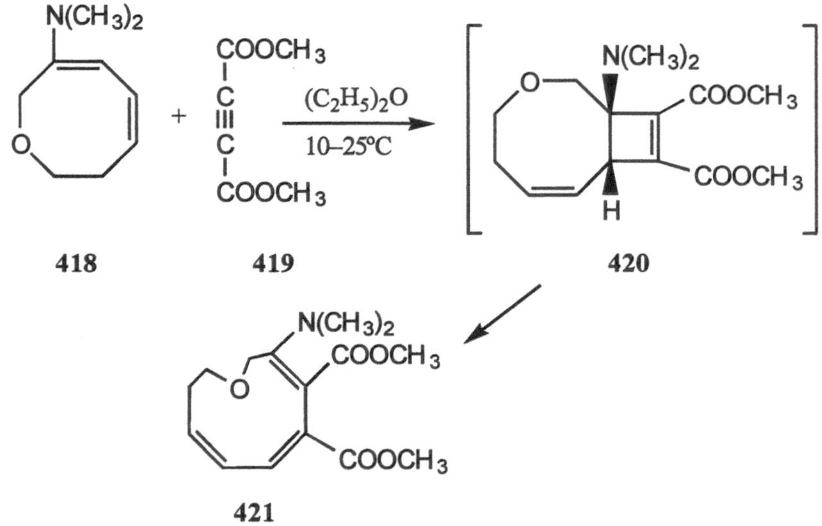

Scheme-192

2.3 Ene Reactions

The reaction of an olefin having an allylic hydrogen atom (the ene) with an activated multiple bond of an enophile to yield an adduct is known as ene reaction (scheme-193). In such reactions, the formation of a new σ-bond between the unsaturated centres of ene and enophile and the migration of an allylic hydrogen to the other terminals of the enophile multiple bond takes place (scheme-193).

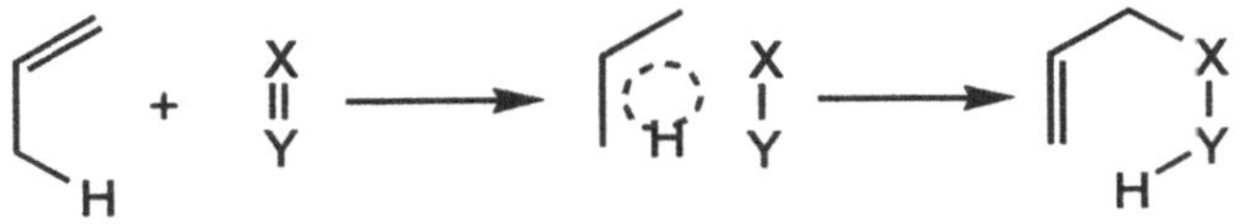

X and Y = C,N,O,S

Scheme-193

Ene reaction shows mechanistic relationship with [4 + 2] cycloaddition, although intermolecular ene reaction does not provide a cyclic product. However, in an intramolecular ene reaction cyclic product is obtained through the formation of a new σ-bond to the allylic carbon atom. Two π-electrons of diene in Diels–Alder reaction are replaced by two electrons of the allylic C–H σ-bond in ene reactions. As the activation energy is higher, higher temperature is required. It has been found that many ene reactions proceed under mild conditions, under the catalytic action of Lewis acid and often with improved stereoselectivity. In most of the reactions HOMO of ene component interacts with LUMO of enophile and catalysts exert their effect by lowering the energy of LUMO of enophile. Intramolecular ene reaction of prenyl ester **422** of thioxoacetic acid, liberated in situ by thermolysis of its adduct with 9,10-dimethylanthracene provides heterocycle **423** (scheme-194)[147]. Cyclic amide **426** is obtained from acylimine **425** which is generated in situ by flash vaccum pyrolysis of acetate **424** (scheme-195)[148].

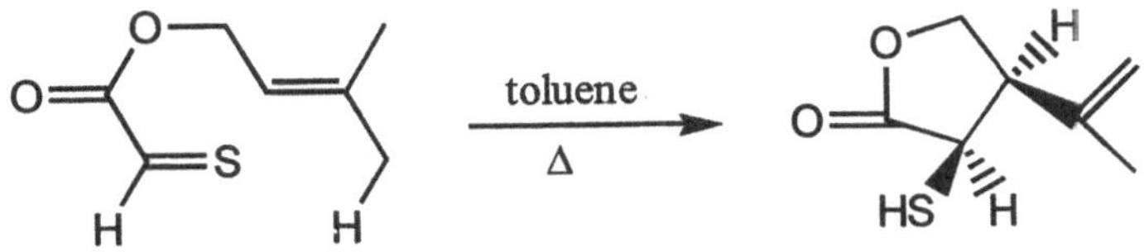

422   **423**

Scheme-194

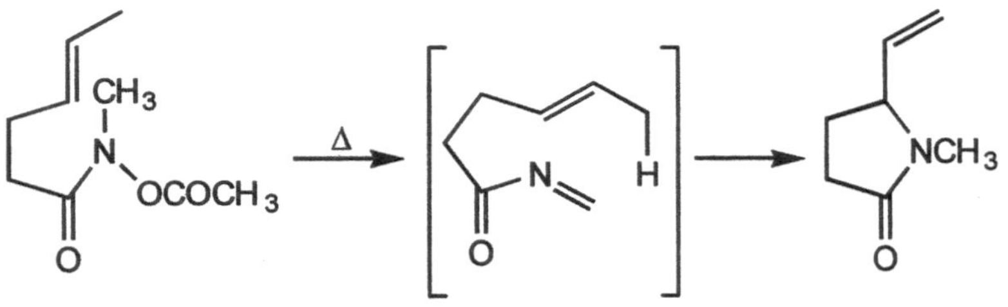

424 **425** **426**

Scheme-195

Activated nitroso compounds comprise an useful group of enophiles, although they are less reactive in ene reactions than in Diels–Alder reactions. Nitroso ketone derivative **428** obtained by the oxidation of hydroxamic acid **427** in the presence of 9,10-dimethylanthracene yields heterocycle **429** (scheme-196)[149].

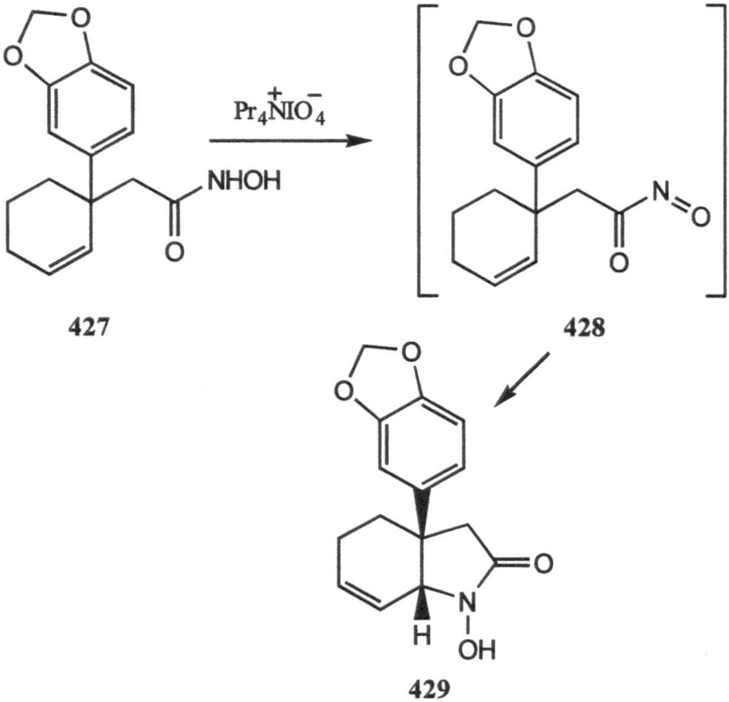

427 **428** **429**

Scheme-196

Acetylenic bonds also readily undergo intramolecular ene reactions e.g., intramolecular cyclization of diyne **430** (scheme-197)[150].

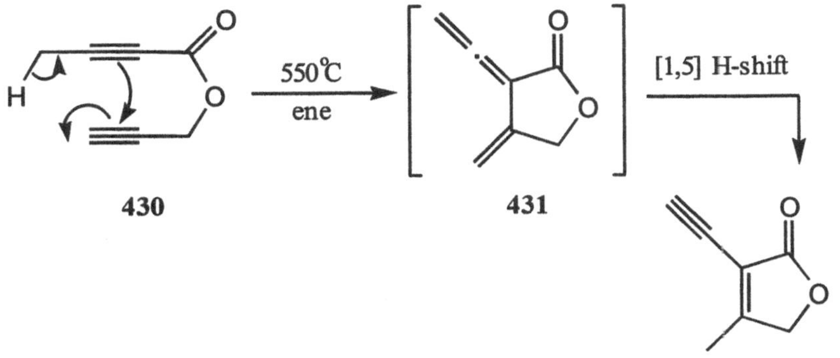

430 **431** **432**

Scheme-197

Ene reactions are highly diastereoselective. A chiral centre present either in ene or enophile is finally incorporated in the product formed.

Cyclization of optically active diene **433** provides pyrrolidine derivative **434** (scheme-198)[151].

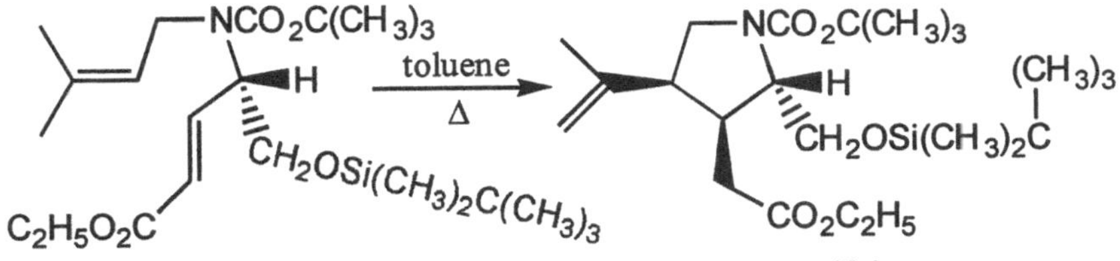

433 **434**

Scheme-198

REFERENCES

1. J. E. Baldwin, *J. Chem. Soc. Chem. Comm.* 734 (1976).
2. P. D. Bartlett and R. H. Rosenwald, *J. Am. Chem. Soc.* **56**, 1990 (1934); P. D. Bartlett, *J. Am. Chem. Soc.* **57**, 224 (1935).

3. W. H. Richardson, C. M. Golino, R. H. Wachs and M. B. Yelvington, *J. Org. Chem.* **36**, 943 (1971).

4. S. Winstein and H. J. Lucas, *J. Am. Chem. Soc.* **61**, 2845 (1939).

5. D. H. R. Barton, D. A. Lewis and J. F. McGhie, *J. Chem. Soc.* 2907 (1957).

6. F. W. Bachelor and R. K. Bansal, *J. Org. Chem.* **34**, 3600 (1969).

7. W. Colteff, *U. S. Patent* 2, 183, 860 (1939), *C. A.*, **34**, 2395 (1940).

8. E. H. Farmer and F. W. Shipley, *J. Chem. Soc.* 1519 (1947).

9. A. Weissperger and H. Bach, *Chem. Ber.* **64**, 1095 (1931); **65**, 631 (1932).

10. S. J. Brois, *J. Org. Chem.* **27**, 3532 (1962).

11. R. Appel and M. Halstenberg, *Chem. Ber.* **109**, 814 (1976).

12. C. Galli, G. Illuminati, L. Mandolini and P. Tamborra, *J. Am. Chem. Soc.* **99**, 2591 (1977).

13. H. E. Baumgarten, *J. Am. Chem. Soc.* **84**, 4975 (1962); *J. Am. Chem. Soc.* **85**, 3303 (1963); J. C. Sheehan and I. Lengyel, *J. Am. Chem. Soc.* **86**, 746 (1964).

14. M. J. Miller, P. G. Mattingly, M. A. Morrison and J. F. Kerwin, *J. Am. Chem. Soc.* **102**, 7026 (1980).

15. J. W. Batty, P. D. Howes and C. J. M. Stirling, *J. Chem. Soc. Perkin Trans.* 1, 65 (1973).

16. F. Fiest, *Ber.* **35**, 1545 (1902); E. Benary, *Ber.* **44**, 493 (1911).

17. W. Madelung, *Ber.* **45**, 1128 (1912).

18. H. Wynberg and H. J. Kooreman, *J. Am. Chem. Soc.* **87**, 1739 (1965).

19. L. F. Miller and R. E. Banburg, *J. Med. Chem.* **13**, 1022 (1970).

20. P. L. Julian et al., *J. Am. Chem. Soc.* **67**, 1203 (1945).

21. R. Robinson, *J. Chem. Soc.* **95**, 2167 (1909); S. Gabriel, *Ber.* **43**, 1283 (1910).

22. Z. H. Skraup, *Monatsh. Chem.* **1**, 316 (1880); **2**, 139 (1881); R. H. F. Manske and M. Kulka, *Org. Reactions* **7**, 59 (1953).

23. E. Roberts and E. E. Turner, *J. Chem. Soc.* 1832 (1927).

24. A. Bischler and B. Napieralski, *Chem. Ber.* **26**, 1903 (1893); W. M. Whaley and T. R. Govindachari, *Org. Reactions* **6**, 74 (1951).

25. G. R. Allen and M. J. Wein, *J. Org. Chem.* **33**, 198 (1968).

26. K. C. Nicolaou, *Tetrahedron* **37**, 4097 (1981).

27. R. R. Webb and S. Danishefsky, *Tetrahedron Lett.* 1357 (1983).

28. M. B. Gasc, A. Lattes and J. J. Perie, *Tetrahedron* **39**, 703 (1983).

29. A. K. Bose, M. S. Manhas and R. B. Ramer, *Tetrahedron* **21**, 449 (1965).

30. B. A. J. Clark, J. Parrick, P. J. West and A. H. Kelly, *J. Chem. Soc. C* 498 (1970).

31. J. A. Montgomery and K. Hewson, *J. Org. Chem.* **30**, 1528 (1965).

32. Y. Kurasawa and A. Takada, *Heterocycles* **14**, 281 (1980).

33. M. A. Kira, M. O. Abdel–Rahman and K. Z. Gadalla, *Tetrahedron Lett.* 109 (1969).

34. M. A. Kira, M. N. Abdul–Enein and M. I. Korkor, *J. Heterocycl. Chem.* **7**, 25 (1970).

35. M. A. Kira, Z. M. Nofal and K. Z. Gadalla, *Tetrahedron Lett.* 4215 (1970).

36. O. Meth–Cohn, B. Narine and B. Tarnowski, *J. Chem. Soc. Perkin Trans.* 1, 1520 (1981)

37. T. Tsuji, *Chem. Pharm. Bull.* **22**, 471 (1974).

38. F. Yoneda, Y. Sakuma and K. Shinozuka and K. Senga, *Heterocycles* **6**, 1179 (1977); F. Yoneda, K. Senga and S. Nishigaki, *Chem. Pharm. Bull.* **21**, 260 (1973).

39. E. Vedejs and G. A. Krarfft, *Tetrahedron* **38**, 2857 (1982).

40. J. Colonge and R. Gelin, *Bull. Soc. Chim. France* 799 (1954).

41. N. R. Easton, D. R. Cassady and R. D. Dillard, *J. Org. Chem.* **28**, 448 (1963).

42. K. E. Schulte and G. Nimke, *Arch. Pharm.* **290**, 597 (1957).

43. L. Lopez and J. Barrans, *Compt. Rend.* **263**, 557 (1966).

44. F. Ya. Perveer and V. M. Demidova, *Zh. Org. Khim.* **1**, 2244 (1965); *C.A.* **64**, 11095*d* (1966).

45. J. Reisch, *Ber.* **97**, 2717 (1964).

46. N. R. Easton, D. R. Cassady and R. D. Dillard, *J. Org. Chem.* **29**, 1851 (1964).

47. S. V. Kessar, *Acc. Chem. Res.* **11**, 283 (1978).

48. A. Mckillop, A. Henderson, P. S. Ray, C. Avendano and E. G. Molinero, *Tetrahedron Lett.* 3357 (1982).

49. U. Schollkopf, *Angew. Chem. Int. Edn. Engl.* **16**, 339 (1977).

50. A. M. Van Leusen and J. Schuat, *Tetrahedron Lett.* 285 (1976).

51. Y. Ito, K. Kobayashi and T. Saegusa, *J. Am. Chem. Soc.* **99**, 3532 (1977).

52. L. Stella, *Angew. Chem. Int. Edn. Engl.* **22**, 337 (1983).

53. P. Kovaic, M. K. Lowery and K. W. Field, *Chem. Rev.* **70**, 639 (1970).

54. K. Friedrich, G. Bock and H. Fritz, *Tetrahedron Lett.* 3327 (1978).

55. K. Isomura, H. Taguchi, T. Tanaka and H. Taniguchi, *Chem. Lett.* 401 (1977); K. Isomura, T. Tanaka and H. Taniguchi, *Chem. Lett.*, 397 (1977).

56. T. L. Gilchrist, C. W. Rees and J. A. R. Rodrigues, *Chem. Commun.* 627 (1979).

57. N. Furukawa, T. Nishio, M. Fukumura and S. Oae, *Chem. Lett.* 209 (1978).

58. G. Just and W. Zehetner, *Tetrahedron Lett.* 3389 (1967).

59. T. L. Gilchrist, C. J. Moody and C. W. Rees, *J. Chem. Soc. Perkin Trans.* 1, 1964 (1975); T. L. Gilchrist, C. J. Moody and C. W. Rees, *J. Chem. Soc. Perkin Trans.* 1, 1871 (1979).

60. R. J. Sundberg, S. R. Suter and M. Brenner, *J. Am. Chem. Soc.* **94**, 513 (1972).

61. F. P. Tsui, Y. H. Chang, T. M. Vogel and G. Zon, *J. Org. Chem.* **41**, 3381 (1976).

62. J. I. G. Cadogen, D. J. Sears, D. M. Smith and M. J. Todd, *J. Chem. Soc. C*, 2813 (1969).

63. J. S. Swenton, T. J. Ikeler and G. LeRoy Smyser, *J. Org. Chem.* **38**, 1157 (1973); F. P. Tsui, T. M. Vogel and G. Zon, *J. Org. Chem.* **40**, 761 (1975); P. A. S. Smith in W. Lwowski (Ed.), *'Nitrenes'*, Wiley–interscience, New York, 1970, chapter 4.

64. R. A. Abramovitch, T. Chellathurai, I. T. McMaster, T. Takaya, C. I. Azogu and D. P. Vanderpool, *J. Org. Chem.* **42**, 2914 (1977).

65. Y. Sakuma and F. Yoneda, *Heterocycles* **6**, 1911 (1977); M. Hamaguchi, *Chem. Commun.* 247 (1978).

66. J. I. G. Cadogan, J. T. Sharp and M. J. Trattles, *Chem. Commun.* 900 (1974).

67. A. Sekiguchi and W. Ando, *Bull. Chem. Soc. Japan* **50**, 3067 (1977); G. R. Champers and M. Jones, *Tetrahedron Lett.* 5193 (1978); W. Ando, A. Sekiguchi, A. J. Rothschild, R. R. Gallucci, M. Jones, T. J. Barton and J. A. Kilgour, *J. Am. Chem. Soc.* **99**, 6995 (1977).

68. W. D. Crow and H. Mc. Nab, *Aust. J. Chem.* **32**, 99, 111 and 123 (1979).

69. R. A. Abramovitch and V. Alexanian, *Heterocycles* **2**, 595 (1974).

70. R. M. Kellogg and S. Wassenaar, *Tetrahedron Lett.* 1987 (1970); W. H. Middleton, *J. Org. Chem.* **34**, 3201 (1969); J. Bulter, S. Wassenaar and R. M. Kellogg, *J. Org. Chem.* **37**, 4045 (1972).

71. J. P. Synder, *J. Am. Chem. Soc.* **96**, 5005 (1974).

72. F. D. Greene and S. S. Hecht, *J. Org. Chem.* **35**, 2482 (1970).

73. L. E. Friedrich and G. B. Schuster, *J. Am. Chem. Soc.* **91**, 7204 (1969).

74. H. D. Becher and K. Gustafssor, *J. Org. Chem.* **42**, 2966 (1977); E. Müller, R. Maeyer, B. Narr, A. Riecker and K. Scheffer, *Justus Liebigs Ann. Chem.* **25**, 645 (1961).

75. E. S. Levchenko and E. M. Dorokhova, *J. Org. Chem.* (USSR) **8**, 2573 (1972).

76. J. C. Pommelet, N. Manisse and J. Chuche, *C. R. Acad. Sci. Paris, Ser. C.* **270**, 1894 (1970).

77. W. Augenstein and F. Krohnke, *Justus Liebigs Ann. Chem.* **697**, 158 (1966).

78. E. J. Smutny, *J. Am. Chem. Soc.* **91**, 208 (1969).

79. A. Padwa, J. Smolanoff and A. Tremper, *Tetrahedron Lett.* 29 (1974); *J. Am. Chem. Soc.* **97**, 4682 (1975).

80. J. E. Baldwin, R. G. Pudussery, A. K. Gureshi and B. Sklarz, *J. Am. Chem. Soc.* **94**, 1395 (1972); **95**, 1945 and 1954 (1973).

81. E. M. Burgess and H. R. Penton, *J. Org. Chem.* **39**, 2885 (1974).

82. J. Bruhn, J. Zsindely, H. Schmid and G. Frater, *Helv. Chim. Acta* **61**, 2542 (1978).

83. W. H. Pirkle, H. Seto and W. V. Turner, *J. Am. Chem. Soc.* **92**, 6984 (1970); W. H. Pirkle and W. V. Turner, *J. Org. Chem.*, **40**, 1617 (1975).

84. A. E. Baydar and G. V. Boyd, *Chem. Commun.* 718 (1976).

85. D. Schuijl–Laros, P. J. W. Schiujl and L. Brandsma, *Rec. Trav. Chim. Pays–Bas* **91**, 785 (1972); L. Brandsma and D. Schuijl–laros, *Rec. Trav. Chim. Pays–Bas* **89**, 110 (1970).

86. E. N. Marvell, G. Caple and I. Shahidi, *J. Am. Chem. Soc.* **92**, 5641 (1970).

87. A. Roedig, H. A. Renk, V. Schaal and D. Scheutzow, *Chem. Ber.* **107**, 1136 (1974).

88. I. Hassan and F. W. Fowler, *J. Am. Chem. Soc.* **100**, 6696 (1978).

89. A. Padwa and L. Gehrlein, *J. Am. Chem. Soc.* **94**, 4933 (1972).

90. F. Yoneda and M. Higuchi, *J. Chem. Soc. Perkin Trans.* 1, 1336 (1977).

91. R. C. Shah and M. B. Ichaporia, *J. Chem. Soc.* 431 (1936); H. P. Ghadiali and R. C. Shah, *J. Ind. Chem. Soc.* **26**, 117 (1949); M. M. Blatter and H. Lukaszweski, *Tetrahedron Lett.* 855 (1964).

92. A. R. Katritzky and J. M. Logowski, *Chemistry of Heterocyclic N-oxides*, Academic press, New York, 1971, pp. 94.

93. W. Schroth and G. Fischer, *Angew. Chem. Int. Edn. Engl.* **2**, 394 (1963).

94. J. C. Jurtz, *Top. Curr. Chem.* **73**, 125(1978).

95. R. Huisgen, *Angew. Chem. Int. Ed. Engl.* **2**, 565, 633 (1963).

96. R. Huisgen, R. Grashey and J. Sauer in S. Patai (Ed.), *The Chemistry of the Alkenes*, Wiley-Interscience, London, 1965, pp. 806.

97. R. B. Woodward and R. Hoffmann, *Angew. Chem. Int. Edn. Engl.* **8**, 781 (1969).

98. I. Fleming, *Frotier Orbitals and Organic Chemical Reactions*, Wiley-Interscience, London, 1976.

99. R. Huisgen, G. Szeimies and L. Mobius, *Chem. Ber.* **99**, 475 (1966).

100. R. Huisgen in A. Padwa (Ed.), *1,3-Dipolar Cycloaddition Chemistry* Vol. **1**, Wiley-Interscience, 1984, pp. 1.

101. R. Huisgen and R. Weinberger, *Tetrahedron Lett.* **26**, 5119 (1985).

102. R. Huisgen, W. Scheer and H. Huber, *J. Am. Chem. Soc.* **89**, 1753 (1967).

103. A. Padwa, L. Fisera, K. F. Koehler, A. Rodriguez and G.S.K. Wong, *J. Org. Chem.* **49**, 276 (1984).

104. S. Takano, Y. Iwabuchi and K. Ogasawara, *J. Am. Chem. Soc.* **109**, 5523 (1987).

105. E. Vedejs and G. R. Martinez, *J. Am. Chem. Soc.* **101**, 6452 (1979).

106. C. Grundmann and P. Grunanger, *The nitrile oxides*, Springer (Berlin) (1971); A. Padwa (Ed.), *1,3-Dipolar Cycloaddition Chemistry* Vol. **2**, Wiley-Interscience, London, 1984, pp. 368.

107. A. P. Kozikowski and M. Adamczyk, *J. Org. Chem.*, **48**, 366 (1983).

108. D. P. Curran, *Tetrahedron Lett.* 3443 (1983).

109. M. Asaoka, M. Abe and H. Takei, *Bull. Chem. Soc. Japan* **58**, 2145 (1985).

110. D. St. C. Black, R. F. Crozier and V. C. Davis, *Synthesis* 205 (1975); J. J. Jufariello in A. Padwa (Ed.), *1,3-Dipolar Cycloaddition Chemistry* Vol. **2**, Wiley-Interscience, London, 1984, pp. 83.

111. A. Padwa, L. Fisera, K. F. Koehler, A. Rodriguez and G. S. K. Wong, *J. Org. Chem.* **49**, 276 (1984).

112. J. J. Tufariello and J. M. Puglis, *Tetrahedron Lett.* **27**, 1263 (1986).

113. A. C. Cope and N. A. Lepel, *J. Am. Chem. Soc.* **82**, 4656 (1960); S. W. Baldwin, J. D. Wilson and J. Aube, *J. Org. Chem.* **50**, 4432 (1985).

114. P. De Shong C. M. Dicken, J. M. Leginus and R. R. Whittle, *J. Am. Chem. Soc.* **106**, 5598 (1984).

115. T. Kametani, S-D. Chu and T. Honda, *J. Chem. Soc. Perkin Trans.* 1, 1593 (1988).

116. A. L. Logothetis, *J. Am. Chem. Soc.* **87**, 749 (1965).

117. A. P. Kozikowski and M. N. Greco, *Tetrahedron Lett.* **23**, 2005 (1982).

118. W. Oppolzer, *Tetrahedron Lett.* 2199 (1970).

119. W. Oppolzer, *Tetrahedron Lett.* 3091 (1970).

120. R. Grashey in A. Padwa (Ed.) *1,3-Dipolar Cycloaddition Chemistry* Vol. **1**, Wiley-Interscience, New York, 1984, pp. 733.

121. M. Regitz and H. Heydt in A. Padwa (Ed.), *1,3-Dipolar Cycloaddition Chemistry* Vol. **1**, Wiley-Interscience, New York, 1984, pp. 393.

122. E. Piers, R. W. Britton, R. J. Keziere and R. D. Smillie, *Canad. J. Chem.* **49**, 2623 (1971).

123. P. D. Bartlett, *Science* **159**, 833 (1968).

124. T. Gilchrist and R. C. Storr, *Organic Reations and Orbital Symmetry*, 2nd Edn., Cambridge University Press, Oxford, 1979.

125. W. T. Brady, *Tetrahedron* **37**, 2949 (1981).

126. D. A. Evans and E. B. Sjogren, *Tetrahedron Lett.* 3783 and 3787 (1985).

127. G. Buchi, J. A. Carlson, J. E. Powell and L. F. Tictze, *J. Am. Chem. Soc.* **92**, 2165 (1970); 95, 540 (1973).

128. G. Jones in A. Padwa (Ed.), *Organic Photochemistry* Vol. 5, Dekker, New York, 1981, pp. 123.

129. S. Toki, K. Shima and H. Sakurai, *Bull. Chem. Soc. Japan* **38**, 760 (1965).

130. J. Sauer, *Angew. Chem. Int. Edn.* **6**, 16 (1967).

131. S. M. Weinreb and J. J. Lewin, *Heterocycles* **12**, 949 (1979).

132. S. J. Danishefsky and C. J. Maring, *J. Am. Chem. Soc.* **107**, 1269 (1985).

133. M. Bednarski and S. Danishefrky, *J. Am. Chem. Soc.* **105**, 3716 (1983).

134. S. M. Weinreb and R. R. Staib, *Tetrahedron* **38**, 3087 (1982).

135. G. W. Kirby, *Chem. Soc. Rev.* **6**, 1 (1977).

136. G. Desimoni and G. Tacconi, *Chem. Rev.* **75**, 651 (1975).

137. D. L. Boger, *Tetrahedron* **39**, 2869 (1983).

138. A. P. Marchand, in S. Patai (Ed.), *The Chemistry of Double Bonded Functional Groups*, Wiley-interscience, London, 1977, pp.533.

139. R. S. Atkinson, B. D. Judkins and N. Khan, *J. Chem. Soc. Perkin Trans.* 1, 2491 (1982).

140. R. F. Heldeweg and H. Hageveen, *J. Am. Chem. Soc.* **98**, 2341 (1976).

141. R. J. Cotter and W. F. Beach, *J. Org. Chem.* **29**, 751 (1964).

142. R. A. Braun, *J. Org. Chem.* **28**, 1383 (1963).

143. E. L. Stogryn and S. J. Brois, *J. Org. Chem.* **30**, 88 (1965).

144. L. A. Paquette, *J. Am. Chem. Soc.* **84**, 4987 (1962).

145. E. Vogel, R. Schubart and W. A. Boll, *Angew. Chem. Int. Edn. Engl.* **3**, 510 (1964).

146. L. A. Paquette and R. W. Begland, *J. Am. Chem. Soc.* **88**, 4685 (1966).

147. S. S. M. Choi and G. W. Kirby, *J. Chem. Soc. Chem. Commun.* 177 (1988).

148. J. M. Lin, K. Koch and F. W. Fowler, *J. Org. Chem.* **51**, 167 (1986).

149. G. E. Keck and R. R. Webb, *J. Am. Chem. Soc.* **103**, 3173 (1981).

150. V. Bilinski, M. Karpf and A. S. Dreiding *Helv. Chim. Acta*, **69**, 1734 (1986).

151. W. Oppolzer and K. Thirring, *J. Am. Chem. Soc.* **104**, 4978 (1982).

CHAPTER 6

THREE-MEMBERED HETEROCYCLES

CONTENTS

1 THREE-MEMBERED HETEROCYCLES WITH ONE HETEROATOM

Three-membered heterocycles are strained molecules because of the bond angle distortion. These heterocycles have high energy contents relative to their acyclic isomers. Three-membered heterocycles generally have shorter C–C bonds than in cyclopropane (thiirane being an exception). The C–X–C bond angles (X = hetero-atom) in oxiranes and aziridines are very close to 60° and the peripheral H–C–H bond angles are near to 118°. In unsaturated three-membered heterocycles, the introduction of the double bond further enhances angle distortion and this inevitably increases the ring strain.

The chemistry of three-membered heterocycles is dominated by the presence of ring strain in these molecules. The ring strain leads to the enhanced reactivity and facilitates ring opening reactions. The stability and the overall reactivity of these heterocycles are not only attributed to the combined effect of bond shortening and angle distortion, but also depend on the presence of heteroatom. In this chapter, three-membered heterocycles containing nitrogen, oxygen and sulfur as heteroatm(s) are included.

1.1 Three-Membered Azaheterocycles

1.1.1 Aziridines (Azacyclopropanes)

1.1.1.1 General

Three-membered saturated heterocycles containing nitrogen as heteroatom are known as aziridines or azacyclopropanes **1**[1,2,2a].

1

Aziridines are biologically interesting heterocycles and their tendency to undergo nucleophilic ring opening reactions, responsible for their alkylating nature and activity, makes them antibiotic and anticancer drugs. Thus the antibiotic and anticancer activity of naturally occurring mitomycin C **2** and the anticancer activity of nitrogen mustard drugs of tetramine **3** type are associated with the presence of aziridine ring. Aziridine ring causes interruption in the cell division by alkylating nucleophilic groups on purine and pyrimidine bases in DNA.

2   **3**

Aziridines are weakly basic as compared with other amines because of the increased s-character of the lone pair on nitrogen. The short carbon–carbon bond (1.48 Å) combined with enhanced π-type characteristics causing delocalizaion of the electrons in the ring is also attributed for the weak basicity of aziridines[3]. Thus, the lone pair of electrons will have less interaction with the conjugative substituents (phenyl) attached to nitrogen[2].

In contrast to the most cyclic and acyclic amines, the bonding constraints of the aziridine ring have the effect of retarding the rate of pyramidal inversion at nitrogen (energy barrier in unstrained secondary amines = 25.10 kJ/mol and energy barrier in aziridine = 71.13 kJ/mol). The higher energy barrier in aziridine can be explained on the basis of the increased angle strain in the planar transition state (the bond angle in planar transition state of aziridine deviates by nearly 60° from the angle in unstrained planar transition state). The substituents attached to nitrogen have pronounced effects on the barriers to nitrogen inversion[4-8]. The electron-withdrawing substituents which are able to delocalize the lone pair of electrons and to stabilize the planar transition state lower the energy barrier or increase the rate of pyramidal inversion. The steric effect of bulky substituents has the same effect since destabilizing non-bonded interactions in the ground state configuration are partially relieved in the transition state i.e. pyramidal form destabilized. However, the electron-withdrawing substituents having unshared electron-pair (NH$_2$, Cl, OCH$_3$) have the effect of dramatically raising the energy barriers to nitrogen inversion and often enable isolation of the enantiomers. The lone pair-lone pair interactions in the planar transition state are considered probably responsible for the increased energy barriers.

1.1.1.2 Synthesis

The synthetic methods for aziridines include :

(i) methods involving intramolecular cyclization reactions and

(ii) methods involving cycloaddition reactions.

1.1.1.2.1 Intramolecular Cyclization Reactions

The ring closure in intramolecular cyclization reactions mainly depends on two factors; ring strain and chain length.

(i) Ring strain : the cyclization is more difficult with greater ring strain in the product of cyclization.

(ii) Chain length : the ring forming ability falls off with an increase in the chain length.

Three-membered rings are relatively easily formed. Although the ring strain in three-membered ring is high, the probability of the ends of the chain being in the right conformation for ring closure is also high.

1.1.1.2.1.1 Gabriel Method

Intramolecular cyclization of β-chloroamines **6**, obtained by treating β-hydroxy-amines **5** with hydrochloric acid, provides aziridines **7**. β-Hydroxyamines **5** can be prepared by the reaction of epoxides (oxiranes) **4** with ammonia or primary amines (scheme-1)[9-13].

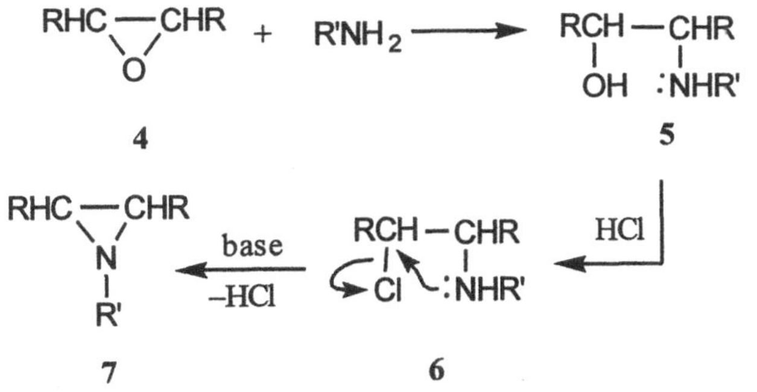

$$
\underset{\textbf{4}}{RHC\overset{\displaystyle \diagdown \,\diagup}{\underset{O}{}}CHR} + R'NH_2 \longrightarrow \underset{\textbf{5}}{\underset{OH\ \ :NHR'}{\overset{\textstyle}{RCH-CHR}}}
$$

$$
\underset{\substack{| \\ R' \\ \textbf{7}}}{RHC\overset{\displaystyle \diagdown \,\diagup}{\underset{N}{}}CHR} \xleftarrow[\text{-HCl}]{\text{base}} \underset{\textbf{6}}{\underset{Cl\ \ :NHR'}{\overset{\textstyle}{RCH-CHR}}} \xleftarrow{\text{HCl}}
$$

Scheme-1

The reaction proceeds with the intramolecular displacement of the halogen atom by an amino group via SN²-mechanism. This reaction is stereospecific and occurs with the inversion at carbon bearing leaving group.

1.1.1.2.1.2 Wenker Method

The reaction of β-amino alcohols (β-hydroxyamines) **5** with sulfuric acid gives
β-amino hydrogen sulfate derivatives **8** which on treatment with a base provide
aziridines **7** (scheme-2).

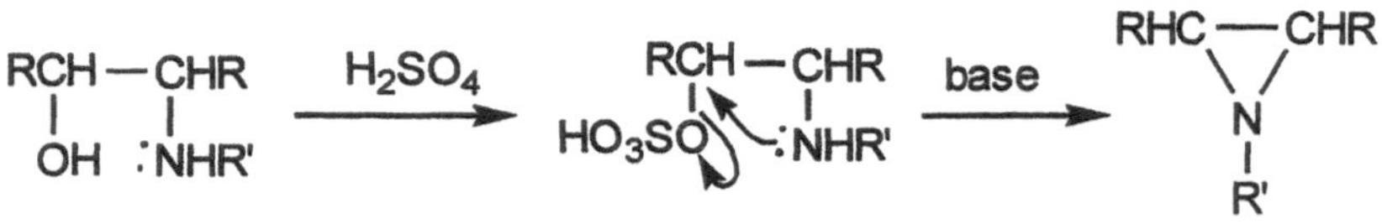

Scheme-2

1.1.1.2.1.3 Chloronitrosation of Tetraalkylalkenes

The methods discussed in the preceding sections are not used to synthesize
2,2,3,3-tetraalkylaziridines because the corresponding β-haloamines required can
not be easily prepared. Closs and Brois[14] developed a method which involves three
steps : (i) chloronitrosation of tetraalkylalkene **9**, (ii) reduction of nitroso chloride
10 and (iii) cyclization of amino derivative **11** with a base (scheme-3).

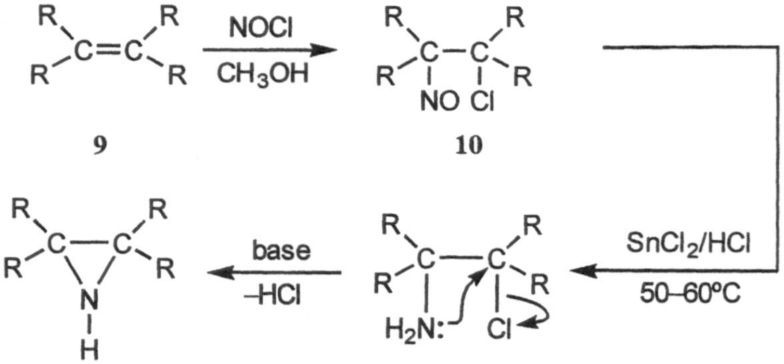

Scheme-3

1.1.1.2.1.4 Hassner Method

It involves stereospecific addition of iodoisocyanate to alkenes **13** via the intermediacy of iodonium ion **14**. The resulting *trans*-diaxial β-iodoisocyanate **15** on treatment with methanol affords the corresponding β-iodocarbamate **16** which with a base undergoes cyclization providing aziridine derivative **17**. The rapid abstraction of a proton is followed by the intramolecular cyclization to N-carbalkoxyaziridine intermediate **16b** which is readily saponified and decarboxylated (scheme-4)[15].

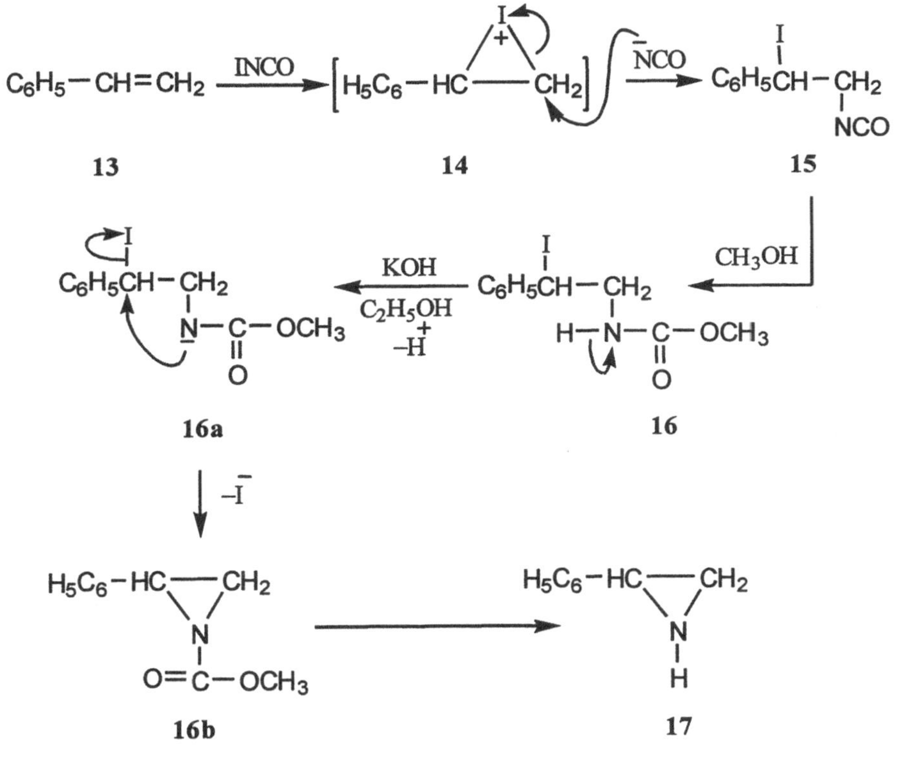

Scheme-4

This method is preferred because of its applicability for the synthesis of substituted aziridines.

1.1.1.2.2 Cycloaddition Reactions

Aziridines are prepared by cycloaddition of a 'one-atom' moiety (such as nitrene) to a 'two-atom' moiety (like alkene).

1.1.1.2.2.1 Nitrene Insertion Method

Direct insertion of a nitrene into an olefinic linkage provides aziridines. When nitrene is photochemically or thermally produced, the yield of aziridine is unsatisfactory because of the formation of side products (scheme-5)[16,17].

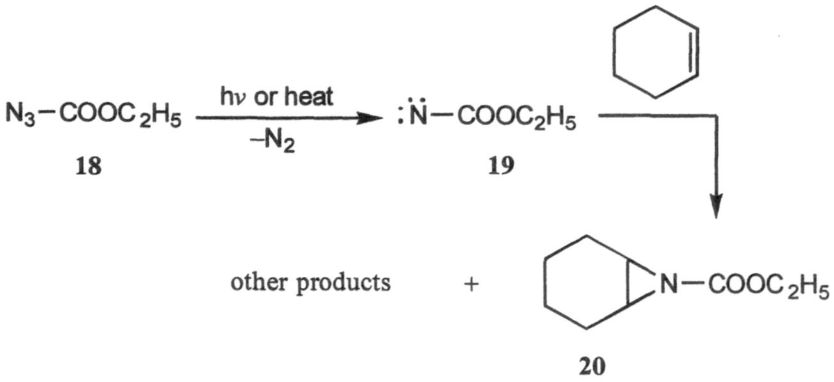

Scheme-5

However, at room temperature, azide **18** reacts with an alkene by 1,3-dipolar addition providing isolable 1,2,3-triazole intermediate **21** which on photolysis yields aziridine in excellent yield (scheme-6)[18].

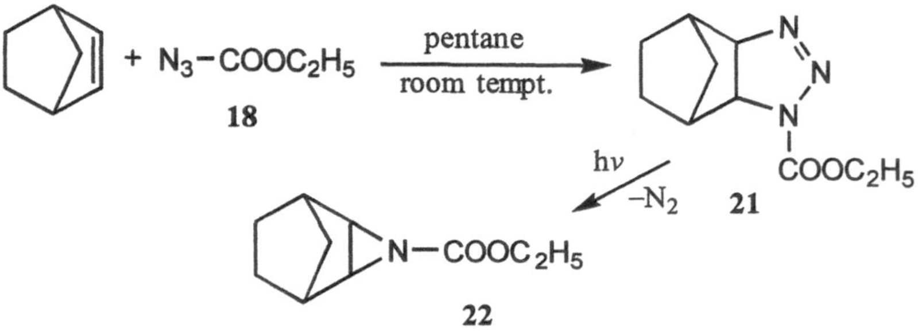

Scheme-6

The reaction is normally stereospecific and the stereochemistry of two carbon unit is maintained in the cyclic product. Thus, two reaction mechanisms are possible in the nitrene insertion. In one mechanism, the azide can lose nitrogen to form diradical nitrene which with alkene undergoes cycloaddition reaction. In the alternative mechanism, which is more general, the azide adds directly to an alkene by 1,3-dipolar addition reaction[19] providing a five-membered heterocyclic compound **24**, which loses nitrogen to yield aziridine (scheme-7).

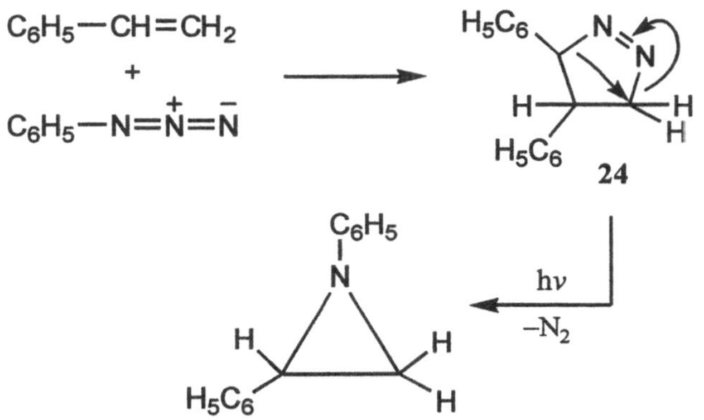

Scheme-7

1.1.1.3 Reactions

Three-membered heterocycles are strained and are, therefore, extremely susceptible to the ring opening reactions due to the release of ring strain.

1.1.1.3.1 Ring Opening Reactions

Aziridine ring is cleaved readily by several reagents providing a variety of functionalized compounds. In the ring opening reactions, carbon–nitrogen bond is most readily cleaved, although carbon–carbon bond is also cleaved in some ring opening reactions.

1.1.1.3.1.1 Nucleophilic Ring Opening Reactions

Aziridine ring undergoes ring opening reactions with the cleavage of carbon–nitrogen bond and the positively charged carbon is then attacked by the nucleophile (scheme-8).

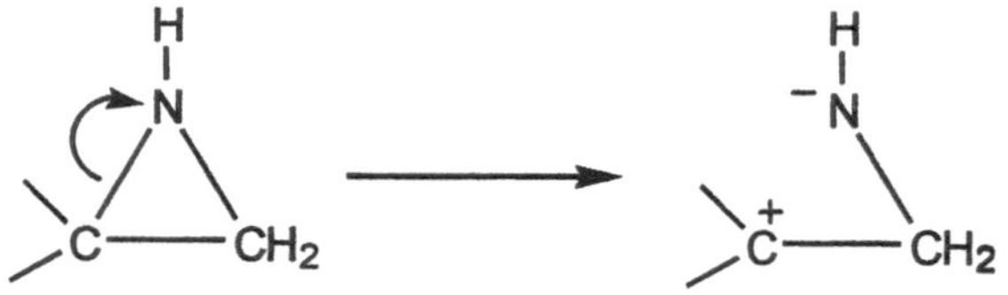

Scheme-8

The facility with which carbon–nitrogen bond of aziridines is cleaved depends upon :

(i)   the electron accepting nature of the substituents at nitrogen atom,

(ii) steric effects of carbon substituents and

(iii) the nature of attacking reagent.

The ring opening becomes progressively easier as the electron accepting ability of the substituent at nitrogen increases and is most rapid with aziridinium cations. Nucleophilic ring opening reactions of N-alkyl- or N-unsubstituted aziridines are often acid catalyzed. The attack of nucleophile on the ring carbon proceeds with the inversion of configuration (attacking at the side opposite to the heteroatom) via SN²-mechanism (scheme-9). Aziridines undergoing nucleophilic ring opening reactions should bear strong electron accepting groups at nitrogen, otherwise strong nucleophiles or severe reaction conditions are required.

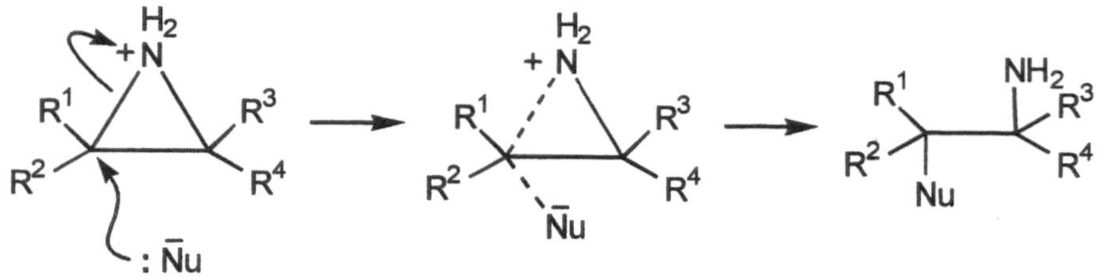

Scheme-9

In asymmetrically substituted aziridines, steric hindrance causes predominantly the attack of nucleophile at the less substituted carbon atom. The reaction is expected to proceed via SN²-type mechanism where the attack of nucleophile is as important as the cleavage of the bond and the mechanism is termed as 'push-mechanism'. However, in some cases the bond cleavage process becomes more important than the process involving nucleophilic attack and the reaction is expected to proceed via SN¹-type mechanism and is termed as 'pull-mechanism' which depends upon :

(i) the presence of carbonium ion stabilizing substituents and

(ii) the ionic nature of the solvent.

The ring opening in asymmetrically substituted aziridines is affected by the carbonium ion stability rather than the steric effects (scheme-10)[20].

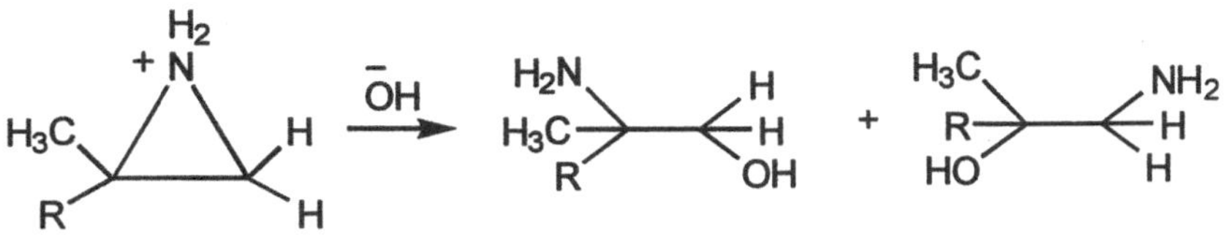

R = H	58%	42%
R = CH$_3$	7%	93%

Scheme-10

The reaction of aziridine with sodiomalonic ester **26** proceeds with the cleavage of the ring and forms reactive intermediate **27** which cyclizes to 3-carboethoxy-pyrrolidinone **28** (scheme-11)[21].

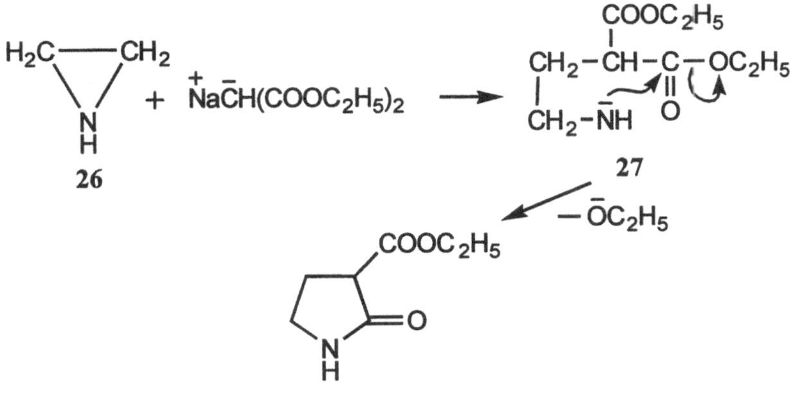

Scheme-11

1.1.1.3.1.2 Electrophilic Ring Opening Reactions

Ring Opening reactions of aziridines are markedly accelerated in acidic media. Aziridine with hydrochloric acid under uncontrolled conditions undergoes ring opening reaction providing a polymeric product **29** (scheme-12).

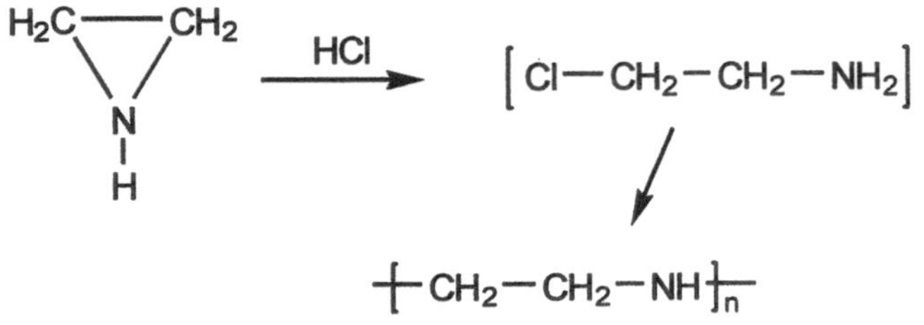

Scheme-12

In the acid catalyzed ring opening reactions, nitrogen is protonated and facilitates pull-mechanism. In the asymmetrically substituted aziridines, the site of attack is governed by the nature of the substituent at the carbon atom and can be explained by the 'push-pull mechanism' (scheme-13)[22,23]. The attack at the hindered carbon atom is caused due to the formation of a stabilized carbonium ion.

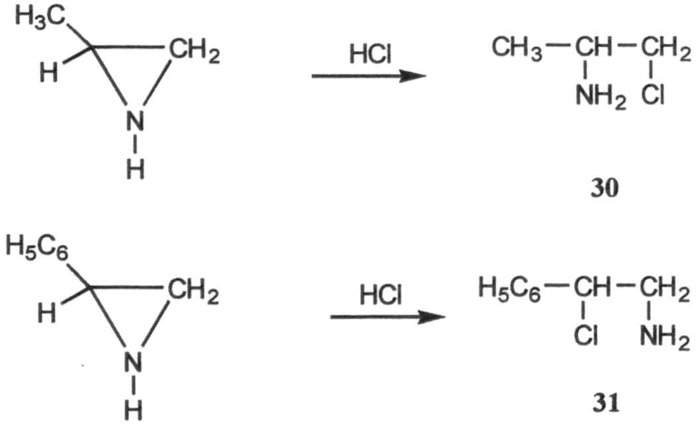

[Effect of stabilized carbonium ion]

Scheme-13

The reaction of aziridine with acetyl chloride proceeds with the electrophilic attack by an acetyl electrophile at the nitrogen atom followed by ring opening (scheme-14).

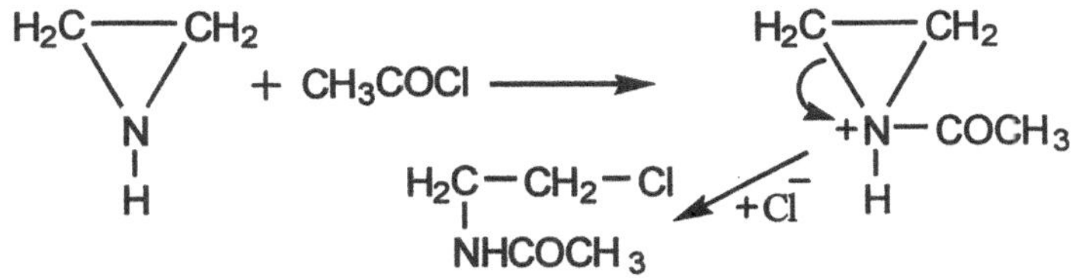

H_2C——CH_2 + CH_3COCl ⟶

32

Scheme-14

1.1.1.3.1.3 Electrocyclic Ring Opening Reactions

N-Substituted aziridines, particularly with conjugatively electron-withdrawing substituents at carbon, undergo electrocyclic ring opening reactions in which carbon–carbon bond is cleaved rather than the carbon–nitrogen bond. In such reactions, the aziridine ring system may be considered analogous to cyclobutene (scheme-15).

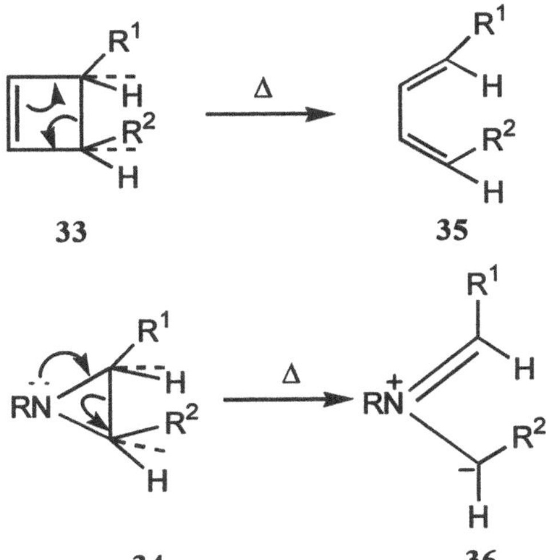

33   35

34   36

Scheme-15

Since these reactions involve 4π-electrons, they will be thermally conrotatory (π-electron system rotates in the same direction). When N-substituted aziridine **37** is pyrolyzed and 1,3-dipolar ring cleaved product **38** obtained is trapped by a dienophile, the stereochemistry of the product **39** is observed to be that expected of a conrotatory ring cleavage (scheme-16).

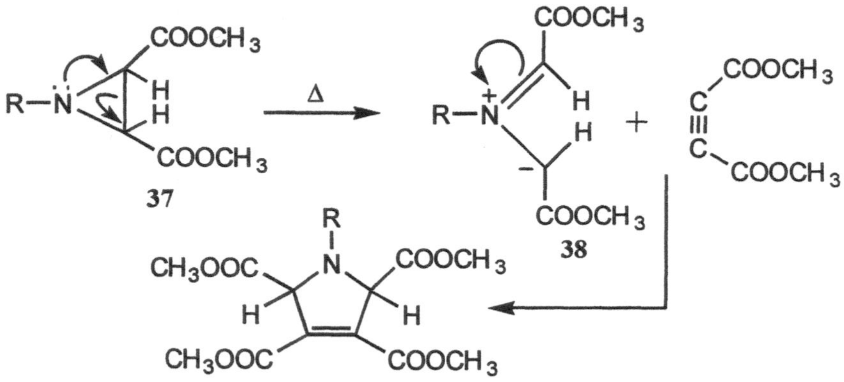

Scheme-16

1.1.1.3.1.4 Other Ring Opening Reactions

Protonated aziridines or quaternary aziridinium salts are exceptionally reactive towards nucleophiles[24]. The reaction of aziridinium salt **40** with methanol proceeds with the ring opening involving cleavage of the carbon–nitrogen bond to form product **41** (scheme-17).

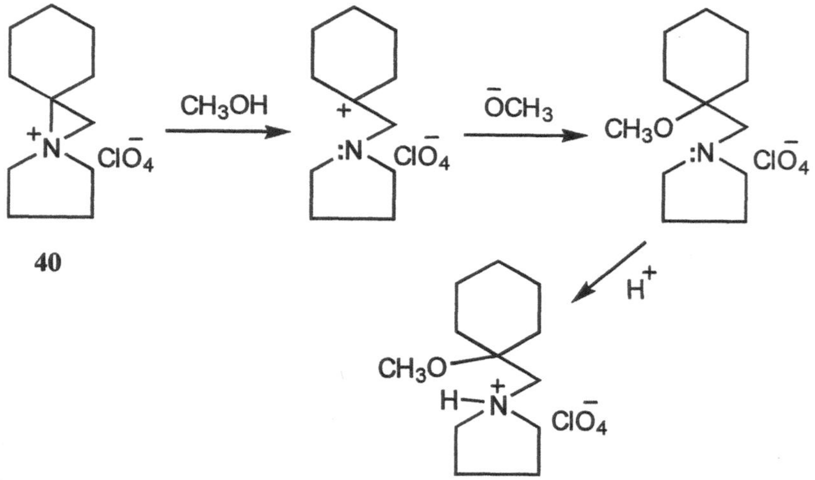

Scheme-17

The condensation of aziridinium salts with aldehydes[25,26], ketones[25,26] and nitriles[27] at moderate temperature results in the expansion of aziridinium ring involving cleavage of the carbon–nitrogen bond (schemes-18 and 19).

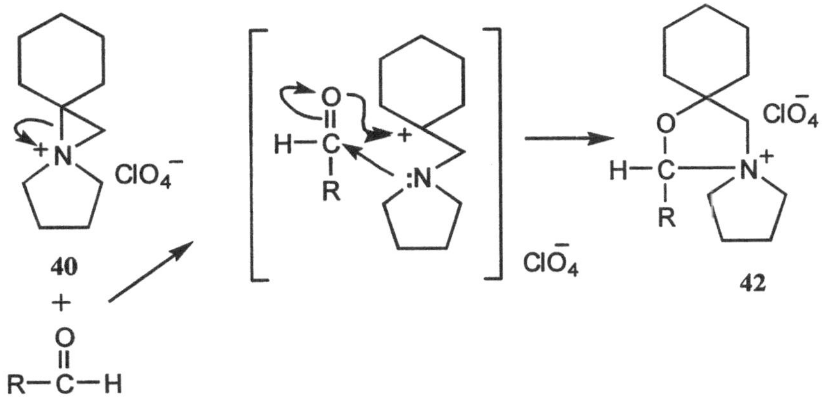

Scheme-18

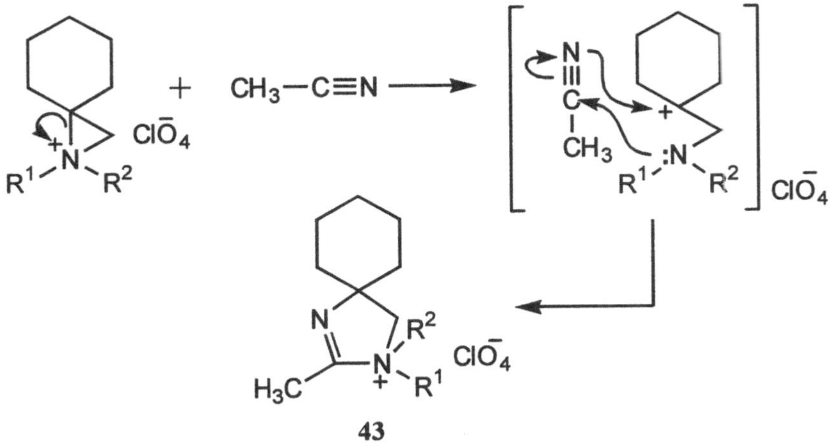

Scheme-19

1.1.1.3.1.5 Fragmentation Reactions

N-Unsubstituted aziridines **44** are stereoselectively deaminated in the reaction with nitrosating agents (nitrosyl chloride, nitrous acid and methyl nitrile). Nitrogen atom is extruded and alkenes are formed via the formation of N-nitrosoaziridines **45** which can be isolated below −20°C (scheme-20)[28,29].

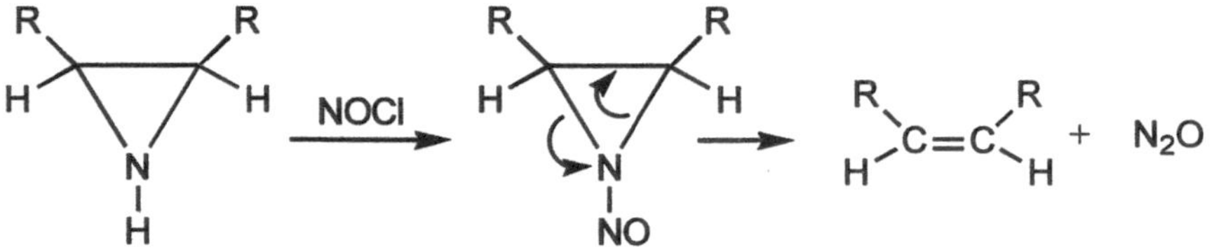

44 45 46

Scheme-20

1.1.1.3.1.6 Rearrangements (Ring Opening Reactions Accompanied by Rearrangements)

1.1.1.3.1.6.1 N-Acylaziridines **47** rearrange thermally or by acid catalysis to oxazolidones **48** (scheme-21)[10]. The rearrangement involves an intramolecular

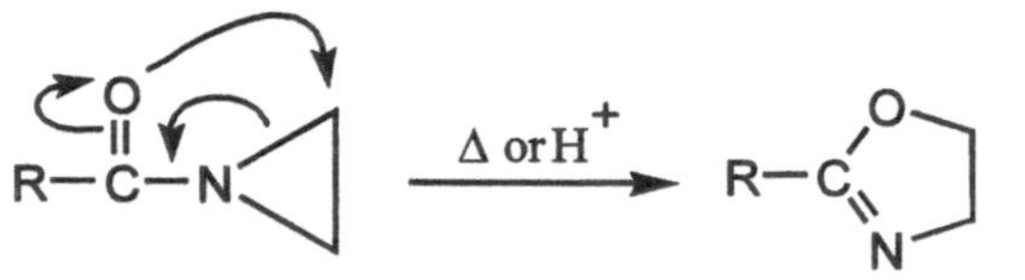

47 48

Scheme-21

attack of carbonyl oxygen at a ring carbon causing cleavage of the carbon–nitrogen bond. The driving force is provided by the relief of strain due to ring opening.

Thioamidoaziridines also undergo similar rearrangement (scheme-22)[30].

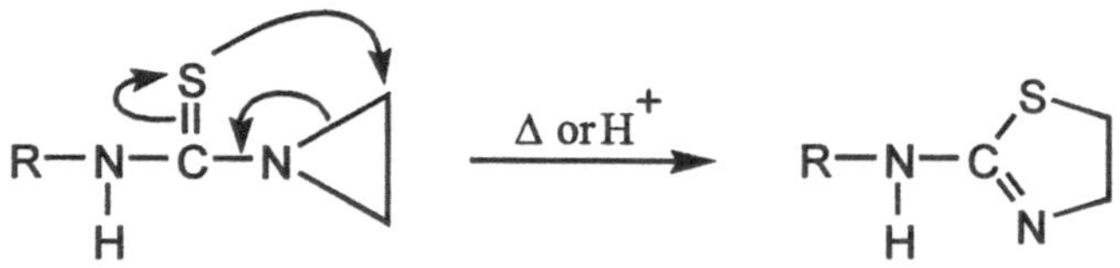

49 50

Scheme-22

1.1.1.3.1.6.2 Pyrolysis of N-acylaziridines **51** causes isomerization to N-allylamides **52** involving intramolecular proton transfer from the side chain carbon to the ring nitrogen with the cleavage of carbon–nitrogen bond (scheme-23)[31].

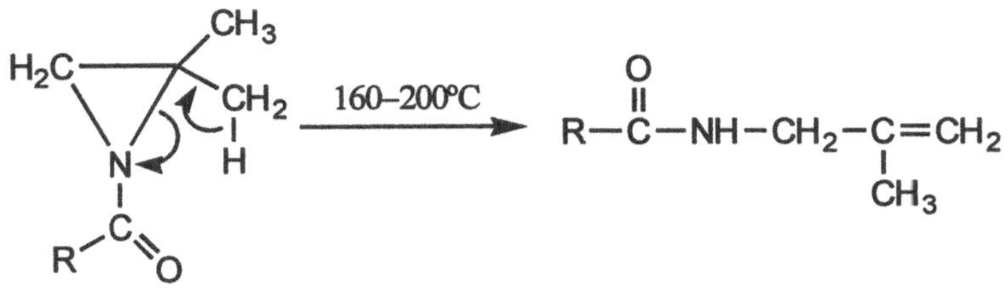

Scheme-23

1.1.1.3.1.6.3 Iodide ion catalyzes the isomerization of aziridines. The process is considered to proceed by the attack of iodide ion at the least hindered carbon in the aziridine ring with the cleavage of carbon–nitrogen bond resulting in cyclization to a five-membered heterocyclic compound **54** (scheme-24)[32].

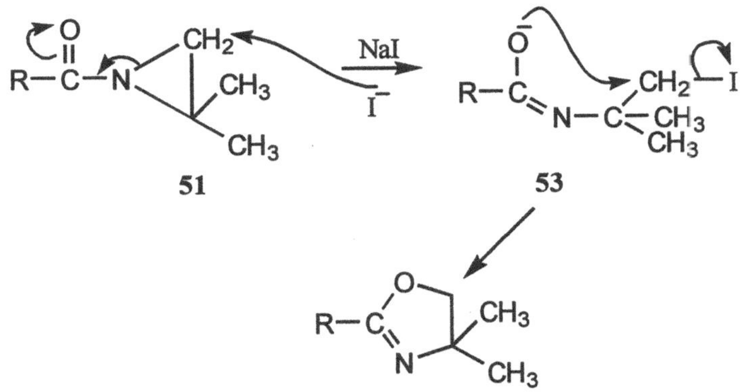

Scheme-24

1.1.1.3.2 Reaction with Carbon Disulfide

Aziridine with carbon disulfide undergoes ring expansion reaction with the formation of thiazoline **55** (scheme-25)[33].

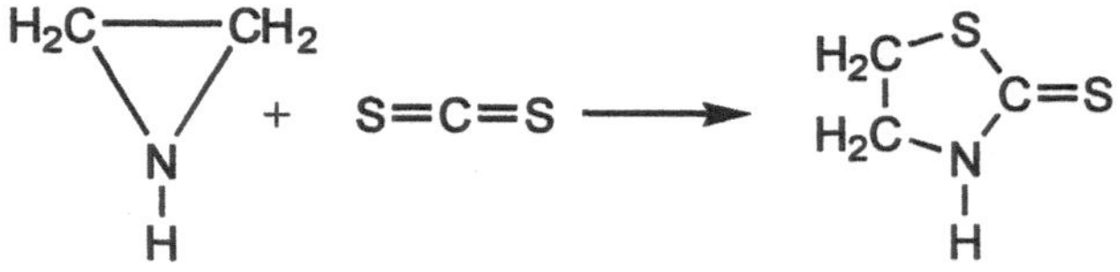

$$H_2C\!-\!\!-\!\!CH_2 \quad + \quad S\!=\!C\!=\!S \quad \longrightarrow \quad \underset{H_2C-N}{\overset{H_2C-S}{\Big\rangle}}\!\!\!\!C\!=\!S$$

Scheme-25

1.1.1.3.3 Friedel–Crafts Reaction

Aziridine undergoes Friedel–Crafts reaction with benzene providing β-phenylethyl-amine **56** (scheme-26).

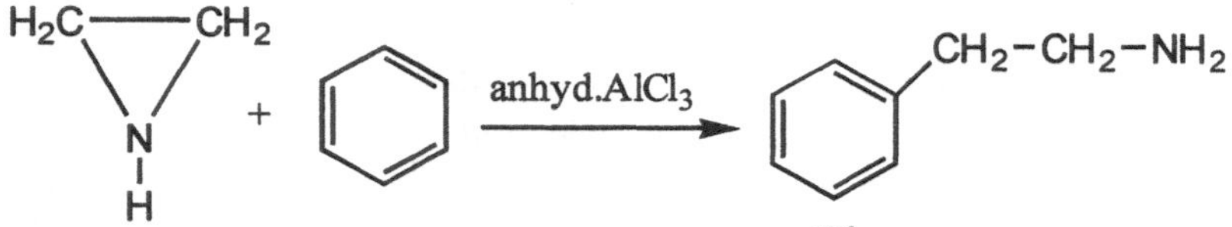

$$H_2C\!-\!\!-\!\!CH_2 \quad + \quad \text{(benzene)} \quad \xrightarrow{\text{anhyd.AlCl}_3} \quad \text{(C}_6\text{H}_5)\text{CH}_2\!-\!CH_2\!-\!NH_2$$

56

Scheme-26

1.1.1.3.4 N-Substitution

The hydrogen atom attached to nitrogen in aziridine is reactive and can be easily substituted (scheme-27).

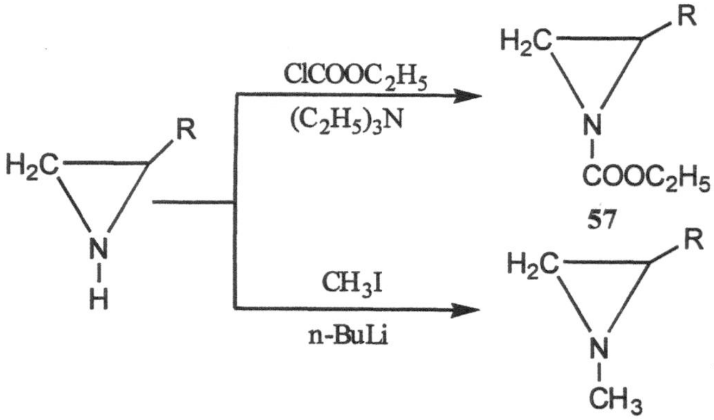

$$\xrightarrow[\text{(C}_2\text{H}_5)_3\text{N}]{\text{ClCOOC}_2\text{H}_5}$$

$$\xrightarrow[\text{n-BuLi}]{\text{CH}_3\text{I}}$$

57

58

Scheme-27

1.1.2 Azirines (Azacyclopropenes)

Introduction of unsaturation in the small rings satisfying the requirement of 120° bond angle for *sp²*-hybridized atoms inevitably increases the ring strain. Unsaturated small ring heterocycles are, therefore, less stable as compared to their saturated counterparts and are difficult to prepare. Azirines[34] can exist in two isomeric forms; 1*H*-azirines (enamines or 2-azirines) **59** and 2*H*-azirines (imines or 1-azirines) **60**.

59   **60**

1*H*-Azirine is isoelectronic with cyclobutadiene having four π-electrons in planar cyclic conjugation. It is an antiaromatic and resonance destabilized as defying the Hückel's aromaticity rule. The instability of 1*H*-azirines **59** is not only due to the large angle strain in the unsaturated three-membered ring, but also due to the potential overlap of the lone pair of electrons at nitrogen with the olefinic π-electrons. The instability, coupled with considerable ring strain, makes the synthesis of 1*H*-azirines very unlikely. However, 1*H*-azirine (2-azirine) **63** is formed as an intermediate in the pyrolysis of triazolines **61** and **62** where both the compounds yield an indentical mixture of 2*H*-azirines **64** and **65** (scheme-28).

61   **63**   **62**

64   **65**

Scheme-28

2*H*-Azirine is more stable than isomeric 1*H*-azirine by 168.62 kJ/mol. 2*H*-Azirines are non-basic because of considerably increased s-character of the C–H bond and of the lone pair on nitrogen atom. The strained nature of 2*H*-azirines is reflected from the abnormal C=N stretching frequency (1800 cm⁻¹) as compared to the normal value (1650 cm⁻¹) of unstrained imines in infrared spectra and the considerably higher coupling constant (^{13}C–H=176 ppm) than in open chain analogs (^{13}C–H=132 ppm) in NMR spectra. 2*H*-Azirines are very strained, but are not antiaromatic.

1.1.2.1 Synthesis

1.1.2.1.1 Thermal and Photochemical Decompositions of Vinylazides

Thermal and photochemical decompositions of vinylazides **66** provide 2*H*-azirines via the formation of vinyl nitrene intermediate **67** (scheme-29)[38-40].

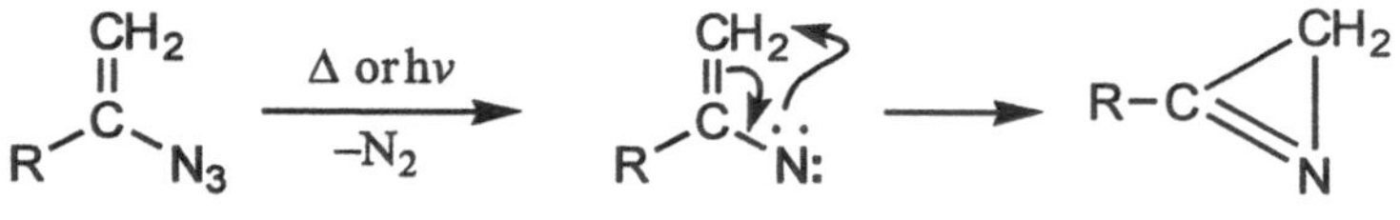

Scheme-29

Vinylazides **66** are prepared by the following sequence of the reactions (scheme-30).

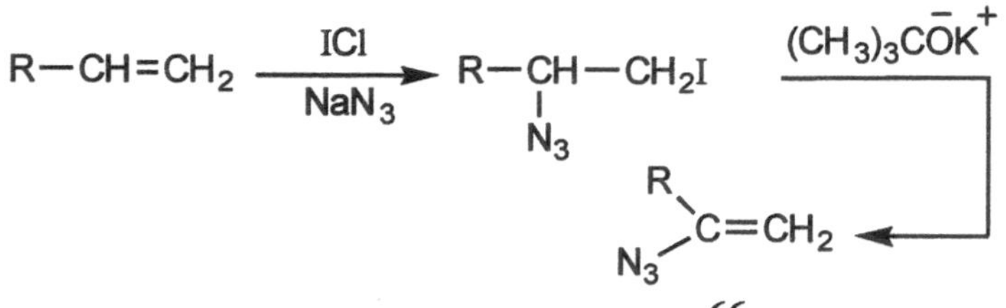

Scheme-30

1.1.2.1.2 Modified Neber Rearrangement

The base catalyzed rearrangement of oxime-p-toluene sulfonates **68** to α-amino ketones **72** via an azirine intermediate **71** is known as Neber rearrangement (scheme-31)[41].

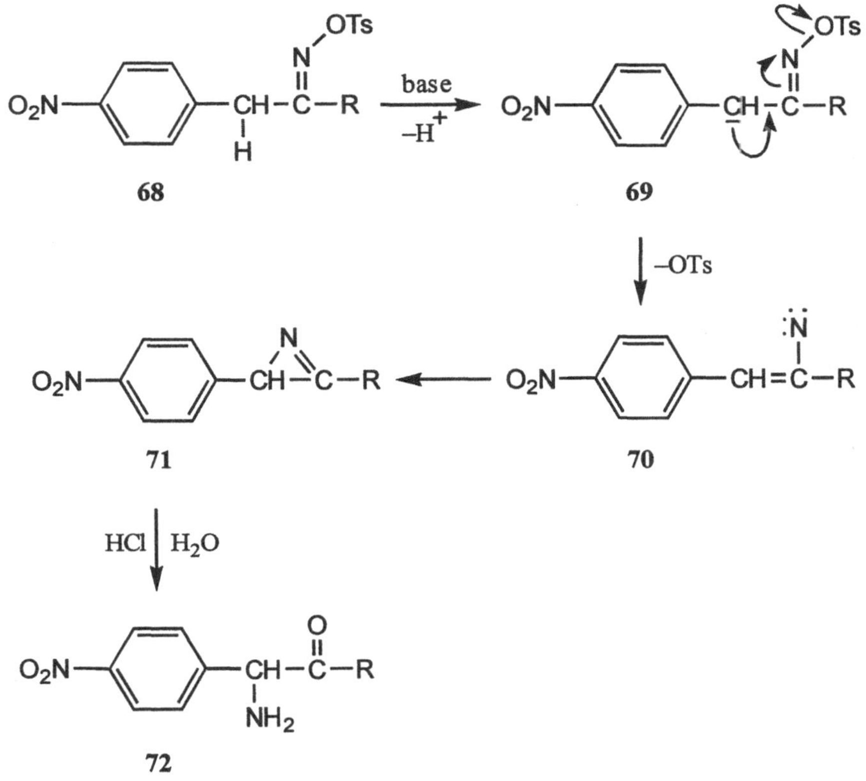

Scheme-31

The rearrangement involves α-proton abstraction by a base and loss of tosylate ion to form α,β-unsaturated nitrene **70** which cyclizes to azirine **71**. The isolation of 2*H*-azirines is difficult and, therefore, a modification has been introduced in which dimethylhydrazone methiodide **72a** with a strong base provides isolable 2*H*-azirines **73** (scheme-32)[42].

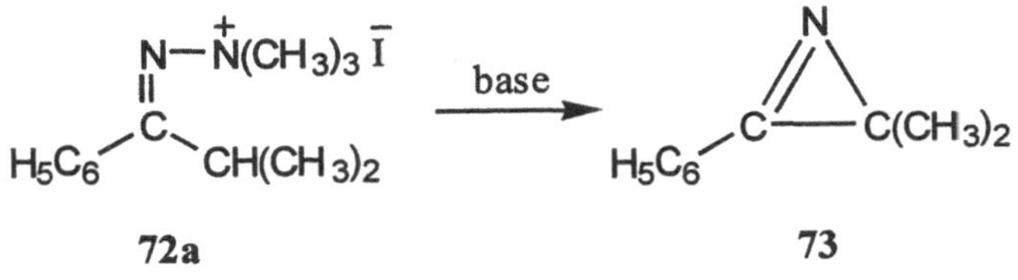

Scheme-32

1.1.2.1.3 Thermal and Photolytic Ring Contractions of Isoxazoles

Pyrolysis of 5-alkoxyisoxazoles **74** involves ring contraction providing isolable 2*H*-azirines **75** (scheme-33)[43].

Scheme-33

Photolysis of 3,5-disubstituted isoxazoles **76** also produces 2*H*-azirines **77** (scheme-34)[44].

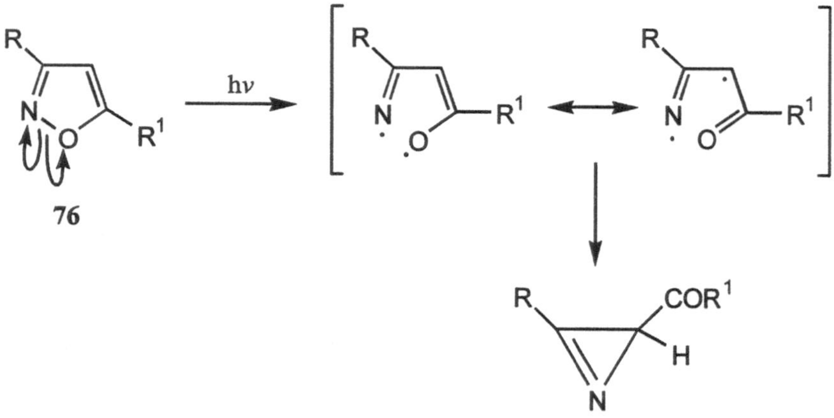

Scheme-34

1.1.2.2 Reactions

The inherent structural features within the ring system, high ring strain, reactive π-bond and a lone pair on nitrogen, are important in determining the chemistry of 2*H*-azirines. The reactions can be categorized into three classes :

(i)   Reactions at C=N bond

(ii) Thermal cleavage of N–C$_2$ bond

(iii) Cleavage of C$_2$–C$_3$ bond.

1.1.2.2.1 Reactions at C=N Bond

The addition of nucleophiles to C=N bond takes place stereospecifically on the least hindered side of the molecule.

1.1.2.2.1.1 Reaction with Lithium Aluminium Hydride

Lithium aluminium hydride reduces C=N bond providing aziridines (scheme-35)[45].

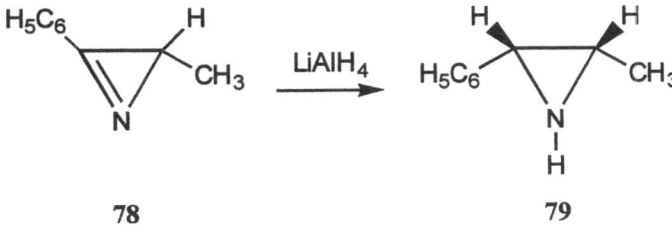

78   79

Scheme-35

1.1.2.2.1.2 Reaction with Grignard Reagents

Azirines undergo *cis*-addition with Grignard reagents (scheme-36).

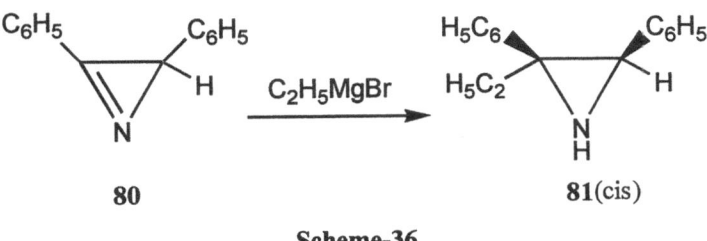

80   81(cis)

Scheme-36

1.1.2.2.1.3 Reaction with Carbanions

The carbanion attacks an imine carbon of 2*H*-azirine and results in an intermediate **83** which undergoes ring cleavage followed by cyclization to yield a five-membered heterocycle **85** (scheme-37)[45,46].

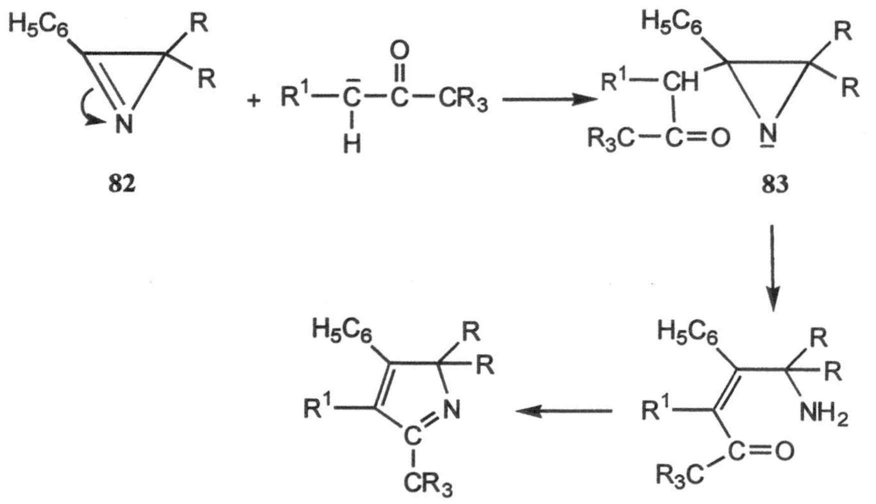

Scheme-37

1.1.2.2.1.4 Reaction with Trichloromethide Anion

Trichloromethide anion adds stereoselectively to 2*H*-azirines to form aziridines **86** which with nitrous acid yield alkenes **88** via the formation of nitroso derivatives **87**; and with a base undergo rearrangement yielding azetidines **90** via the formation of bicyclobutane derivatives **89** (scheme-38)[45,47].

1.1.2.2.1.5 Cycloaddition

2*H*-Azirines undergo cycloaddition with dienes **91** and heterodienes **94** providing cycloaddition products **92** and **95**, respectively[48]. The C=N bond of strained 2*H*-azirines shows increased reactivity than the corresponding bond in unstrained imines (scheme-39).

1.1.2.2.2 Thermal Cleavage of N–C$_2$ Bond

Thermolysis of azirines containing conjugative substituent at C-2 results in the formation of five-membered rearranged product **97** as a major product with the cleavage of N–C$_2$ bond (scheme-40)[49].

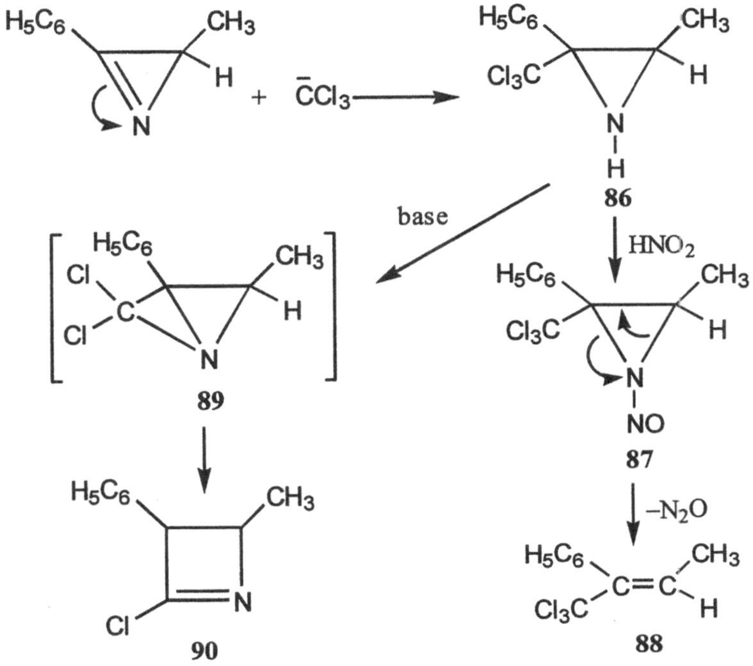

Scheme-38

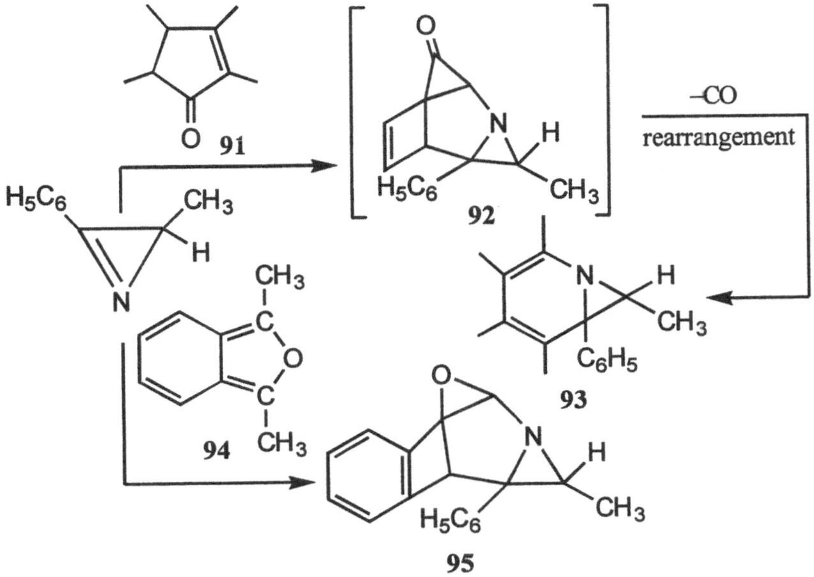

Scheme-39

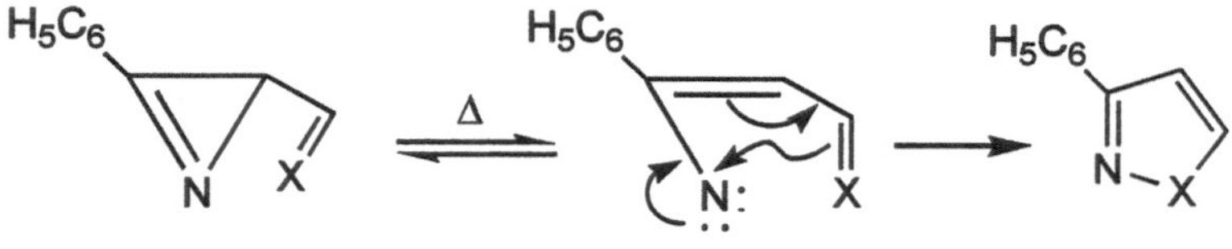

96   **97**

X= O, NPh

Scheme-40

1.1.2.2.3 Photolytic Cleavage of C₂–C₃ Bond

Photolysis of azirines, in contrast to thermolysis, usually results in the cleavage of C_2–C_3 bond forming nitrile ylides which can be trapped by a number of dipolarophiles (scheme-41)[49].

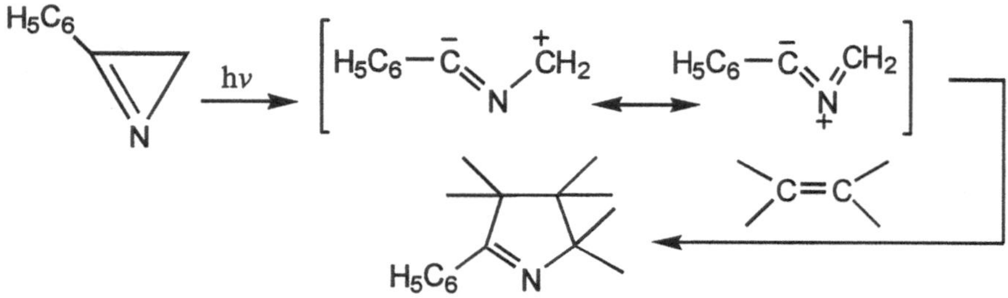

99

Scheme-41

In the absence of dipolarophile, the intermediate nitrile ylide, formed by azirine containing stabilizing group at C-3, undergoes intramolecular cyclization yielding an isolable product **100** (scheme-42).

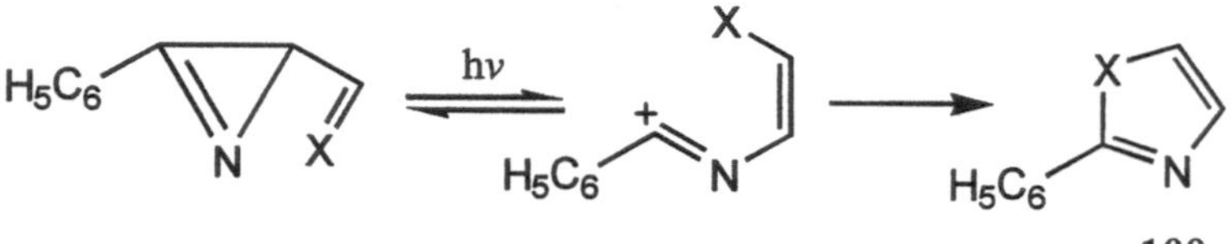

100

Scheme-42

1.2 Three-Membered Oxaheterocycles

1.2.1 Oxiranes

1.2.1.1 General

Three-membered saturated heterocycles containing oxygen as a heteroatom are known as oxiranes **101**[50,51]. These are also designated as epoxides or α,β-epoxyethanes.

101

The C–C bond distance in oxirane (1.472 Å) is between the normal C–C (1.54 Å) and C=C (1.34 Å) bond distances. The C–O–C bond angle is 61° which is much smaller than that of dimethyl ether (111.5°). The H–C–H bond angle (116°) in oxirane lies between those expected of sp^3 (109.5°) and sp^2 (120°) hybridized carbon atoms[52]. Thus, oxirane has been considered to have hybridization between sp^3 and sp^2. The strain energy has been estimated to be 113.80 kJ/mol. This ring system is present in many biologically important compounds such as **102** and **103**.

Leukotriene
102

(+)-Disparlure
103

1.2.1.2 Synthesis

1.2.1.2.1 Epoxidation of Alkenes (Insertion of Oxygen into C=C Bond)

1.2.1.2.1.1 Peracid Oxidation[54]

It involves cycloaddition of one-atom moiety from peracid to two-atom moiety of an alkene. *m*-Chloroperbenzoic acid[55] is the most convenient oxidant for the

oxidation of C=C bond in alkenes. This is an electrophilic-addition reaction and is
called Prilezhaev reaction (scheme-43)[56].

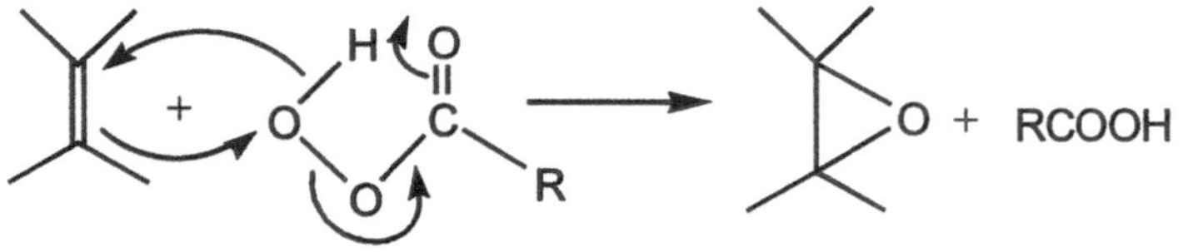

Scheme-43

Epoxidation is highly dependent upon the electron density at the double bond.
The substituents increasing electron density at the double bond enhance
epoxidation (scheme-44)[57].

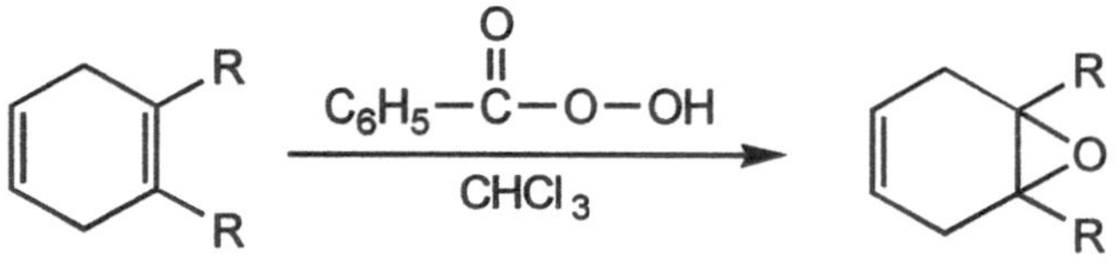

Scheme-44

The reaction is stereospecific and retains the stereochemistry of alkenes
(scheme-45).

1.2.1.2.1.2   Epoxidation with Alkaline Hydrogen Peroxide

The epoxidation of alkenes, conjugated with electron-withdrawing groups (carbonyl
or cyano), is best accomplished with alkaline hydrogen peroxide[58]. The nucleophilic
addition proceeds by Michael addition of hydroperoxide anion to C=C bond
followed by intramolecular displacement of hydroxide ion (scheme-46)[59]. It is not
stereoselective and differs from the peracid oxidation as only one isomer of oxirane
is formed (scheme-47)[60].

1.2.1.2.1.3   Epoxidation with tert-Butyl Hydroperoxide

Alkenes can also be epoxidized with tert-butyl hydroperoxide and triton B
preventing the hydrolysis of reactive functional groups (scheme-48)[60].

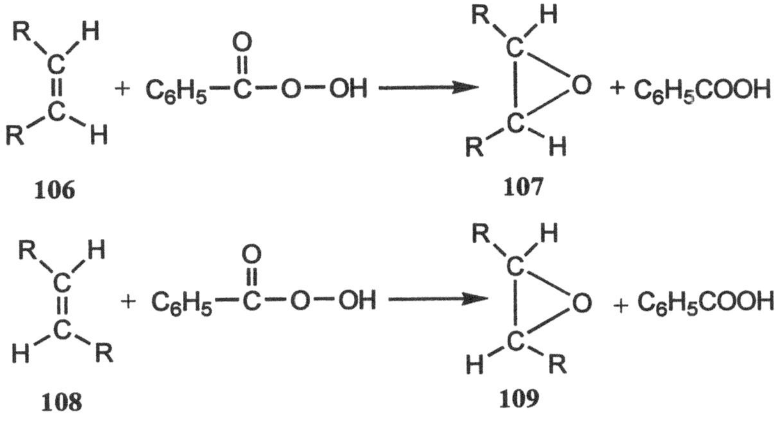

Scheme-45

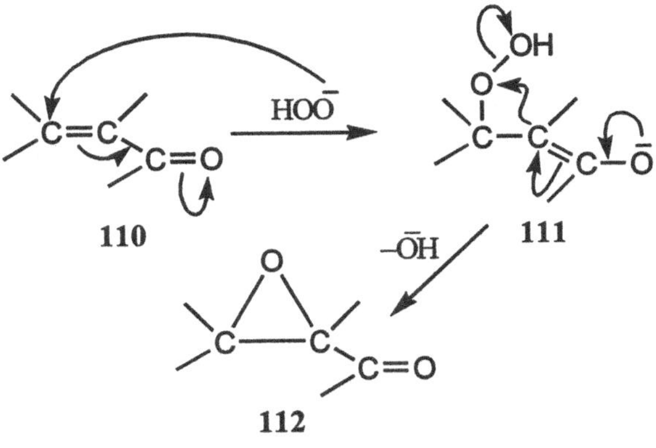

Scheme-46

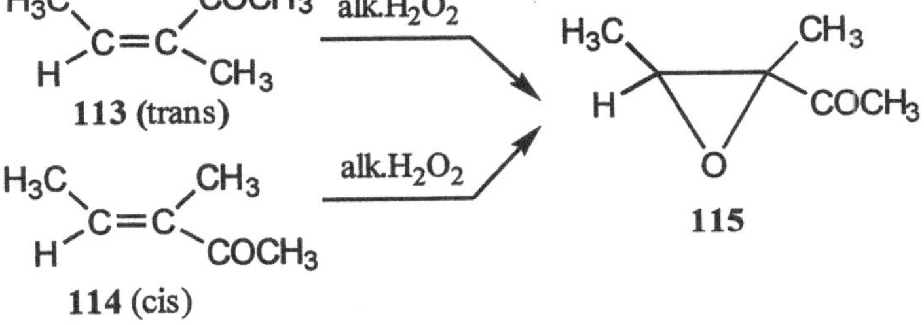

Scheme-47

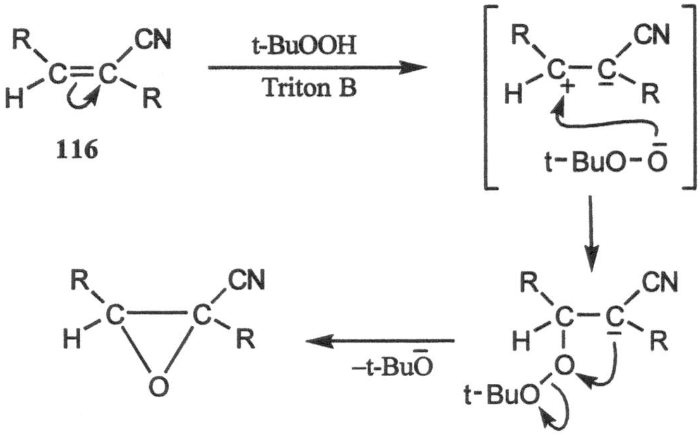

Scheme-48

1.2.1.2.2 Nucleophilic Alkylation of Carbonyl Compounds (Methylene Insertion Reactions)

1.2.1.2.2.1 Reaction of Sulfur Ylides (Dimethyloxo- and Dimethylsulfonium Methylide) with Carbonyl Compounds

The reaction of dimethyloxosulfonium methylide **119** with carbonyl compounds results in the transfer of a methylene group from the sulfur ylide to the carbonyl group providing oxiranes[61,62]. It involves nucleophilic attack from the less hindered side to form carbon–carbon bond (scheme-49).

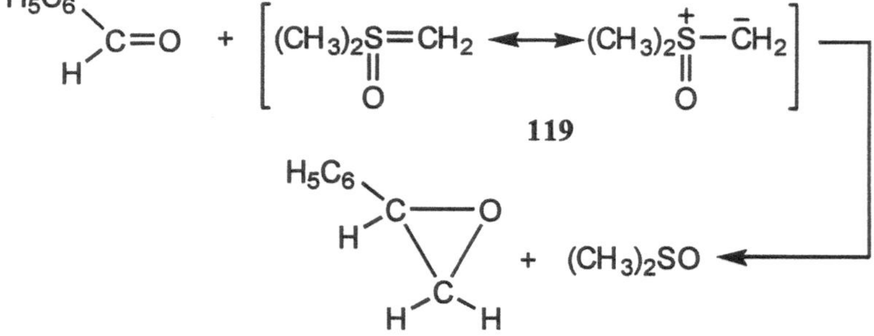

Scheme-49

Dimethylsulfonium methylide **121** also acts as a source of methylene group and transfers it to the carbonyl group forming oxirane, but usually attcks from the more hindered side (scheme-50)[62]. The difference in the side of attack has been

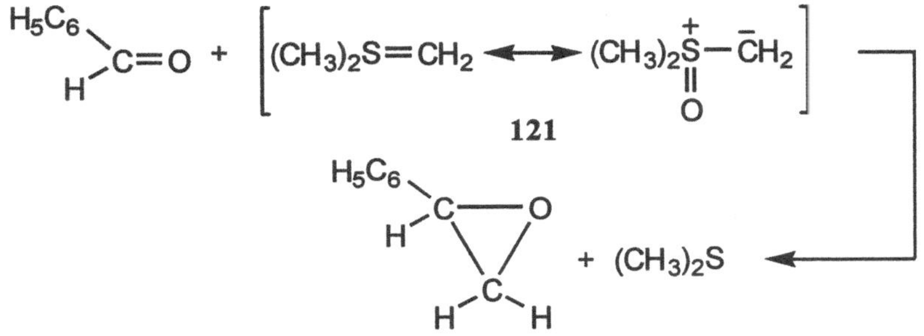

Scheme-50

attributed to the greater stability and lesser reactivity of **119** relative to **121**. The generally accepted mechanism for the reaction of sulfur ylides with aldehydes or ketones is represented in (scheme-51).

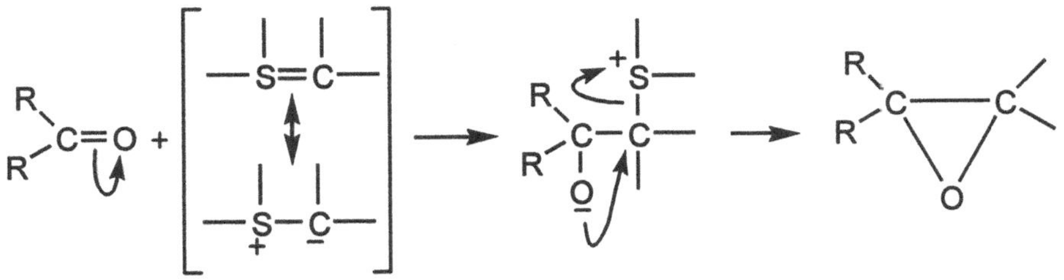

Scheme-51

However, sulfur ylides **119** and **121** have different behaviour towards their reaction with α,β-unsaturated ketones **122**. Dimethyloxosulfonium ylide **119** gives exclusively a cyclopropane **124**, while dimethylsulfonium ylide **121** provides oxirane **126** (scheme-52)[61,63].

1.2.1.2.2.2 Reaction of Diazoalkanes with Aldehydes and Ketones

The reaction of diazoalkanes, most commonly diazomethane, with aldehydes and ketones produces oxiranes **129** alongwith an aldehyde or ketone containing one more carbon (scheme-53)[64]. The separation of oxirane is difficult from the mixture.

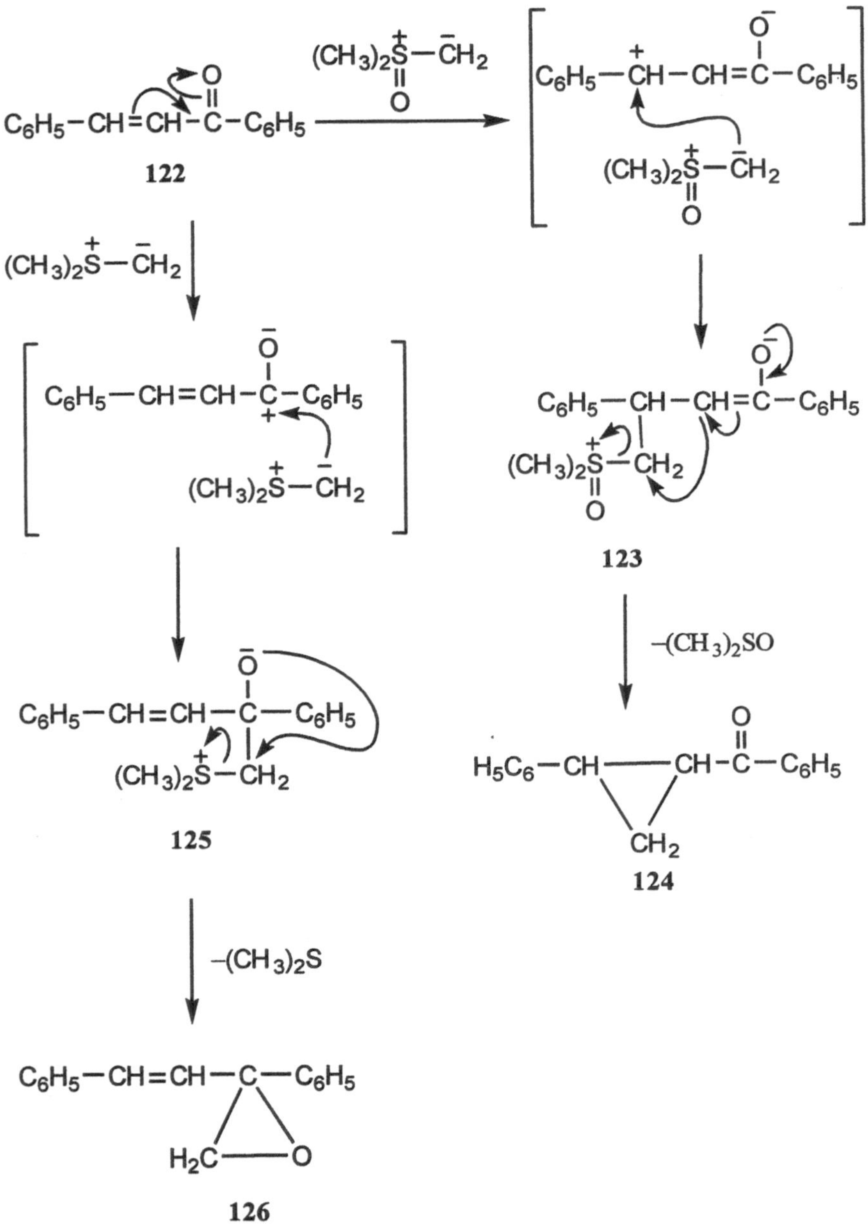

C6H5−CH=CH−C−C6H5
122
(CH3)2S−CH2
C6H5−CH−CH=C−C6H5
(CH3)2S−CH2
C6H5−CH=CH−C−C6H5
(CH3)2S−CH2
C6H5−CH−CH=C−C6H5
(CH3)2S−CH2
123
−(CH3)2SO
H5C6−CH−CH−C−C6H5
CH2
124
C6H5−CH=CH−C−C6H5
(CH3)2S−CH2
125
−(CH3)2S
C6H5−CH=CH−C−C6H5
H2C−O
126

Scheme-52

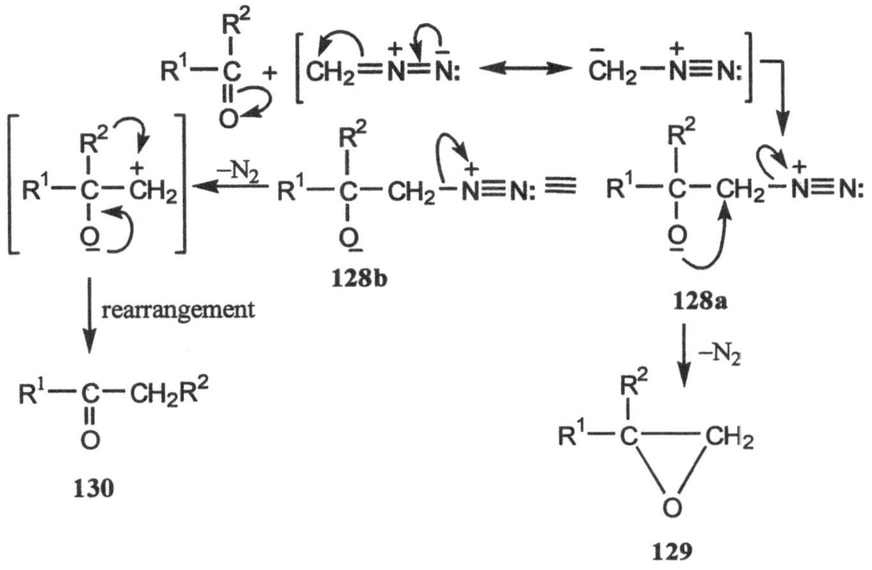

Scheme-53

1.2.1.2.3 Intramolecular Cyclization

Intramolecular cyclization (SN²) of *trans*-halohydrines **131** in the presence of a base provides oxiranes **132**[65]. Halohydrines are prepared by the reaction of alkenes with hypohalous acid (scheme-54).

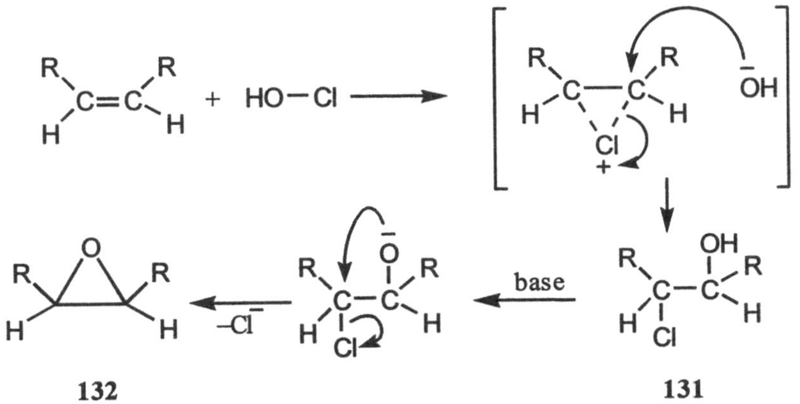

Scheme-54

The hypohalogenation is stereospecific and proceeds by an electrophilic attack of halide ion from the less hindered side providing cyclic halonium ion intermediate, which is attacked by the hydroxide ion at the site of greatest incipient carbonium ion stabilization to give *trans*-halohydrines. This reaction differs from the peracid oxidation in which oxygen is inserted from the less hindered side, while in this reaction oxygen insertion takes place from the more hindered side.

1.2.1.2.4 Condensation Reactions

1.2.1.2.4.1 Darzen Reaction

The condensation of aldehydes or ketones with α-halo esters or α-halo ketones containing α-hydrogen atom in the presence of a base provides oxiranes **133** (scheme-55)[65-67]. The reaction is stereoselective and produces *trans*-oxiranes.

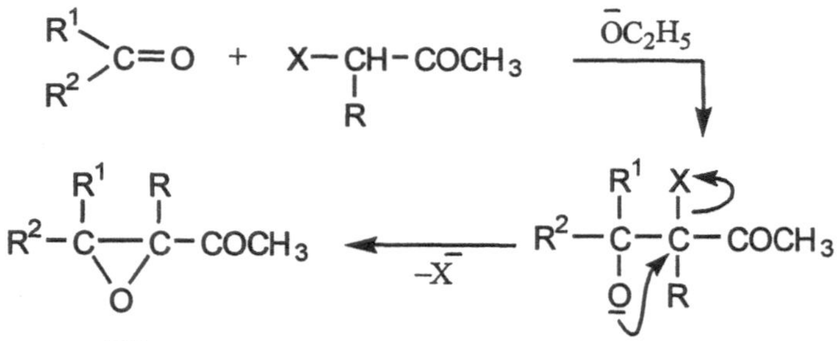

Scheme-55

1.2.1.2.4.2 Reaction of α-Halo Ketones with Grignard Reagents

The reaction of α-halo ketones with Grignard reagent forms halohydrines which on dehydrohalogenation provide oxiranes (scheme-56)[68].

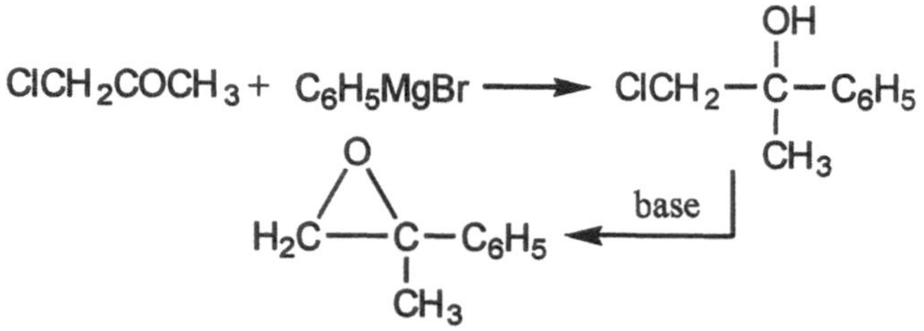

Scheme-56

1.2.1.2.4.3 Reaction of Aromatic Aldehydes with Phosphorus Triamides

The reaction of aromatic aldehydes with phosphorus triamides affords oxiranes **138** (scheme-57)[69]. An electronegative group at the aromatic ring enhances the formation of oxiranes, but the electron-releasing substituents favour the formation of intermediate adducts **135**.

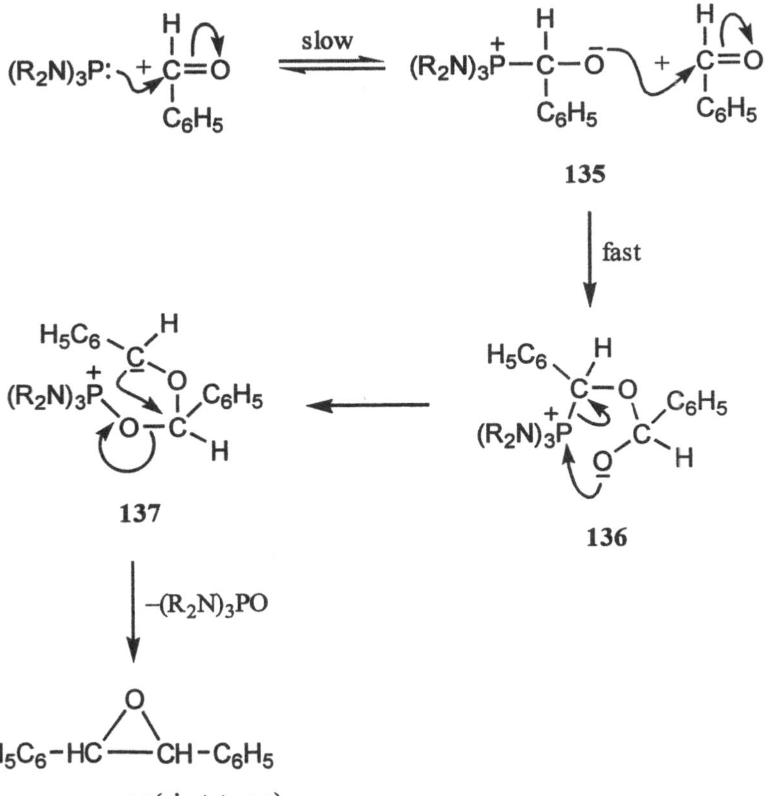

Scheme-57

1.2.1.3 Reactions

1.2.1.3.1 Ring Opening Reactions

In oxiranes, the carbon–oxygen bond is most readily cleaved due to the enhanced reactivity attributable to the high degree of ring strain.

1.2.1.3.1.1 Nucleophilic Ring Opening Reactions

Nucleophile attacks at the side opposite to the heteroatom with the inversion of configuration (scheme-58)[70].

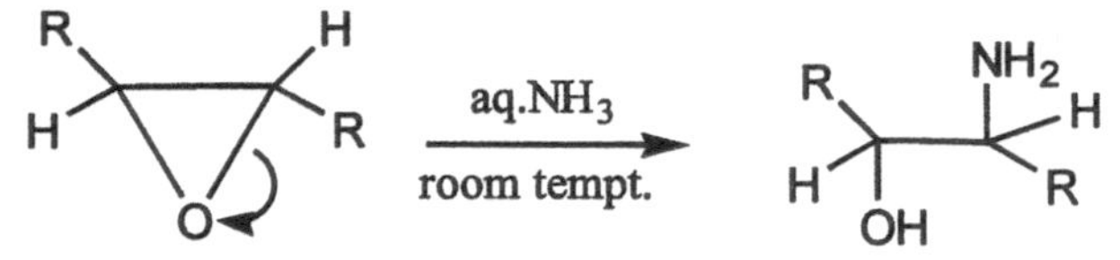
R
H
H
R
aq.NH3
room tempt.
NH2
R
H
H
R
OH

141

Scheme-58

In asymmetrically substituted oxiranes **141**, the steric hindrance causes attack of nucleophile at the less hindered carbon in the ring (scheme-59)[71]. However, the

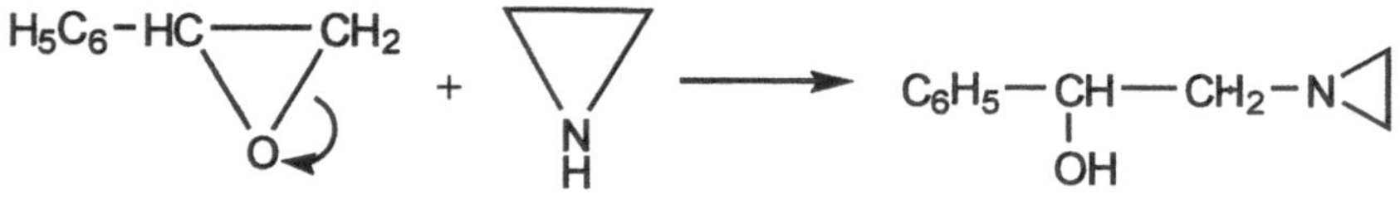
H5C6-HC
CH2
O
+
N
H
C6H5—CH—CH2-N
OH

141 **142**

Scheme-59

reaction is considerably influenced by the nature of the solvent and reagent[72] (scheme-60). The steric hindrance directs the attack of nucleophile at the less

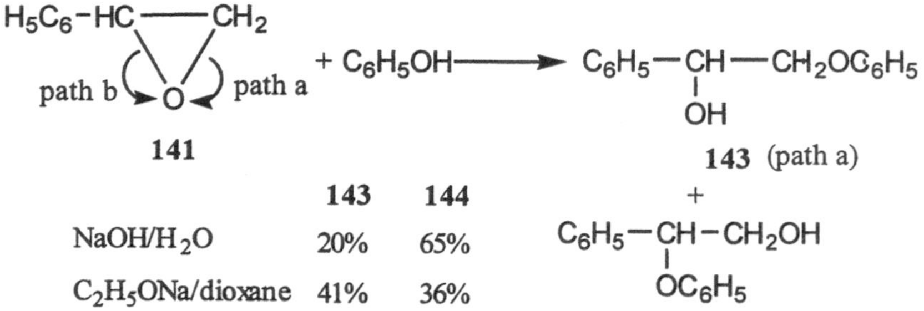
H5C6-HC
CH2
path b
O
path a
+ C6H5OH
C6H5—CH—CH2OC6H5
OH
+
C6H5—CH-CH2OH
OC6H5

141

143 (path a)

144 (path b)

	143	**144**
NaOH/H_2O	20%	65%
C_2H_5ONa/dioxane	41%	36%

Scheme-60

hindered carbon atom, but the bond breaking is facilitated by the resonance stabilization of carbonium ion and causes attack of nucleophile at the hindered carbon, which is enhanced by the solvent of high ionizing power (scheme-61).

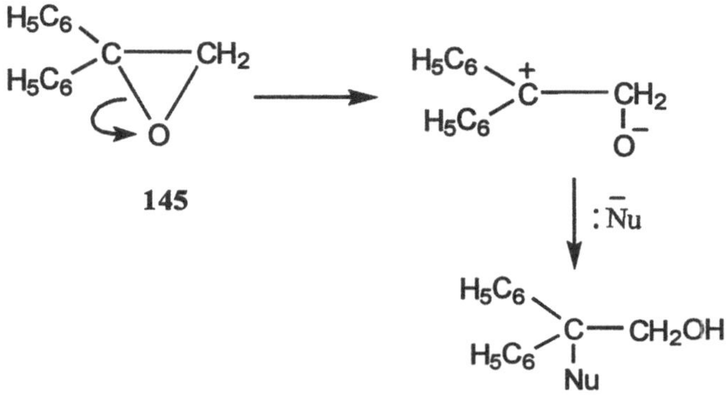

145

146

Scheme-61

The reduction of asymmetrical oxiranes **141** with lithium aluminium hydride provides highly substituted alkanols **147**. However, reduction in the presence of aluminium chloride yields less substituted alkanols **148** (scheme-62)[73,74].

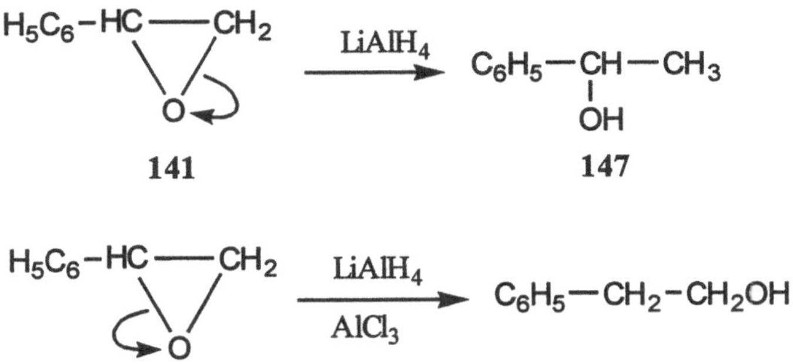

141   **147**

141   **148**

Scheme-62

The reaction of oxiranes with Grignard reagents provides alkanols of the long carbon chain (scheme-63)[75].

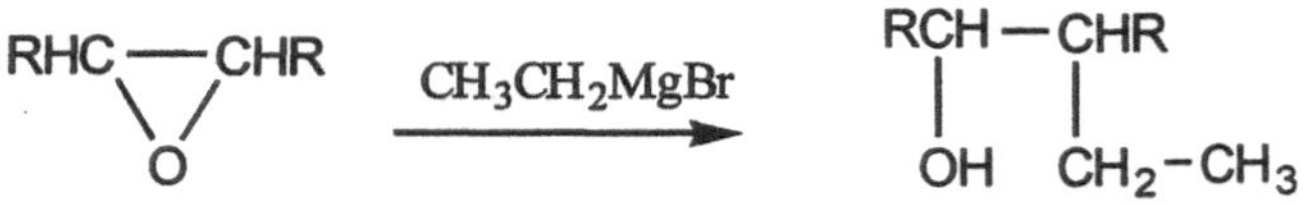

149 150

Scheme-63

However, gem-disubstituted oxiranes **151** form rearranged alkanols **152** (the alkyl group from the Grignard reagent and the hydroxyl group appear on the same carbon atom of the oxirane) (scheme-64). In such cases the oxirane is isomerized to an aldehyde or ketone before reacting with Grignard reagent.

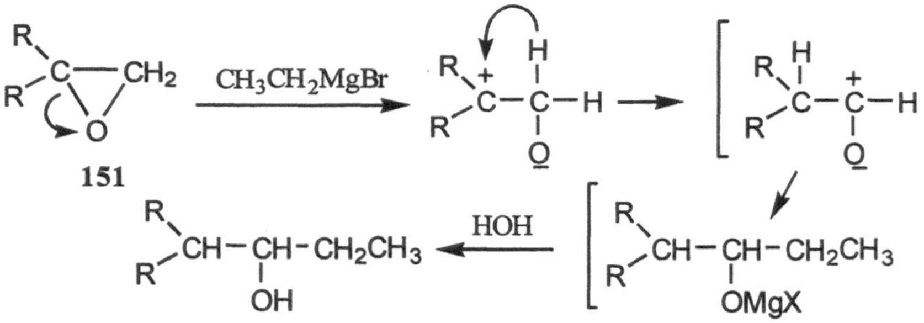

151

152

Scheme-64

The reaction of a Grignard reagent with vinylic oxirane **153** gives a mixture containing normal product and the predominating allylic rearranged product **154** (scheme-65)[76].

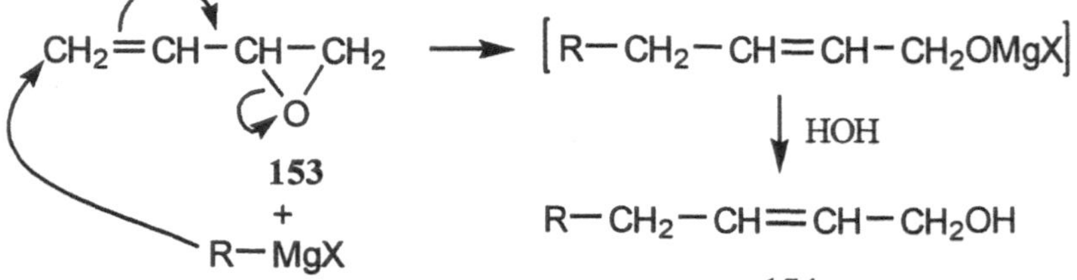

153

154

Scheme-65

The phosphorus ylides are more reactive than their corresponding compounds. The phosphorus ylides are stabilized by resonance if substituted with an electron-withdrawing substituent (–COR, –CN, –COOR, –CHO) at the α-position.

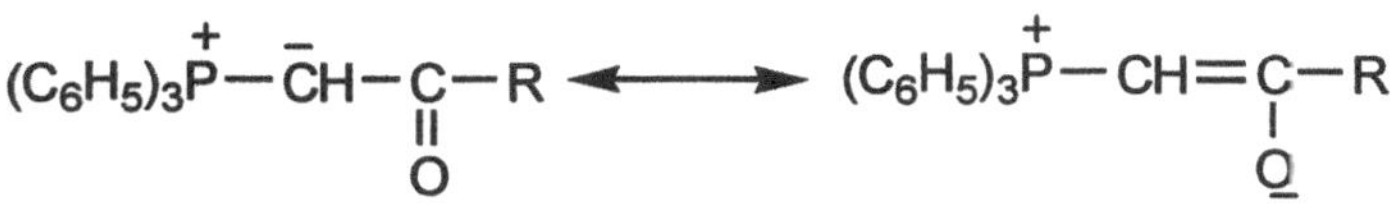

$$(C_6H_5)_3\overset{+}{P}-\overset{-}{C}H-\underset{\underset{O}{\|}}{C}-R \quad\longleftrightarrow\quad (C_6H_5)_3\overset{+}{P}-CH=\underset{\underset{\underline{O}}{|}}{C}-R$$

155a   155b

The reaction of oxiranes with phosphorus ylides of phosphoranes **155**, phosphonates **156** and phosphinates **157** converts oxiranes to cyclopropanes **158** (scheme-66). The reaction proceeds with the inversion of configuration[77-80].

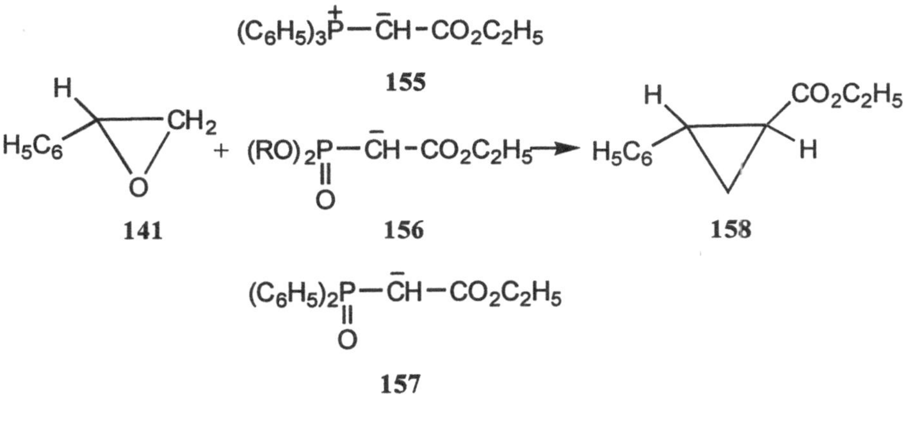

$$(C_6H_5)_3\overset{+}{P}-\overset{-}{C}H\text{-}CO_2C_2H_5$$

155

141   156   158

$$(C_6H_5)_2\underset{\underset{O}{\|}}{P}-\overset{-}{C}H-CO_2C_2H_5$$

157

Scheme-66

The reaction involves SN² attack of the carbanion at the less hindered carbon atom with the cleavage of ring. The zwitterion (betain) **159** formed cyclizes to the five-membered phosphorus containing ring **160** which undergoes ring contraction to provide cyclopropane **158** (scheme-67).

1.2.1.3.1.2 Electrophilic Ring Opening Reactions

Electrophilic reagents react with oxiranes most readily with the cleavage of bond between oxygen and the least substituted carbon atom. These reactions are considerably influenced by the solvents as well as by the reagents (scheme-68).

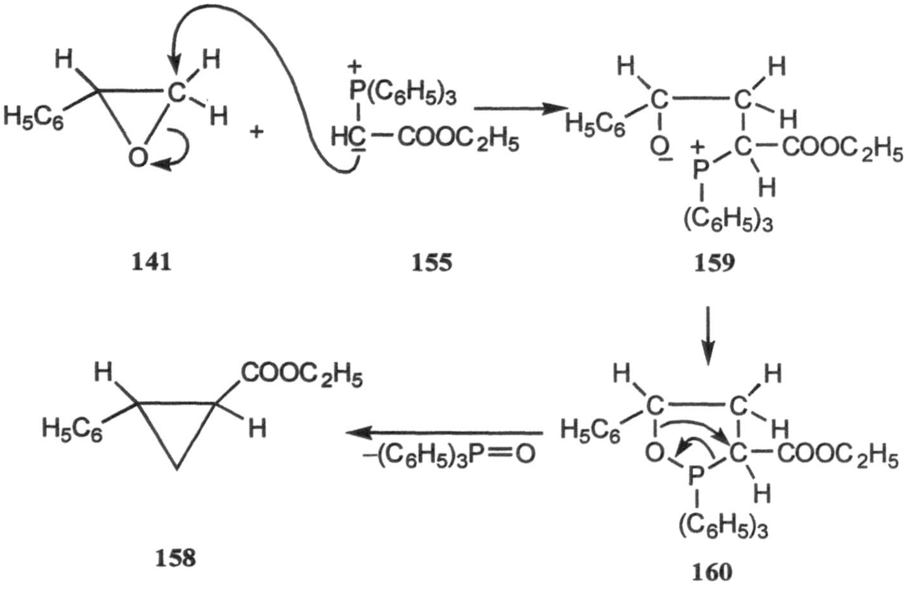

Scheme-67

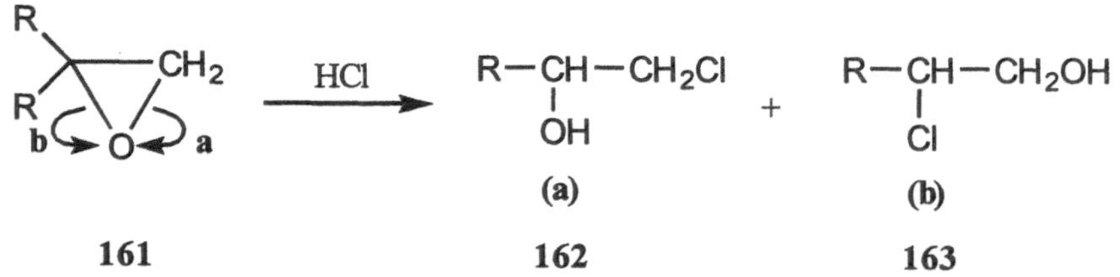

Scheme-68

The nature of the substituents present on the oxirane ring exerts a powerful effect on the direction of the ring opening. The presence of carbonium ion stabilizing groups directs the cleavage of the ring in the direction of stabilized carbonium ion. The direction is reversed if the carbonium ion is destabilized due to the presence of electron-withdrawing group(s) (scheme-69)[81,82].

The reaction of oxirane with an acid chloride proceeds by the electrophilic attack at the heteroatom to produce an intermediate onium salt **167** which, being highly reactive, is readily attacked by a chloride ion at the carbon atom with the cleavage of the ring (scheme-70)[83].

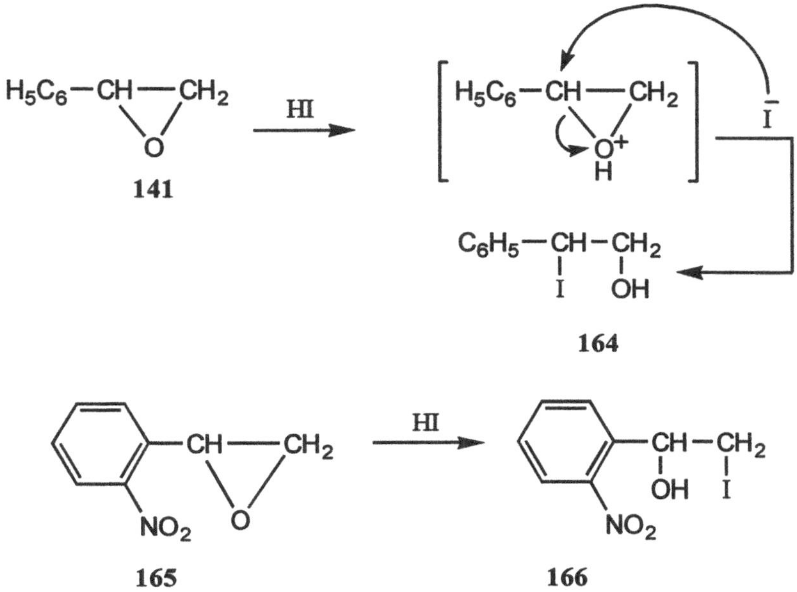

Scheme-69

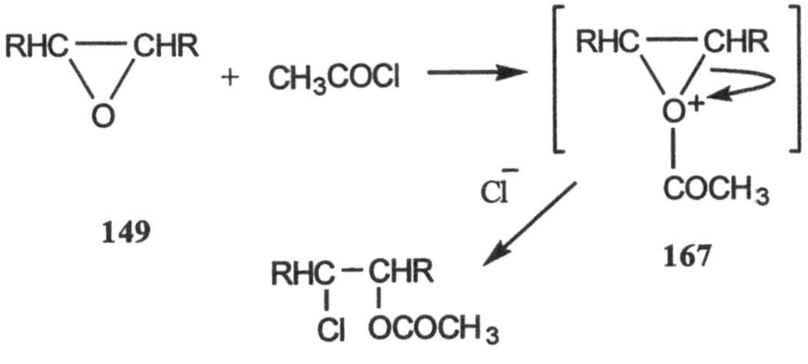

Scheme-70

1.2.1.3.1.3 Other Ring Opening Reactions

1.2.1.3.1.3.1 Reaction with Carbonyl Compounds

Oxiranes **149** react with carbonyl compounds cleaving carbon–oxygen bond and undergo ring expansion to form dioxalanes **169** (scheme-71)[84].

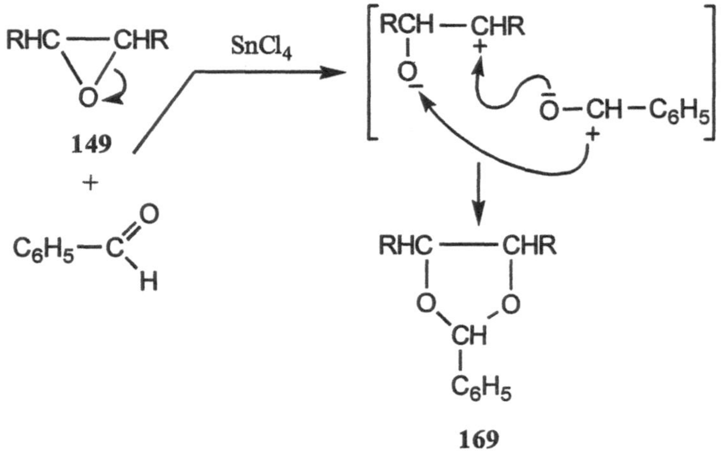

Scheme-71

1.2.1.3.1.3.2 Reaction with Weak Nucleophiles

Oxiranes **170** with weak nucleophiles undergo elimination reactions with the cleavage of carbon–oxygen bond and the abstraction of a proton from the less substituted carbon to provide allyl alcohols **171** (scheme-72)[85,86].

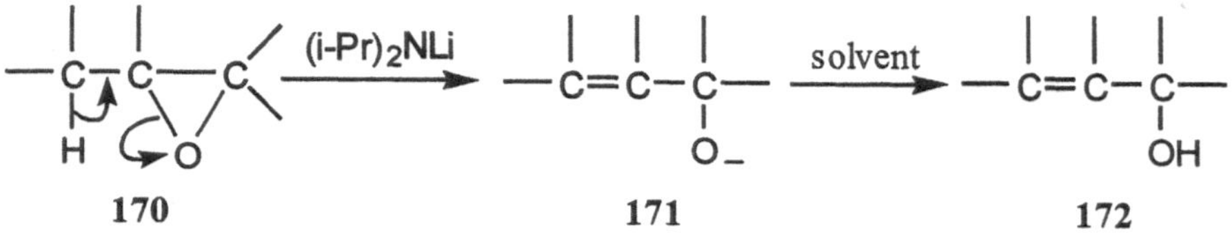

Scheme-72

1.2.1.3.2 Thermal Reactions

The reaction of oxiranes **149** with triphenyl- or trialkylphosphines at elevated temperature proceeds with the extrusion of heteroatom (oxygen) providing stereoselective alkenes **172** and **173** (scheme-73)[87-88]. The reaction involves the nucleophilic attack of tertiary phosphine at the oxirane carbon atom giving a betaine **174** which after the rotation of C–C bond by 180° is followed by four centered elimination to form an alkene **173** of opposite configuration predominantly with the elimination of tertiary phosphine oxide (scheme-74).

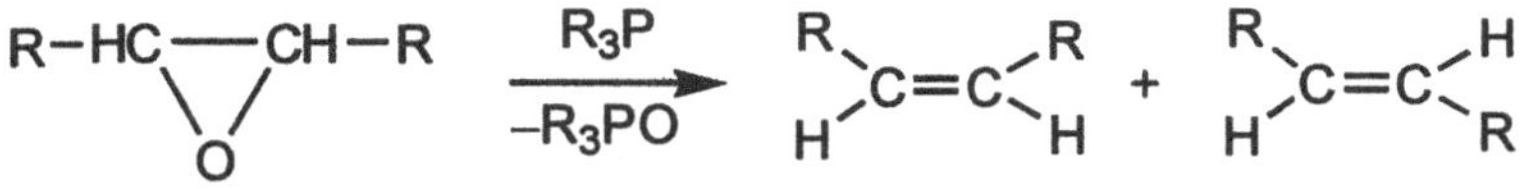

149	172	173
cis-oxirane	19%	81%
trans-oxirane	72%	28%

Scheme-73

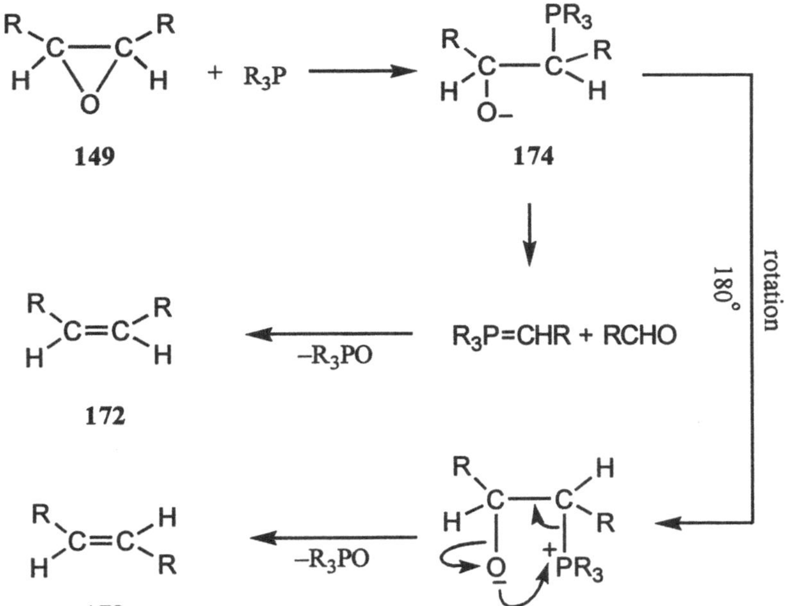

Scheme-74

1.2.1.3.3 Rearrangements

Oxiranes **175** rearrange most readily to the carbonyl compounds **176** in the presence of an acid involving the formation of carbonium ions as intermediates with the cleavage of carbon–oxygen bond (scheme-75).

Oxiranes with some strong bases do not undergo nucleophilic ring opening reactions, but rearrange to alkenes (scheme-76)[89].

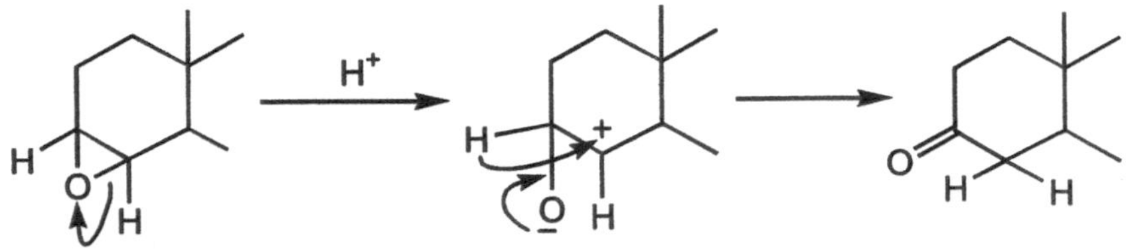

175 **176** **177**

Scheme-75

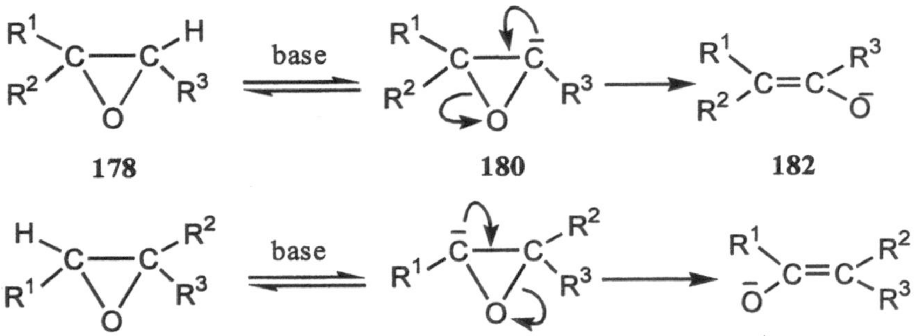

178 **180** **182**

179 **181** **183**

Scheme-76

Oxiranes **184** undergo thermal isomerization to the carbonyl compounds **186**. Similar transformation can also be accomplished by boron trifluoride–ether or anhydrous magnesium bromide–ether (scheme-77)[89,90].

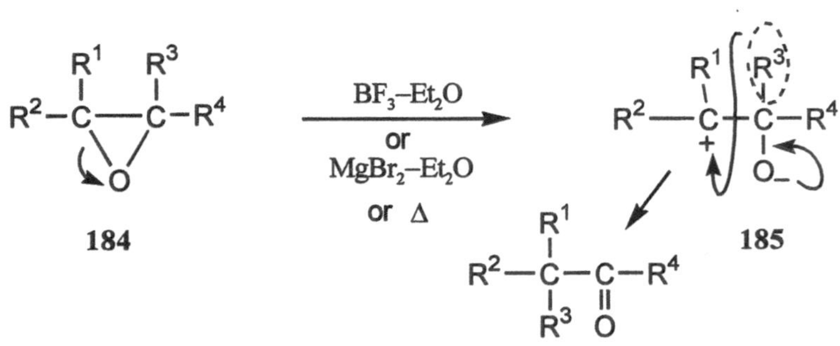

184 **185**

186

Scheme-77

Oxiranes **188** on treatment with certain transition metal catalysts rearrange to aldehydes or ketones (scheme-78)[91-93].

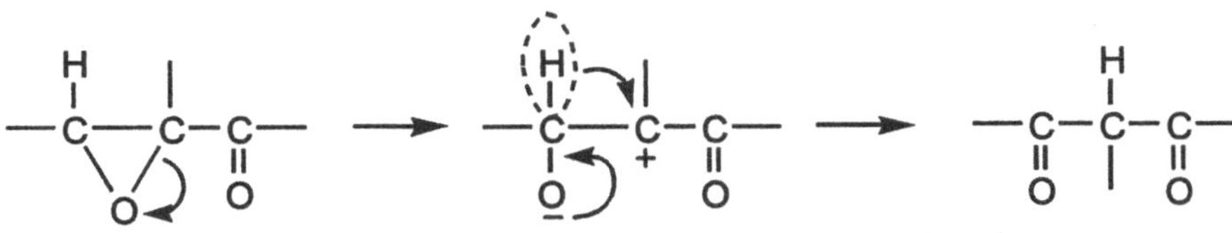

Scheme-78

1.2.2 Oxirenes (Oxacyclopropenes)

Three-membered unsaturated heterocycles containing oxygen as a heteroatom are known as oxirenes. Oxirene **191** is a resonance destabilized antiaromatic heterocycle with four π-electrons like 2-azirine. Oxirene can not be isolated and exists as an unstable intermediate because of the destabilizing electronic conjugation.

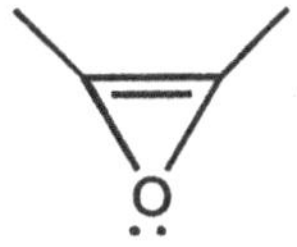

191

Molecular orbital calculations also suggest that it is less stable than its acyclic analog[95]. Oxirenes have been reported as reaction intermediates in the oxidation of alkynes **192** with peracids (scheme-79)[96] and in the photochemical Wolf rearrangement of α-diazoketones **195** (scheme-80)[97].

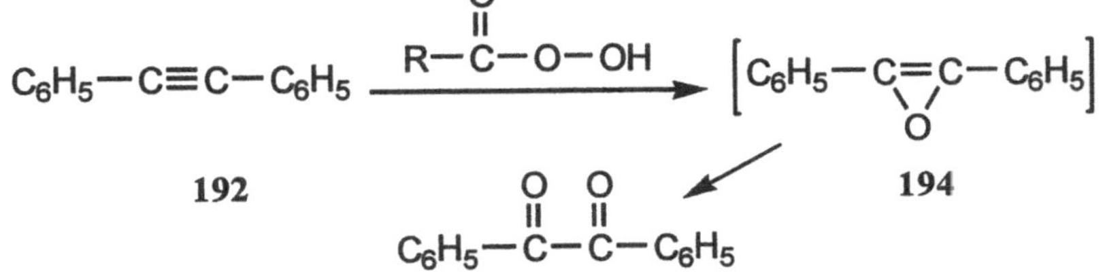

Scheme-79

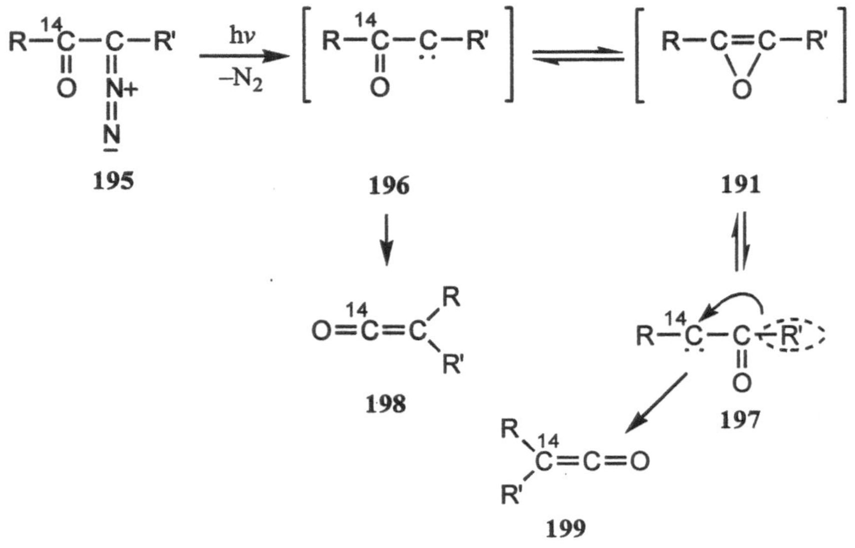

R—$\overset{14}{\text{C}}$—C—R' $\xrightarrow[-N_2]{h\nu}$ [R—$\overset{14}{\text{C}}$—C—R'] ⇌ [R—C=C—R']

195   196   191

O=$\overset{14}{\text{C}}$=C⟨R,R'

198

R—$\overset{14}{\text{C}}$—C⟵R'
197

$\overset{R,R'}{\text{C}}$$\overset{14}{=}$C=O
199

Scheme-80

1.3 Three-Membered Thiaheterocycles

1.3.1 Thiiranes

1.3.1.1 General

Thiiranes are three-membered sulfur containing saturated heterocycles[98]. These are also known as thiacyclopropanes or episulfides 200. Thiirane has the lowest strain energy (83.26 kJ/mol) among the three-membered heterocyclic rings. The dipole moment of thiirane (1.66D) is lower than that of oxirane (1.88D) and the difference can be attributed to the smaller polarity of the C–S bond than that of the C–O bond. The carbon–carbon bond length (1.429Å), intermediate between the normal C–C (1.54Å) and C=C (1.34Å) bond langths suggests its partial double bond character. The C–S bond length (1.829Å) is almost of the same order as in dialkyl sulfides (1.810Å). The H–C–H and C–S–C bond angles are 116° and 48.5°, respectively. In infrared spectra, the higher C–H vibrational frequency (1475 cm^{-1}) reflects the strain in thiirane ring. The chemistry of thiirane is similar to that of the other three-membered saturated heterocycles, but specificity is associated with the presence of a sulfur atom. Thiirane, because of the ability of sulfur atom to expand its valence shell, exists into two oxidized forms, 201 and 202.

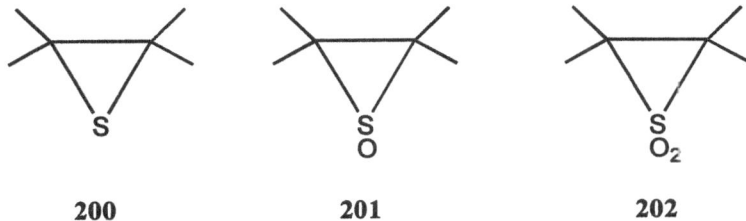

200	**201**	**202**

1.3.1.2 Synthesis

1.3.1.2.1 From Oxiranes

The most widely used method involves the reaction of oxiranes with thiocyanate ions[99]. The reaction mechanism is believed to involve the formation of a cyclic intermediate **204** (scheme-81).

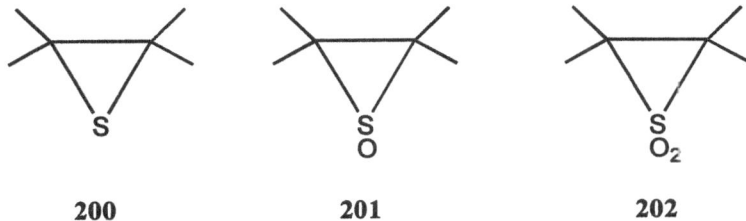

Scheme-81

The reaction of oxiranes with thiourea also provides thiiranes[100]. The reaction proceeds by a similar mechanism involving cyclic intermediate **207** (scheme-82).

Oxiranes can be directly converted to thiiranes when treated with phosphine sulfide such as Ph_3PS or with 3-methylbenzothiazole-2-thione **208** and trifluoroacetic acid (scheme-83)[101].

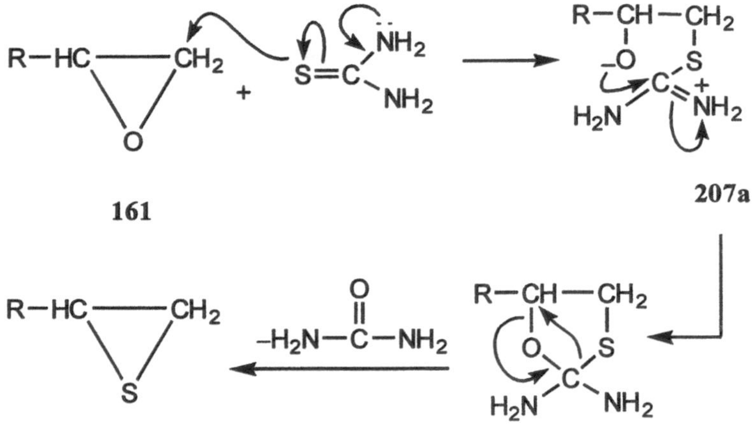

R−HC──CH₂ + S=C⟨NH₂ / NH₂ ⟶ R−CH──CH₂ | O S | ... C⁺ / H₂N NH₂

161

207a

R−HC──CH₂ / S ← −H₂N−C(=O)−NH₂ ← R−CH──CH₂ | O S | C / H₂N NH₂

206

207

Scheme-82

—C—C— / O →(Ph₃PS or / benzothiazole-2-thione (CH₃, N, S) + CF₃COOH)→ —C—C— / S

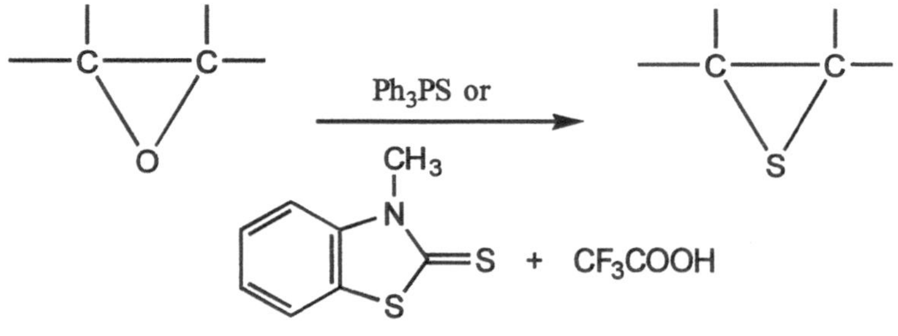

208

Scheme-83

1.3.1.2.2 Intramolecular Cyclization Reactions

The intramolecular cyclization of 2-halo mercaptans **211** in the presence of a base provides thiiranes. 2-Halo mercaptans **211** are obtained from 2-hydroxy mercaptans **210** which, in turn, are prepared from 1,2-halohydrines **209** (scheme-84).

β-Halo disulfides **212**, formed by the addition of arenethiosulfenyl chlorides or trimethylsilylsulfinyl bromide to the compounds containing carbon−carbon double bond, are converted to thiiranes when treated with sodamide or sodium sulfide (scheme-85)[102,103].

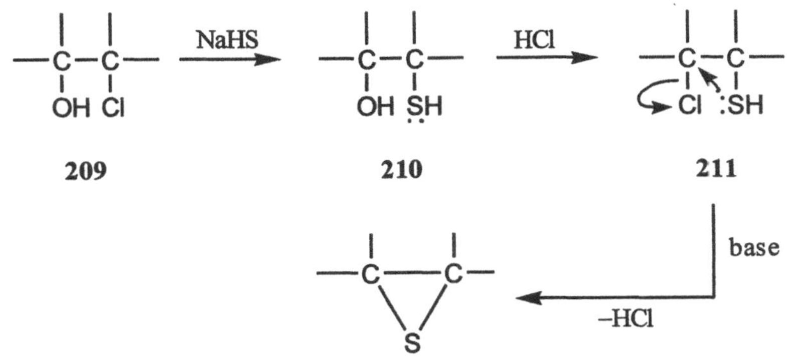

Scheme-84

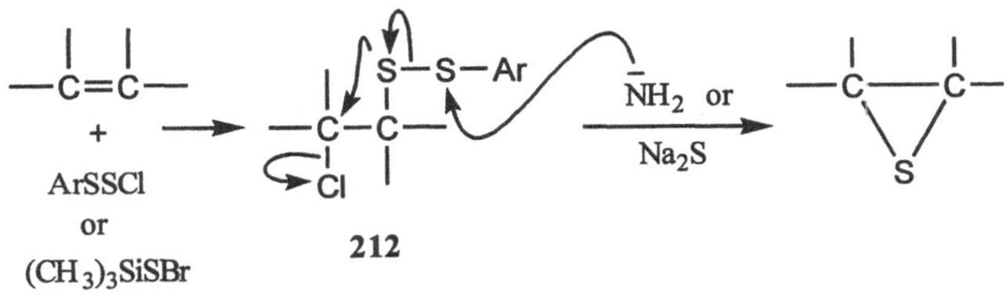

Scheme-85

1.3.1.2.3 Methylene Insertion Reactions

1.3.1.2.3.1 Reaction of Thioketones with Sulfur Ylides

The reaction of thioketones with dimethyloxosulfonium methylide **119** results in the transfer of methylene group from ylide to thioketone providing thiiranes (scheme-86)[104].

1.3.1.2.3.2 Reaction of Thioketones with Diazoalkanes

The reaction of thioketones with diazoalkanes involves the insertion of methylene group into the carbon–sulfur double bond affording thiiranes[105,105]. The reaction proceeds with the formation of 1,3,4-thiadiazoline cycloadduct **214** which extrudes nitrogen to yield thiirane **216** via thiocarbonyl ylide intermediate **215** (scheme-87)[107].

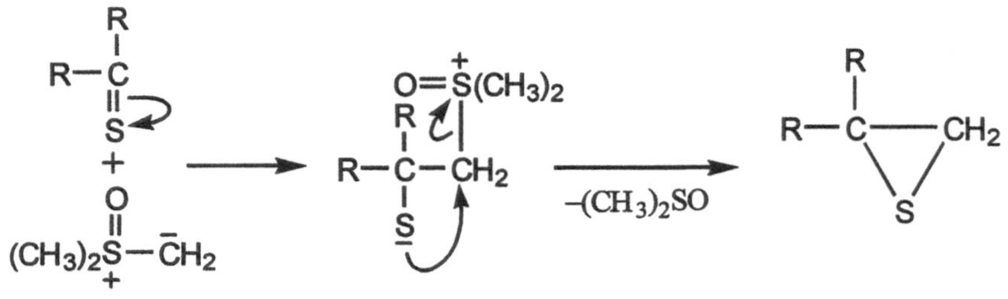

R—C(R)=S ... 119

O=$\overset{+}{S}(CH_3)_2$

R—$\overset{R}{\underset{S^-}{C}}$—CH$_2$ 213

$-(CH_3)_2SO$

R—$\overset{R}{C}$—CH$_2$ (thiirane, S)

(CH$_3$)$_2\overset{+}{S}$—$\bar{C}H_2$

119   **213**

Scheme-86

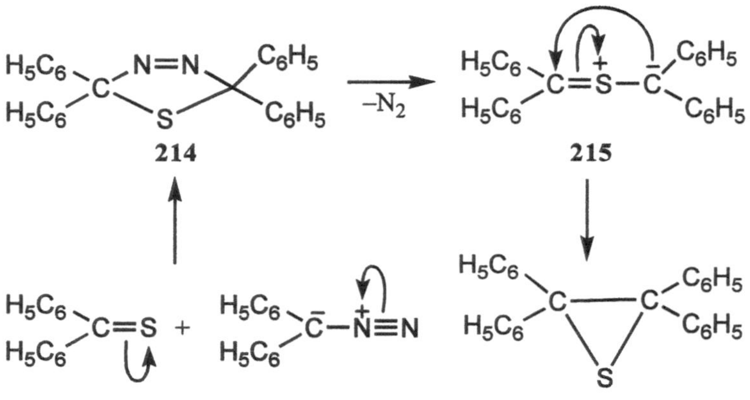

H$_5$C$_6$, H$_5$C$_6$—C—N=N—C(C$_6$H$_5$)(C$_6$H$_5$)—S   **214**

$-N_2$

H$_5$C$_6$, H$_5$C$_6$—C=$\overset{+}{S}$—$\bar{C}$—C$_6$H$_5$, C$_6$H$_5$ **215**

H$_5$C$_6$, H$_5$C$_6$—C=S + H$_5$C$_6$, H$_5$C$_6$—$\bar{C}$—$\overset{+}{N}$≡N

H$_5$C$_6$, H$_5$C$_6$—C—C—C$_6$H$_5$, C$_6$H$_5$ (S)   **216**

Scheme-87

1.3.1.2.4   Reaction of Cyclic Carbonates with Thiocyanates

The reaction of cyclic ethylene carbonate **217** with alkali thiocyanates provides thiiranes involving cyclic intermediate **219** (scheme-88)[108].

1.3.1.3   Reactions

1.3.1.3.1   Ring Opening Reactions

Thiiranes also undergo ring opening reactions. Their reactivity towards nucleophiles is similar or slightly greater than that of oxiranes. Thiiranes are less reactive towards electrophiles due to the lower electron density at sulfur than at oxygen in oxiranes.

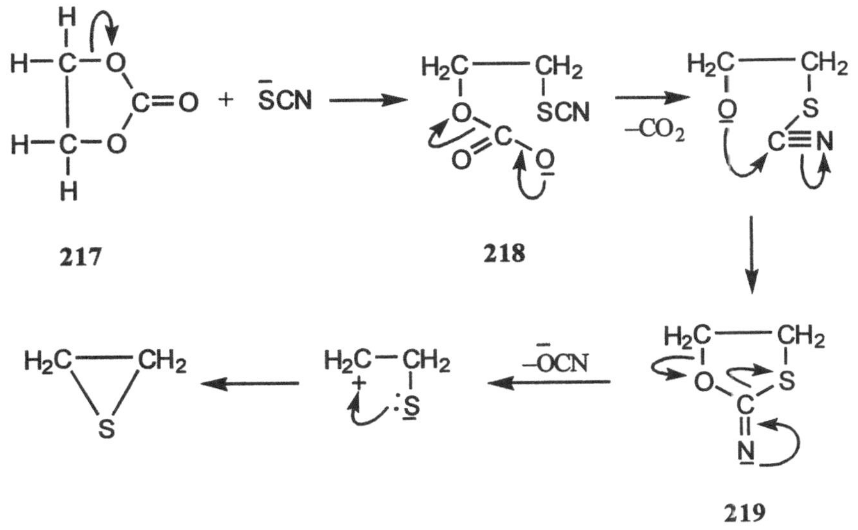

Scheme-88

1.3.1.3.1.1 Nucleophilic Ring Opening Reactions

Nucleophilic ring opening reactions in thiiranes proceed stereospecifically with the inversion of configuration at the site of attack i.e. opposite to the heteroatom leading to the corresponding mercaptide ion or to the mercaptan compound depending on whether sulfur atom is being protonated before or concurrently with the ring opening (schemes-89 and 90).

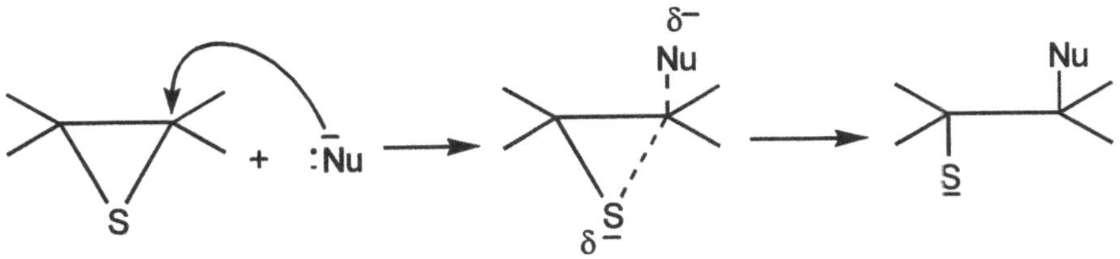

Scheme-89

In asymmetrically substituted thiiranes, the nucleophile attacks at the less substituted carbon atom from the opposite side of the sulfur atom (heteroatom) (scheme-91).

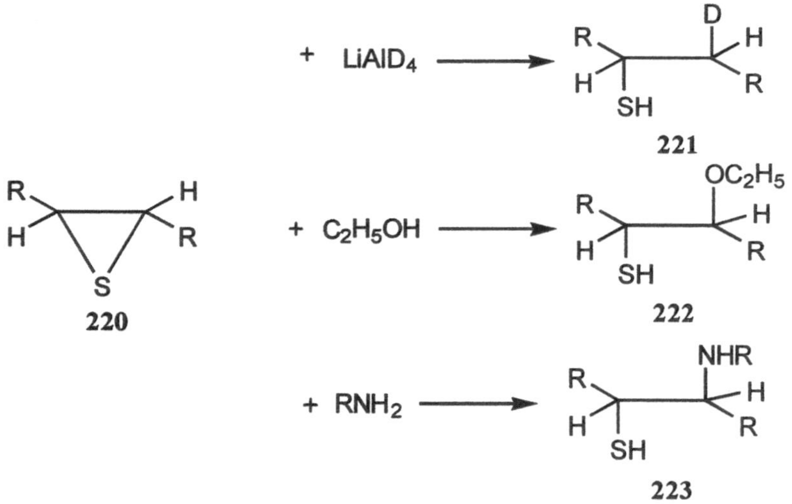

Scheme-90

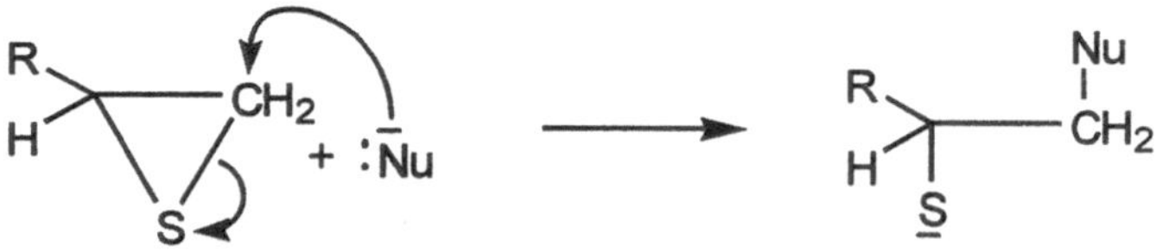

Scheme-91

In the presence of carbonium ion stabilizing substituents and ionic solvents, the reaction proceeds by SN¹-mechanism and the nucleophile attacks at the alternative site (scheme-92).

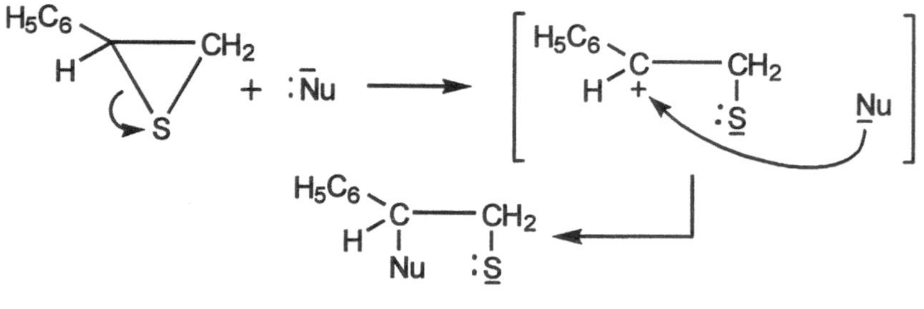

Scheme-92

1.3.1.3.1.2 Electrophilic Ring Opening Reactions

The reaction of thiiranes **224** with hydrochloric acid results in the ring cleavage with the formation of halo mercaptans **225** (scheme-93)[109].

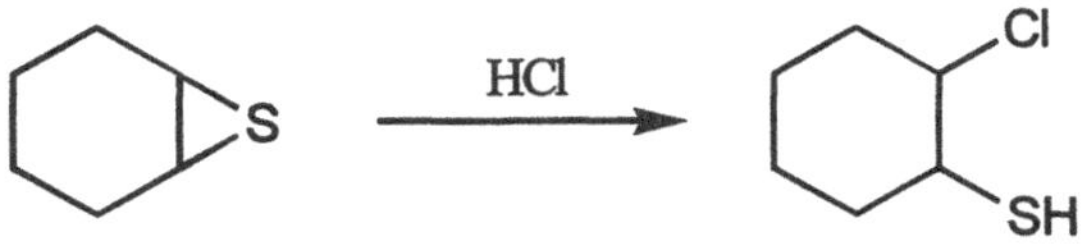

Scheme-93

The reaction of thiiranes with acetyl chloride is similar to that with oxiranes and proceeds by the electrophilic attack at the sulfur atom to produce extremely reactive intermediate **226** which is readily attacked by the nucleophile at the carbon atom with the cleavage of ring (scheme-94)[110].

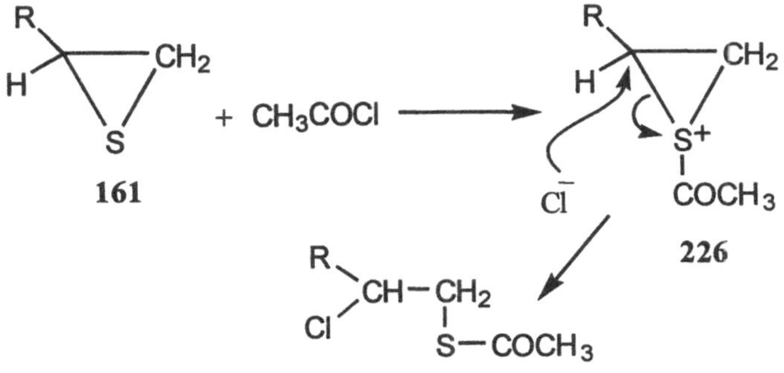

Scheme-94

The reaction of asymmetrical thiiranes with acetic anhydride in pyridine is contrary to that observed with acetyl chloride and proceeds with the direct attack of acetate ion at the less substituted carbon atom (scheme-95)[110].

1.3.1.3.2 Desulfurization (Extrusion of Sulfur Atom)

The reaction of thiiranes **230** with tertiary phosphines or phosphites results in the desulfurization of thiirane ring with the formation of the corresponding alkenes

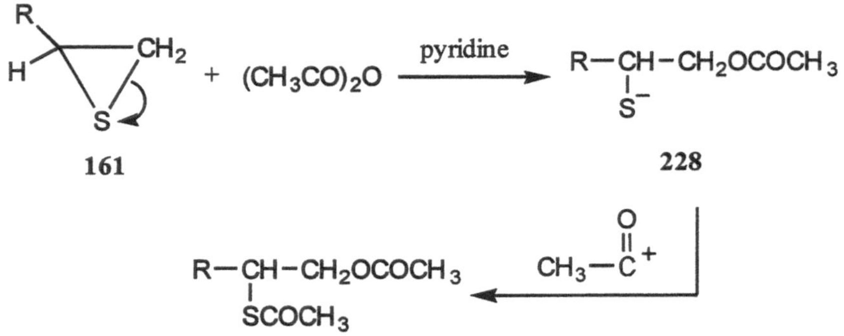

161

228

229

Scheme-95

(scheme-96)[111-113]. The reaction is stereospecific and proceeds by the concerted mechanism involving nucleophilic attack of phosphorus at the sulfur atom.

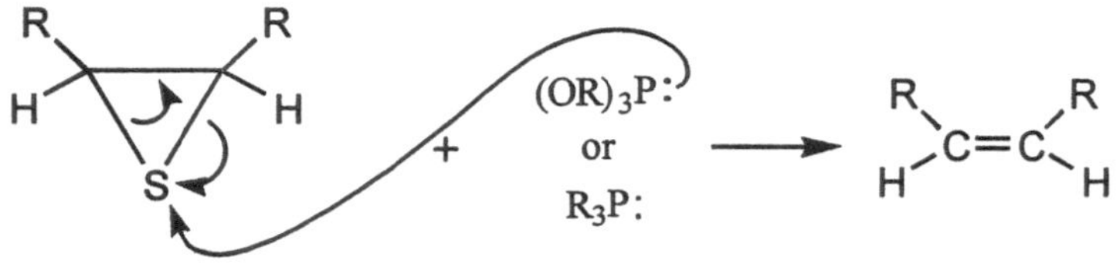

230

231

+

$(OR)_3PS$

or

R_3PS

Scheme-96

The reaction of thiiranes **230** with methyl iodide involves the extrusion of sulfur atom and proceeds with the formation of an intermediate (episulfonium salt) **231** to yield stereoselective alkenes (scheme-97)[114].

Thiiranes **235**, particularly with aryl and other conjugative groups, are desulfurized by heat with the formation of alkenes (scheme-98)[115].

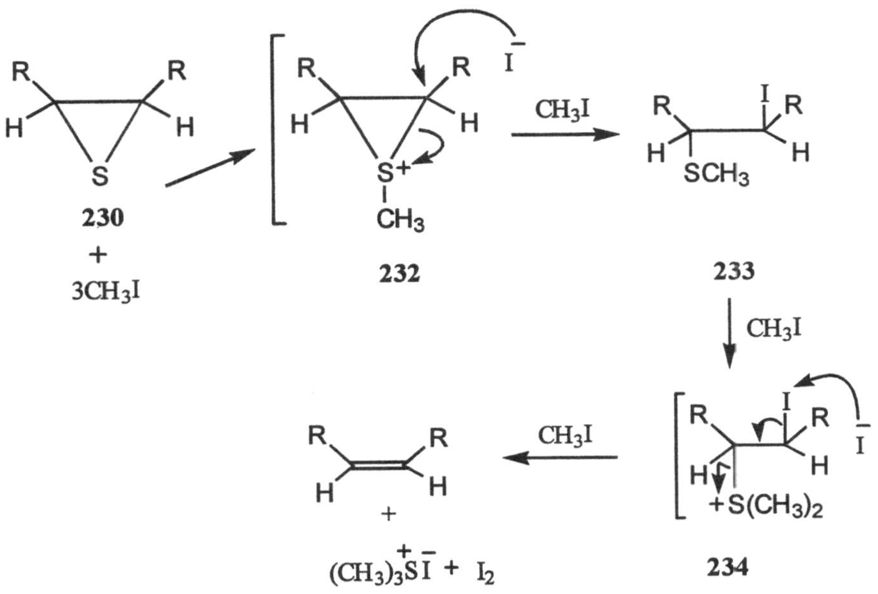

Scheme-97

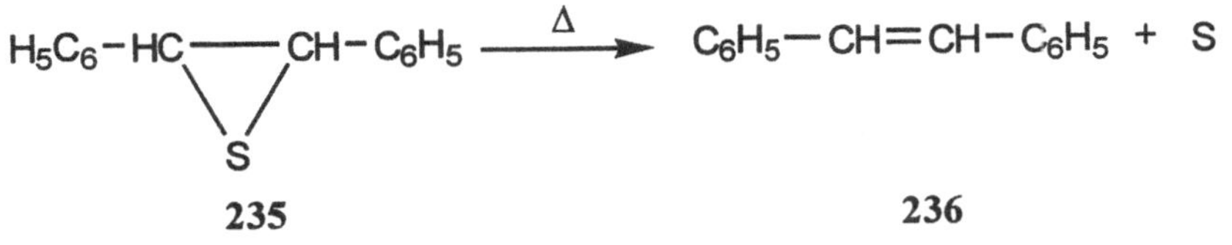

Scheme-98

1.3.1.3.3 Photochemical Reactions

The ultraviolet irradiation of thiirane **235a** in the presence of tetracyanoethylene (TCNE) results in the formation of tetrahydrothiophene **237** involving photochemical [3+2] cycloaddition via 1,3-dipolar cycloaddition of a thiocarbonyl ylide intermediate **236a** (scheme-99)[116]. But similar treatment of thiirane **235a** with fumaronitrile involves desulfurization providing alkene (scheme-100).

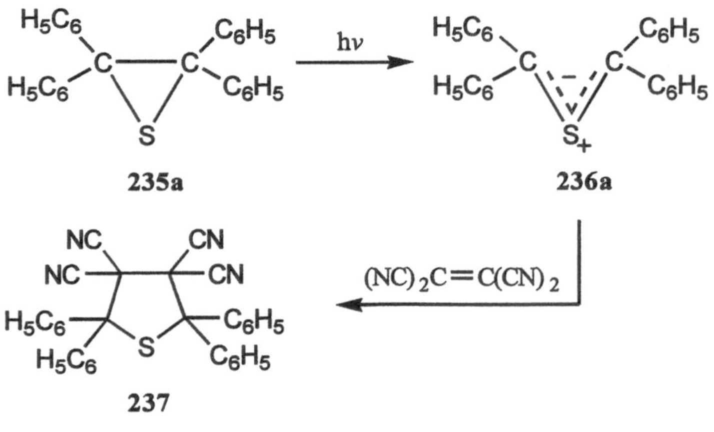

H₅C₆ ... C—C ... C₆H₅ hν → H₅C₆ ... C—C ... C₆H₅
H₅C₆ ... C₆H₅   H₅C₆ ... C₆H₅
          S   S₊

235a   **236a**

NC   CN
NC——   ——CN   (NC)₂C=C(CN)₂
H₅C₆——   ——C₆H₅   ←
H₅C₆   S   C₆H₅

237

Scheme-99

H₅C₆ ... C—C ... C₆H₅   furmaronitrile   C₆H₅—C=C—C₆H₅
H₅C₆ ... C₆H₅   ──────────→   H₅C₆   C₆H₅
          S   hν, –S

235a

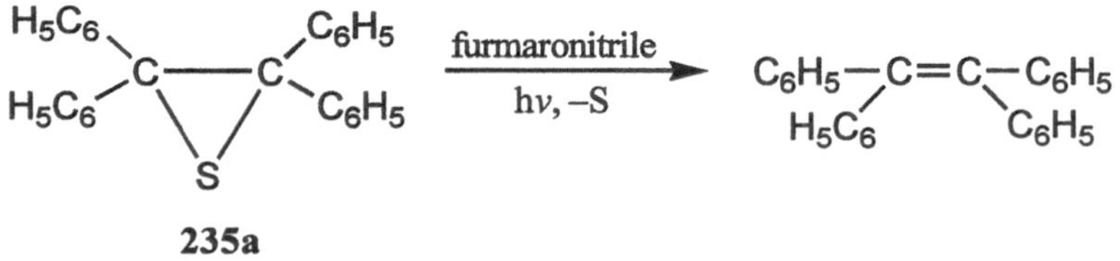

Scheme-100

1.3.2 Thiirenes (Thiacyclopropenes)

Three-membered unsaturated heterocycles containing sulfur as a heteroatom are called thiirenes or thiacyclopropenes. Thiirene **238** is isoelectronic with cyclobutadiene and belongs to the class of antiaromatic heterocycles containing

238

4π-electrons. Thiirenes, because of having resonance destabilized 4π-electron system, are less stable than their acyclic analogs and, therefore, have been observed as transient intermediates. The photochemical decomposition of 1,2,3-

thiadiazole **239** gives thioketene via thiirene as an intermediate **238** (scheme-101)[117]. The instability is reduced with the formation of thiirenium salts **240**,

hν
λ = 238–280 mm
$-N_2$
$CH_2=C=S$ + $CH\equiv C-SH$

239 **238**

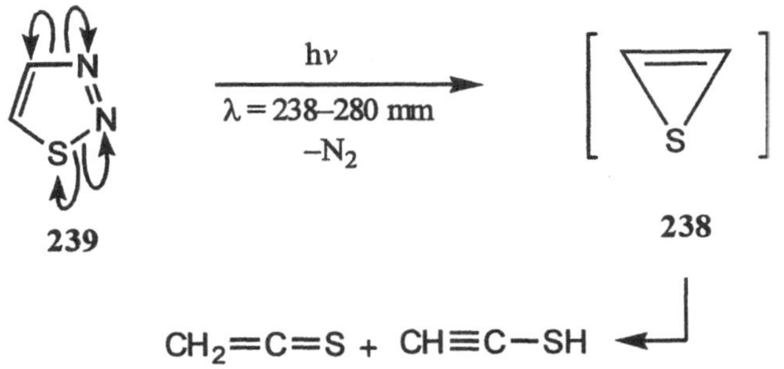

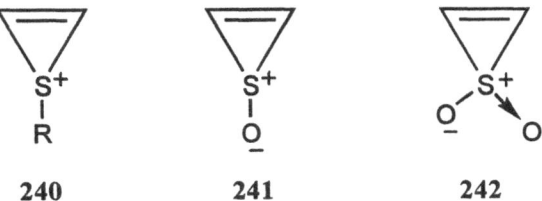

Scheme-101

thiirene-1-oxides **241** and thiirene-1,1-dioxides **242** because free pair of electrons is not present and resemble cyclopropene, 2π-electron ring system. S-Methyl

240 **241** **242**

derivatives **243**, prepared by the reaction of methanesulfenyl chloride with di-tert-butylacetylene, is stable at room temperature (scheme-102)[118]. CNDO/2 Calculations[119] have indicated that thiirene-1,1-dioxide **247** is more stable than thiirene-1-oxide **245** because of π–π (carbon and oxygen) stabilizing interactions.

$(CH_3)_3C-C\equiv C-C(CH_3)_3$
+
CH_3SCl
$(CH_3)_3C$
$C(CH_3)_3$
CH_3 Cl^-

243

Scheme-102

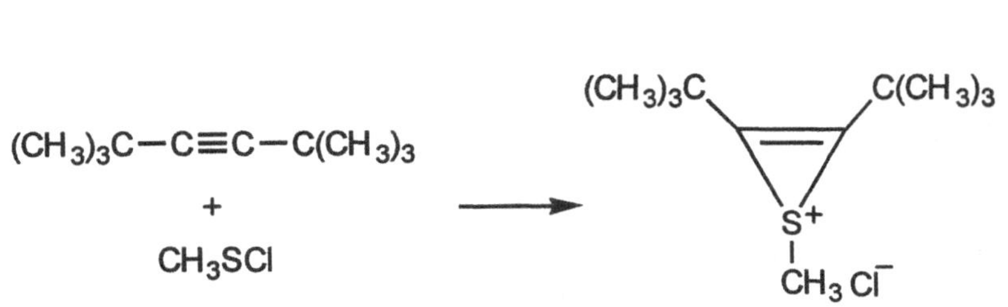

Thiirene-1-oxide **245** is prepared by the reaction of α,α'-dibromodibenzyl sulfoxide **244** with triethylamine[120]. The reaction[121] of α,α-dichlorobenzyl sulfone **246** with an excess of base, triethylenediamine in dimethyl sulfoxide, at room temperature yields 2,3-diphenylthiirene-1,1-dioxide **247** (scheme-103).

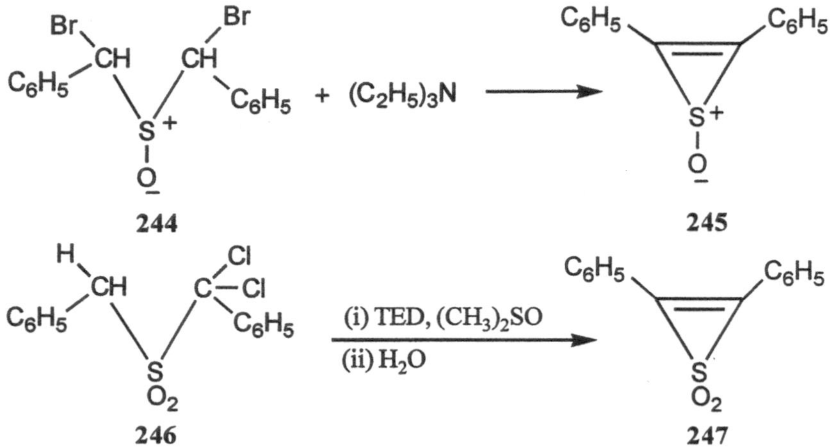

Scheme-103

2 THREE-MEMBERED HETEROCYCLES WITH TWO HETEROATOMS

Three-membered heterocycles with two heteroatoms are highly reactive and possess certain unusual properties because of the ring strain. These heterocycles, similer to three-membered heterocycles with one heteroatom, can also be prepared easily as compared to the larger rings because of the facile ring closure. These are interesting not only from their reactions point of view, but also from theoretical stand point of view.

2.1 Diaziridines

2.1.1 General

Three-membered saturated heterocycles with one carbon atom and two nitrogen heteroatoms are known as diaziridines **248**[122,123].

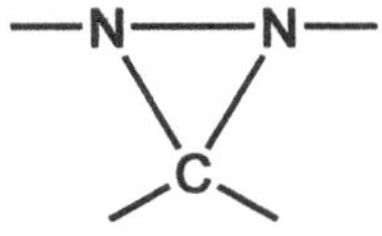

248

Diaziridines are weak bases and form salts. The solubility in aqueous mineral acids decreases with increasing the chain length of the alkyl substituents. The nitrogen-inversion in diaziridine is retarded because of the increased rotational energy barrier resulting from the lone pair-lone pair interactions on the nitrogen atoms and thus diaziridine can be resolved into enantiomers.

2.1.2 Synthesis

2.1.2.1 Reaction of Ketones with Ammonia or Primary Amines

Diaziridines are prepared by the reaction of ketones with ammonia or primary amines in the presence of an aminating agent (chloramine or hydroxylamine-O-sulfonic acid) (scheme-104)[124,125].

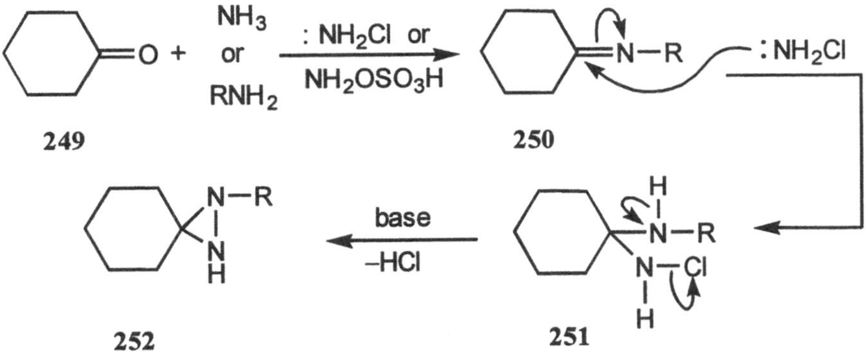

Scheme-104

2.1.2.2 Reaction of Diazirines with Grignard Reagents

The reaction of diazirines **253** with Grignard reagent provides diaziridines **255** (scheme-105)[126].

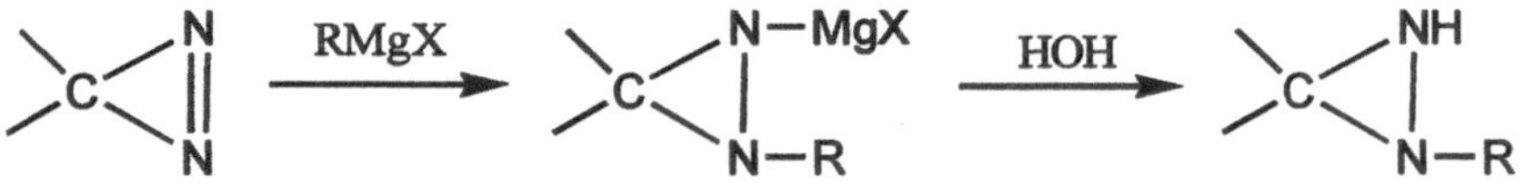

253 254 255

Scheme-105

2.1.2.3 Photochemical Method

Diaziridines **257** can also be prepared by the electrocyclic ring closure of stabilized azomethine imides **256** in the presence of ultraviolet light. The reaction is reversed when diaziridine **257** is heated (scheme-106)[127].

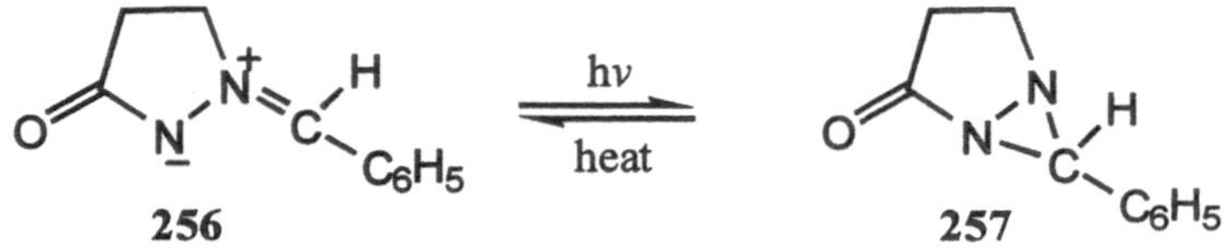

256 257

Scheme-106

Photolysis of Δ^2-tetrazolines **258** also results in the formation of diaziridines **259** with the extrusion of N_2 (scheme-107)[128].

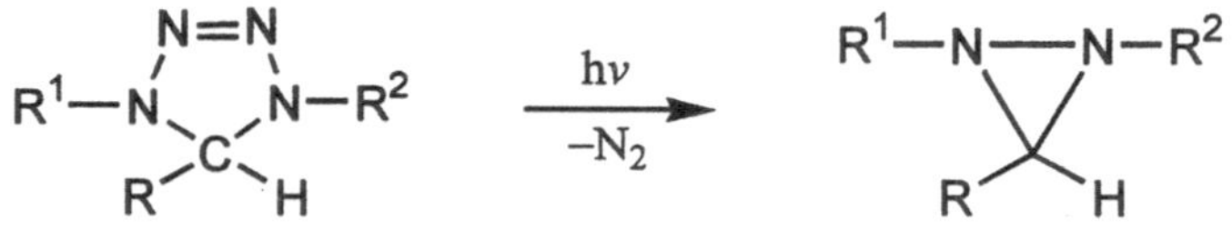

258 259

Scheme-107

2.1.3 Reactions

Diaziridines are more stable than oxiranes. The diaziridine ring is stable to alkaline reagents, but hydrolyzed readily in acid solution. However, diaziridines undergo reactions involving the cleavage of C–N and N–N bonds and the retention of the ring.

2.1.3.1 Cleavage of N–N Bond (Reduction)

Diaziridines are reduced to two molecules of amines with the cleavage of N–N bond (scheme-108)[129].

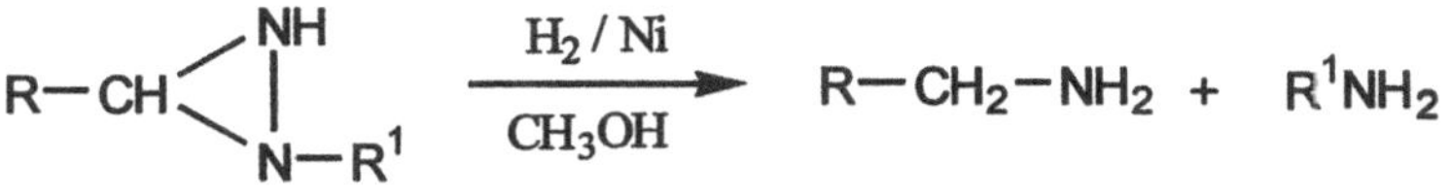

R—CH (NH, N—R¹) 260 $\xrightarrow[CH_3OH]{H_2/Ni}$ R—CH$_2$—NH$_2$ + R¹NH$_2$

                                               261   262

Scheme-108

The diaziridines with at least one unsubstituted nitrogen are reduced by lithium aluminium hydride with the cleavage of N–N bond to the products different from those obtained in the catalytic hydrogenation (scheme-109)[129].

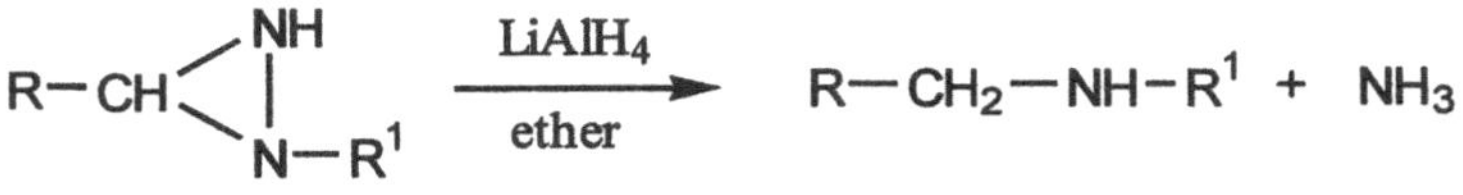

R—CH (NH, N—R¹) 260 $\xrightarrow[ether]{LiAlH_4}$ R—CH$_2$—NH—R¹ + NH$_3$

                                                 263

Scheme-109

2.1.3.2 Cleavage of C–N Bond (Acid Hydrolysis)

The acid hydrolysis of diaziridines causes the cleavage of C–N bond and is considerably accelerated by an increase in the number of substituents on the carbon atom. However, the presence of substituents on the nitrogen atom does not cause any appreciable effect (scheme-110).

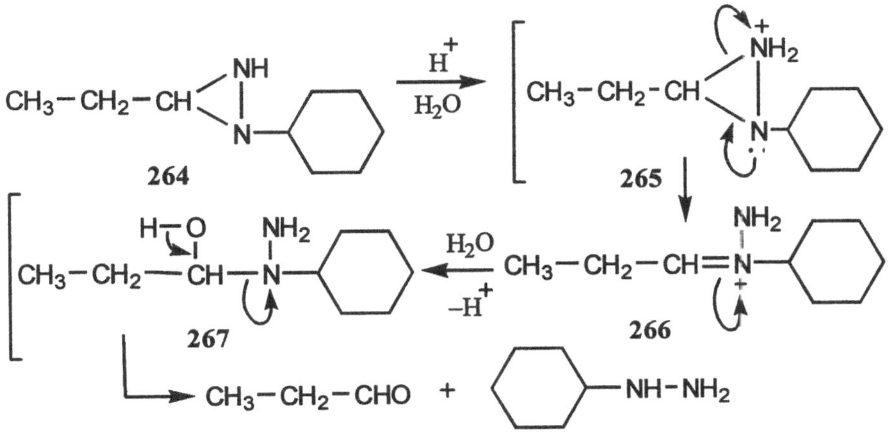

Scheme-110

2.1.3.3 Retention of Ring

The diaziridines **270** containing at least one N–H group undergo reactions similar to those of the secondary amines. The condensation of diaziridines with chloral proceeds with the retention of the ring (scheme-111).

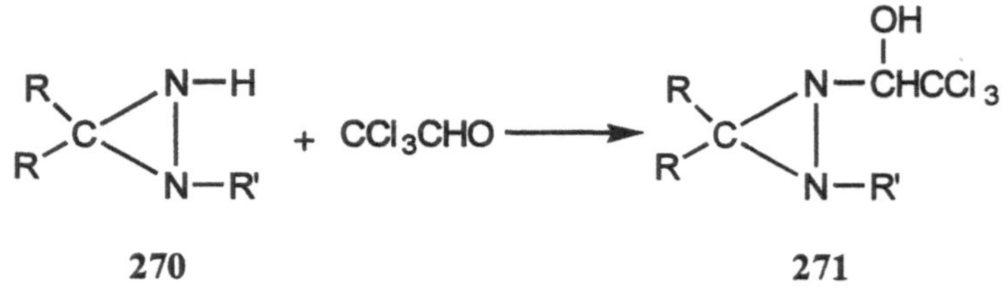

Scheme-111

2.2 Diazirine

2.2.1 General

Three-membered unsaturated heterocycles with two nitrogen heteroatoms are known as diazirines[123]. Diazirines **253** are the cyclic isomers of diazoalkanes, but differ considerably in properties.

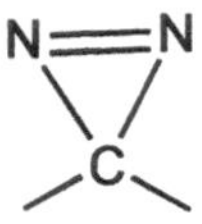

253

2.2.2 Synthesis

2.2.2.1 Reaction of tert-Alkylazomethine with Dichloramine

The reaction of tert-alkylazomethine **272** with dichloramine results in the formation of diazirine **253** involving addition of dichloramine to formaldehyde imine (scheme-112)[130,131].

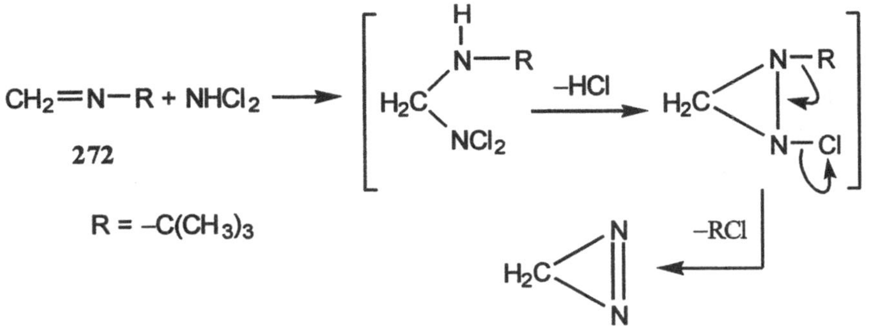

272

R = −C(CH3)3

253

Scheme-112

2.2.2.2 Reaction of Amidines with Sodium Hypochlorite

3-Chlorodiazirines **277** are prepared by the reaction of alkyl- or arylamidines with sodium hypochlorite in the presence of DMSO (scheme-113).

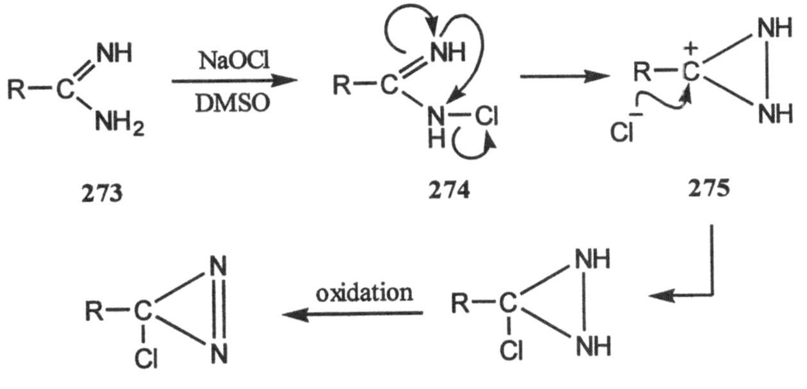

273 274 275

277 276

Scheme-113

2.2.2.3 Oxidation of Diaziridines

The oxidation of diaziridines **278** with oxidizing agents such as silver oxide, alkaline permanganate and mercuric oxide provides diazirines **279** (scheme-114)[132].

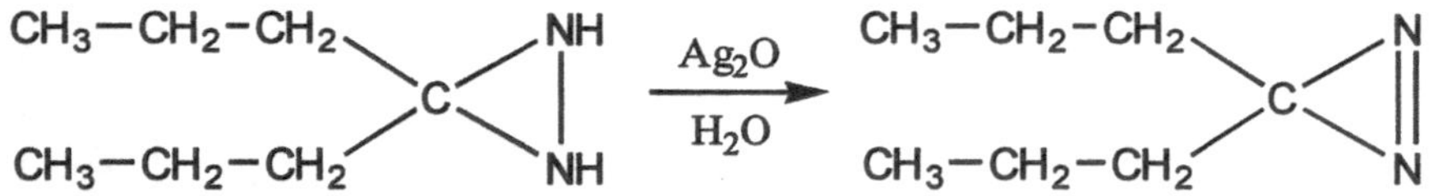

Scheme-114

However, aryltrifluoromethyldiazirines **281**, which can not be prepared by the usual oxidation methods, are synthesized by the oxidation with DMSO–oxalyl chloride (scheme-115)[133].

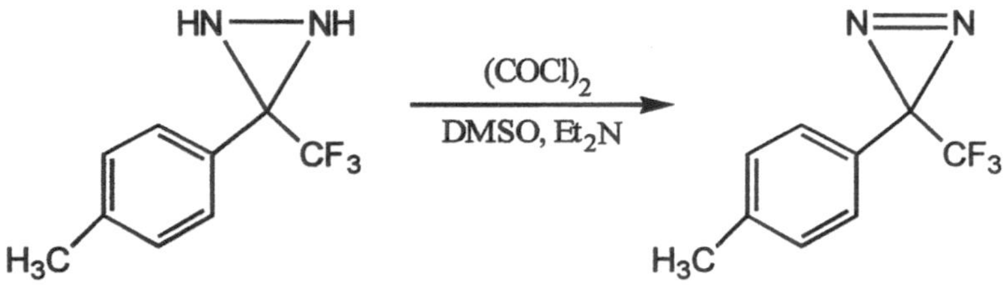

Scheme-115

2.2.3 Reactions

Diazirines are more stable and less reactive than diazoalkanes. They do not react with acids and alkalies; and with other electrophiles at room temperature. However, they react with Grignard reagents and undergo thermolytic and photolytic decompositions.

2.2.3.1 Thermal and Photochemical Decompositions

Thermal and photochemical decompositions of diazirines proceed with the formation of carbene intermediates which can be trapped by alkenes or undergo rearrangement reactions (scheme-116)[134].

2.2.3.2 Reaction with Grignard Reagents (Retantion of Ring)

Diazirines react with Grignard reagents without the cleavage of the ring providing dazirines (scheme-117)[132].

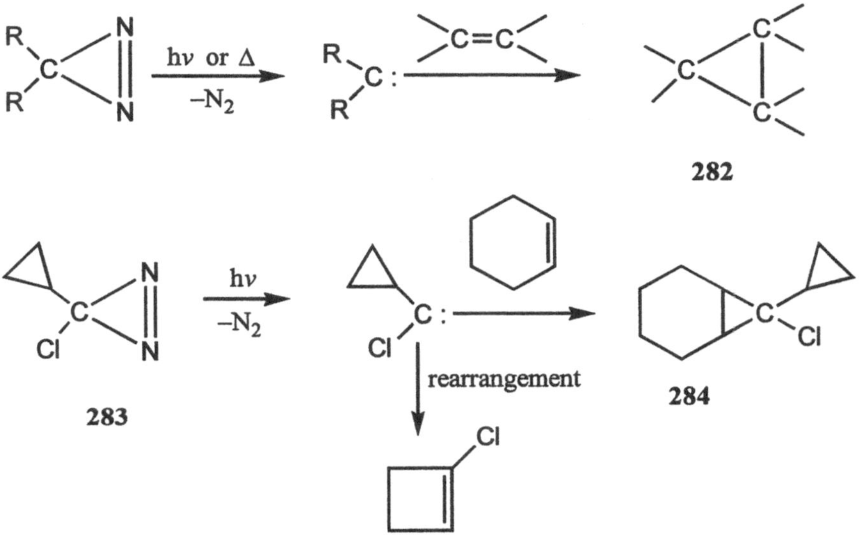

282

283

284

285

Scheme-116

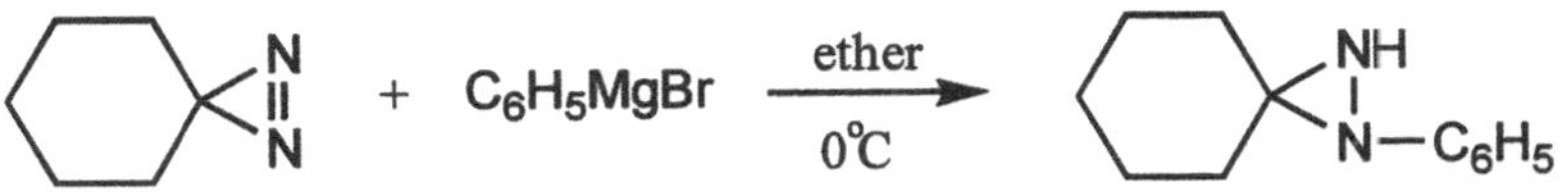

Scheme-117

2.3 Oxaziridines (Oxaziranes)

2.3.1 General

Three-membered saturated heterocycles with carbon, nitrogen and oxygen atoms are known as oxaziridines or oxaziranes **286**[134]. Oxaziridines are non-basic and do not form salts with acids. Oxaziridine shows remarkable conformational stability about nitrogen because of with considerably higher nitrogen inversion barrier. Thus, the isomers which differ only in their configuration about nitrogen are resolvable and configurationally stable[135].

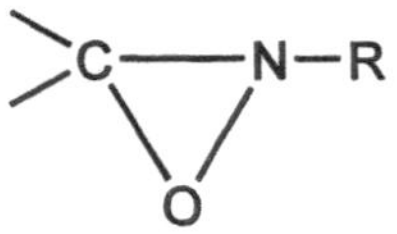

286

2.3.2 Synthesis

2.3.2.1 Peracid Oxidation of Imines or Schiff Bases (Insertion of Oxygen Atom into C=N Bond)

It involves direct insertion of oxygen atom into C=N bond and can be easily accomplished by the oxidation of imines (schiff bases) with peracids (scheme-118)[136]. The reaction is of wide applicability as the required imines can be easily prepared by the reaction of primary amines with ketones and aldehydes (scheme-119). The mechanism of peracid-imine reaction is similar to that of the epoxidation of alkenes (scheme-120).

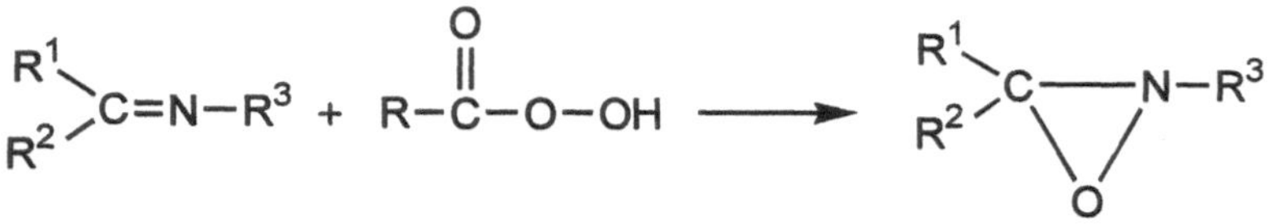

287   **286**

Scheme-118

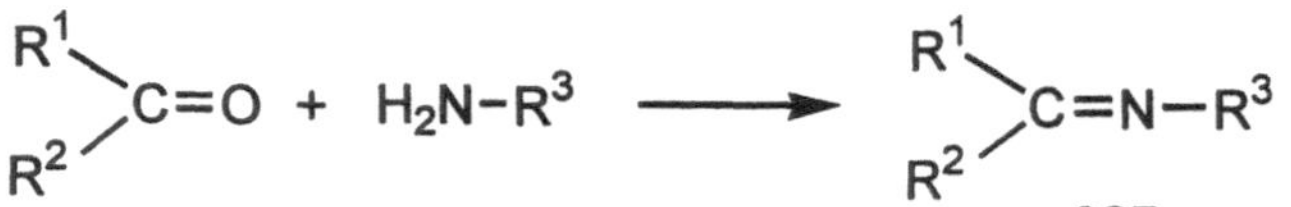

287

Scheme-119

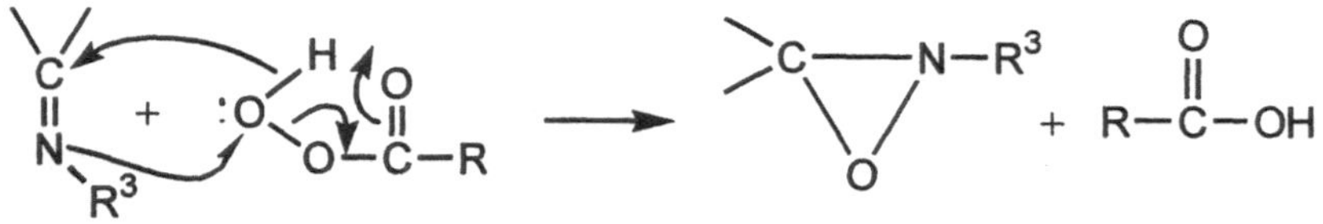

Scheme-120

2.3.2.2 Reaction of Aldehydes and Ketones with Hydroxylamine-O-sulfonic Acids and Chloramines (Insertion of Nitrogen Atom into C=O Bond)

The reaction of aldehydes and ketones with hydroxylamine-O-sulfonic acids or chloramines involves the insertion of nitrogen atom into C=O bond yielding oxaziridines (scheme-121)[137-139].

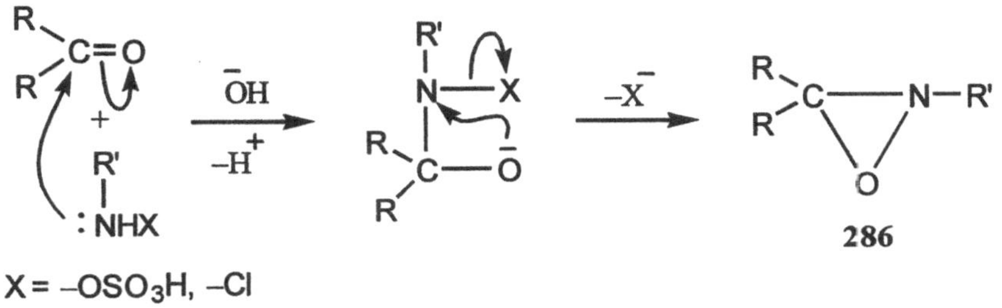

X = –OSO₃H, –Cl

Scheme-121

The reaction proceeds by 1,2-addition of nitrogen-containing component to the carbonyl group, followed by the intramolecular cyclization.

2.3.2.3 Photoisomerization of Nitrones

Photoisomerization of nitrones **288** to oxaziridines **289** provides another route for the synthesis of oxaziridines (scheme-122)[140,141]

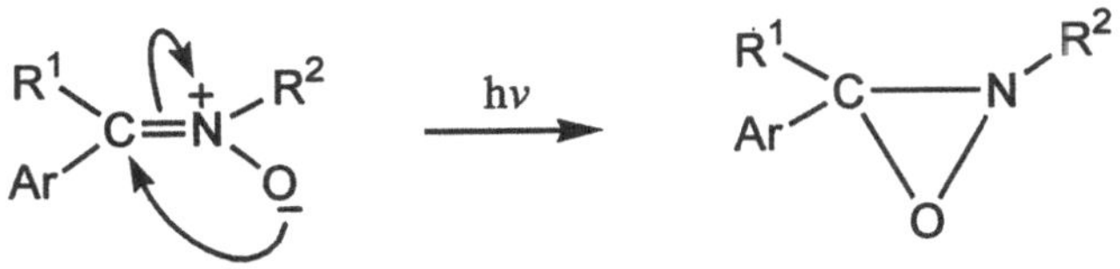

288   **289**

Scheme-122

2.3.3 Reactions

Oxaziridines have high strain energy and as such undergo ring opening reactions involving the cleavage of C–O, C–N and N–O bonds. However, their stability varies with the nature and the number of substituents.

2.3.3.1 Reaction with Reducing Agents

Oxaziridines undergo reductive cleavage with lithium aluminium hydride and with catalytic hydrogen providing the corresponding imines and secondary amines respectively (scheme-123)[142,143].

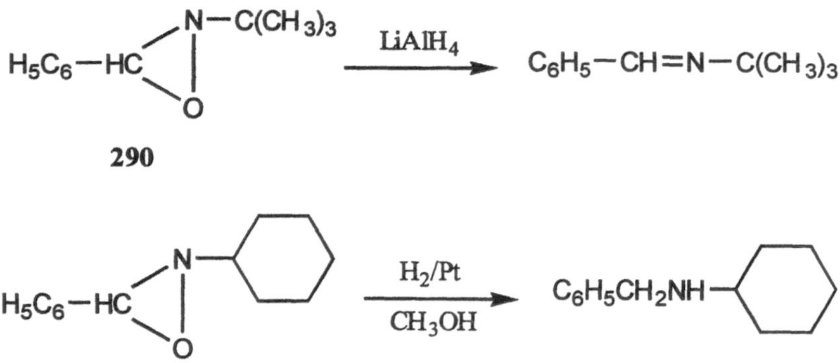

Scheme-123

2.3.3.2 Acid Hydrolysis

Acid hydrolysis of oxaziridines proceeds with the cleavage of C–N or N–O bond depending on the nature of substituents. 3-Aryloxaziridines **290** undergo acid hydrolysis involving cleavage of the C–N bond with the formation of aromatic aldehydes and alkyl hydroxylamine (scheme-124)[142,143].

The reaction involves protonation of oxygen to form stabilized benzylic carbonium ion **293** with the cleavage of C–O bond. The carbonium ion **293** is then attacked by the hydroxyl ion, followed by the C–N bond fission forming aromatic aldehyde and hydroxylamine. In case of alkyloxaziridines **295**, the reaction proceeds via the formation of electron-deficient nitrogen species with the cleavage of N–O bond. The migration of hydride ion occurs from the neighbouring carbon atom providing aldehydes and ammonia (scheme-125). However, if hydrogen is not available on the neighbouring carbon atom, the migration of alkyl group to the nitrogen atom occurs with the formation of an amine (scheme-126)[144,145].

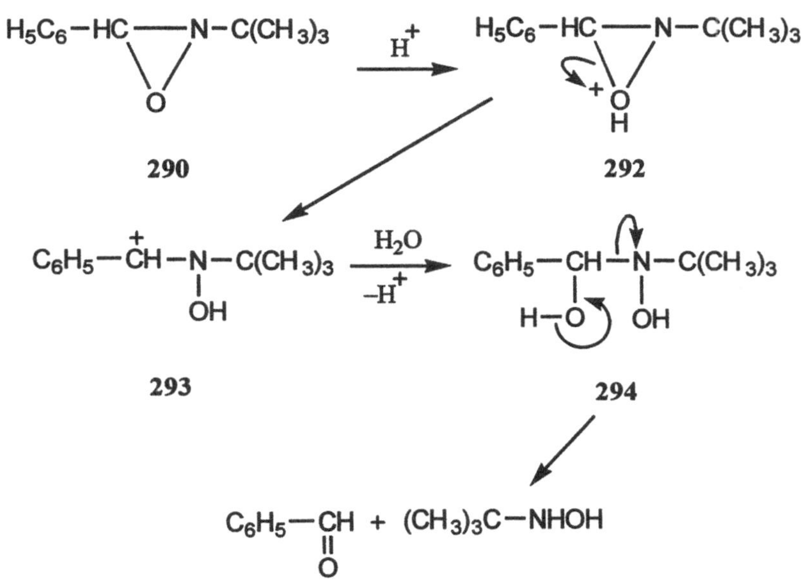

H₅C₆—HC————N—C(CH₃)₃ $\xrightarrow{\ H^+\ }$ H₅C₆—HC————N—C(CH₃)₃

$$\underset{\textbf{290}}{\overset{O}{\diagup}} \qquad\qquad \underset{\textbf{292}}{\overset{\overset{+}{O}}{\underset{H}{\diagup}}}$$

$$\underset{\textbf{293}}{C_6H_5-\overset{+}{C}H-\underset{\underset{OH}{|}}{N}-C(CH_3)_3} \xrightarrow[{-H^+}]{H_2O} \underset{\textbf{294}}{C_6H_5-CH-N-C(CH_3)_3}$$

$$\underset{\overset{\|}{O}}{C_6H_5-CH} \ + \ (CH_3)_3C-NHOH$$

Scheme-124

$$\underset{\textbf{295}}{H_2C————N-CH_2-(CH_2)_2-CH_3} \xrightarrow{\ H^+\ } \left[\underset{\textbf{296}}{H_2C————N-CH_2-(CH_2)_2-CH_3} \right.$$

$$\left[\underset{\textbf{298}}{CH_2-N-\overset{+}{C}H-(CH_2)_2-CH_3} \xleftarrow{\ H-O\ } \underset{\textbf{297}}{CH_2-\overset{+}{N}-CH-(CH_2)_2-CH_3} \right.$$

$$\downarrow {-H^+}$$

$$H_2C=O + CH_3-(CH_2)_2-CH=NH \xrightarrow[HOH]{H^+} CH_3(CH_2)_2CHO + NH_3$$

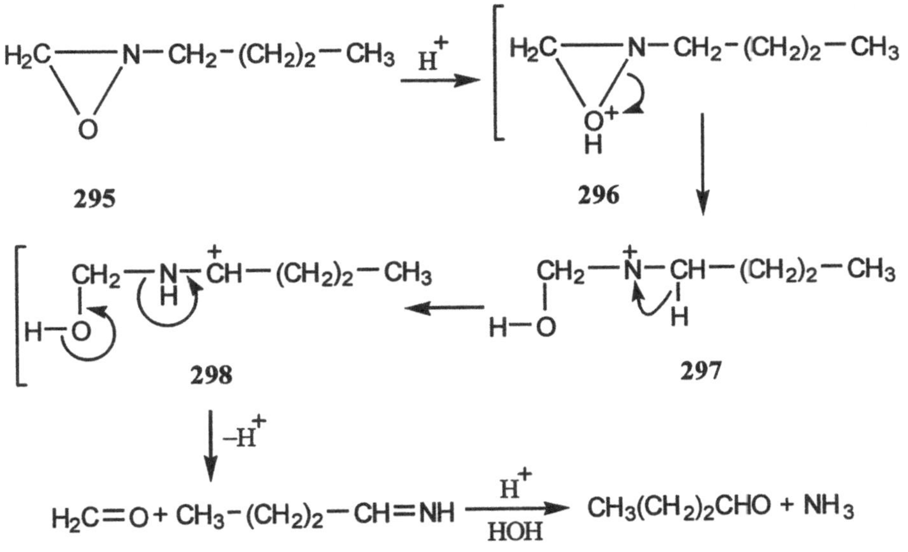

Scheme-125

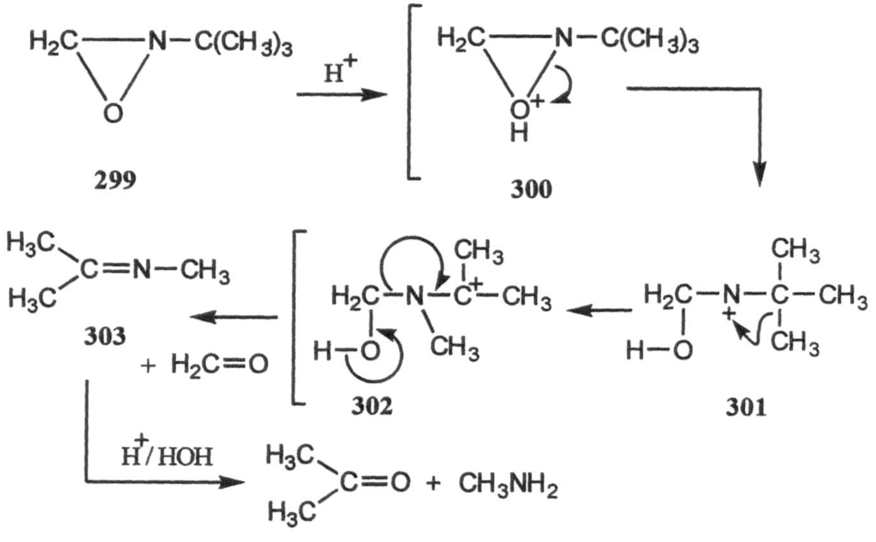

Scheme-126

2.3.3.3 Reaction with Basic Reagents

The oxaziridine ring itself is unreactive towards basic reagents, but suitably substituted oxaziridines are decomposed with the N-O bond fission (scheme-127)[146-148].

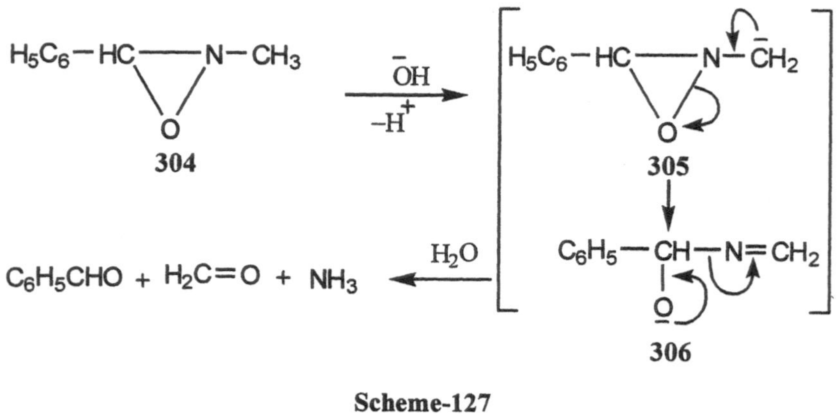

Scheme-127

The reaction proceeds with the formation of carbanion **305** which undergoes degradation to form carbonyl compound.

2.3.3.4 Thermal Rearrangements

Thermal rearrangements of oxaziridines depend upon the substituents at the position-3.

2.3.3.4.1 Oxaziridines **290**, substituted with an aryl group at the position-3, isomerize to nitrones **288** involving C–O bond fission (scheme-128)[142].

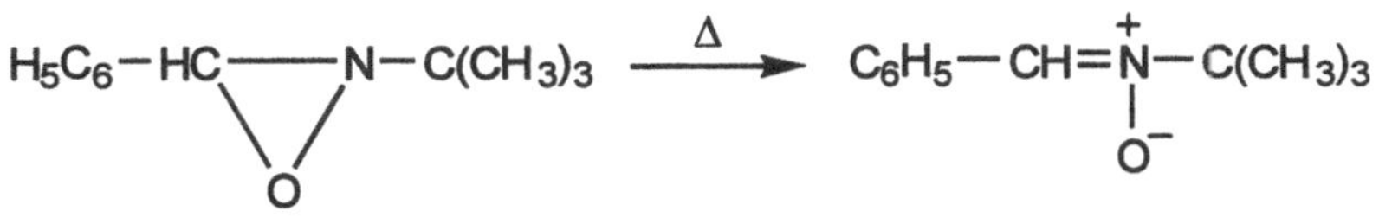

$$H_5C_6-HC \overset{\qquad}{\underset{O}{\diagdown\diagup}} N-C(CH_3)_3 \quad \xrightarrow{\Delta} \quad C_6H_5-CH=\overset{+}{\underset{\underset{O^-}{|}}{N}}-C(CH_3)_3$$

290   **288**

Scheme-128

2.3.3.4.2 The oxaziridines with an alkyl group at the ring carbon rearrange to amides with the cleavage of N–O bond (scheme-129).

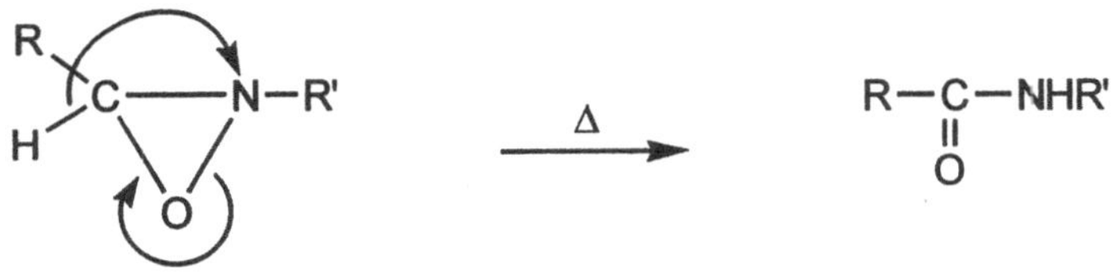

$$R-\overset{O}{\overset{\|}{C}}-NHR'$$

Scheme-129

2.3.3.4.3 The oxaziridines derived from the cyclic ketones undergo ring expansion during thermal rearrangement (scheme-130).

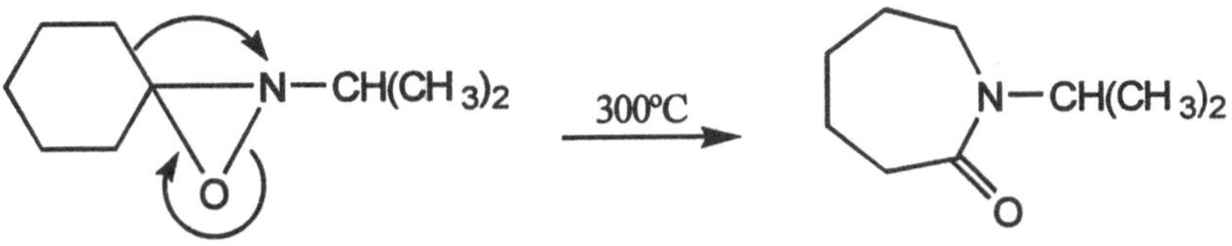

$$\xrightarrow{300°C}$$

308   **309**

Scheme-130

2.3.3.5 Photochemical Rearrangements

Photochemical rearrangement proceeds with the migration of a group from carbon to nitrogen involving cleavage of N–O bond to form amides (scheme-131)[149].

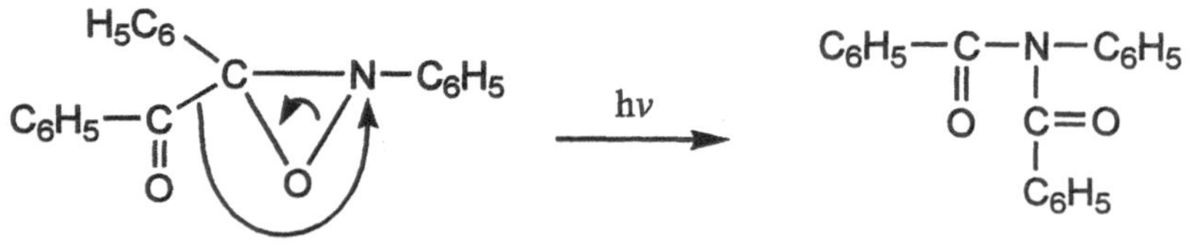

310   311

Scheme-131

Bicyclic oxaziridines **312** on photolysis undergo ring expansion. The reaction has been considered to proceed via free radical mechanism with the cleavage of N–O bond (scheme-132)[150-153].

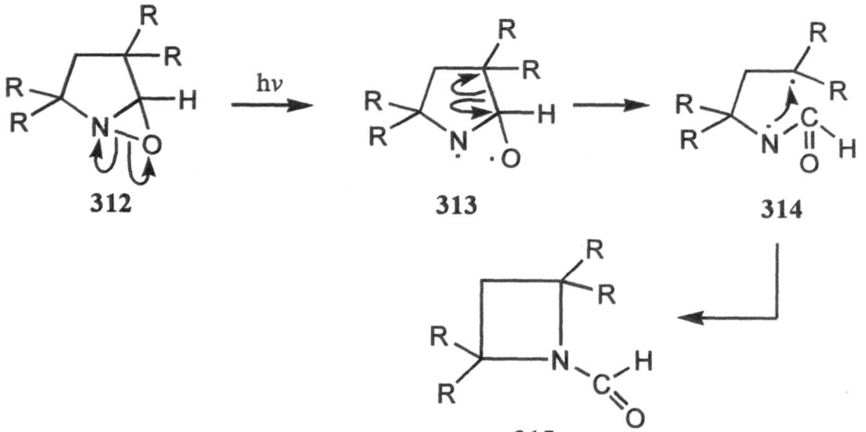

312   313   314

315

Scheme-132

REFERENCES

1. J. A. Deyrup in A. Hassner (Ed.), *Small Ring Heterocycles*, Part I, Wiley-Interscience, New York, 1983, pp. 1.

2   O. C. Dermer and G. E. Ham, *Ethyleneimine and Other Aziridines*, Academic Press, New York, 1969; D. Tanner and C. Birgersson, *Tetrahedron Lett.* 2533 (1991)

2a. A. Padwa and A. D. Woolhouse in A. R. Katritzky and C. W. Rees (Eds.), *Comprehensive Heterocyclic Chemistry* Vol. 7, Pergamon Press, Oxford, 1984, pp. 47; A. Padwa and R. L. Chinn in H. Suschitzky and E. F. V. Scriven (Eds.), *Progress in Heterocyclic Chemistry* Vol. 1, Pergamon Press, Oxford, 1989, pp. 90; A. Padwa and F. R. Kinder, Vol. 2, 1990, pp. 29; Vol 3, 1991, pp. 52; A. Padwa and S. S. Murphree, Vol. 4, 1992, pp. 44; Vol. 5, 1993, pp. 65; Vol. 6, 1994, pp. 68.

3. A. de Meijere, *Angew. Chem. Int. Edn. Engl.* **18**, 809 (1979).

4. R. E. Carter and T. Darkenberg, *Chem. Commun.* 582 (1972).

5. J. B. Lambert, *Top. Stereochem.* **6**, 52 (1971).

6. S. J. Brois, *J. Am. Chem. Soc.* **92**, 1079 (1970).

7. S. J. Brois, *J. Am. Chem. Soc.* **90**, 506 (1968).

8. S. J. Brois, *Tetrahedron lett.* 5997 (1968).

9. S. Gabriel, *Chem. Ber.* **21**, 1049 (1888).

10. S. Gabriel and R. Stelzner, *Chem. Ber.* **28**, 2929 (1895).

11. P. E. Fanta in A. Weissberger (Ed.), *The Chemistry of Heterocyclic Compounds* **XIX**, Part I, Wiley-Interscience, New York, 1964, pp. 524.

12. K. Hafner and C. Konig, *Angew. Chem. Int. Edn.* **2**, 96 (1963).

13. H. Quast, *Heterocycles*, **14**, 1677 (1980); J. R. Pfister, Synthesis, 969 (1984).

14. G. L. Closs and S. J. Brois, *J. Am. Chem. Soc.* **82**, 6068 (1960).

15. F. W. Fowler, A. Hassner and L. A. Levy, *J. Am. Chem. Soc.* **89**, 2077 (1967).

16. K. Hafner and C. Konig, *Angew. Chem.* **75**, 89 (1963).

17. W. Lwowski and T. W. Mattingly, Jr., *J. Am. Chem. Soc.* **87**, 1947 (1965) and references cited therein.

18. P. Scheiner, *J. Org. Chem.* **30**, 7 (1965).

19. E. Huisgen, *Angew. Chem. Int. Edn.* **2**, 565 and 633 (1963).

20. J. F. Bunnett, R. L. McDonald and F. P. Olsen, *J. Am. Chem. Soc.* **96**, 2855 (1974).

21. H. Stamm, *Angew. Chem.* **74**, 694 (1962).

22. S. Gabriel and H. Ohle, *Chem. Ber.* **50**, 804 (1977).

23. F. Wolfheim, *Chem. Ber.* **47**, 1440 (1914).

24. N. J. Leonard, J. V. Paukstelis and L. E. Brady, *J. Org. Chem.* **29**, 3383 (1964).

25. J. S. Doughty, C. L. Lazell and A. R. Collett, *J. Am. Chem. Soc.* **72**, 2866 (1950).

26. N. J. Leonard, and L. E. Brady, *J. Org. Chem.* **28**, 2850 (1963).

27. N. J. Leonard and L. E. Brady, *J. Org. Chem.* **30**, 817 (1965).

28. R. D. Clark and G. K. Helmkamp, *J. Org. Chem.* **29**, 1316 (1964).

29. C. L. Bumgardner, K. S. McCallum and J. P. Freeman, *J. Am. Chem. Soc.* **83**, 4417 (1961).

30. A. S. Deutsch and P. E. Fanta, *J. Org. Chem.* **21**, 892 (1956).

31. P. E. Fanta and A. S. Deutsch, *J. Org. Chem.* **23**, 72 (1958).

32. H. W. Whitlock, Jr. and G. L. Smith, *Tetrahedron Lett.* 1389 (1965).

33. L. B. Clapp and J. W. Watjen, *J. Am. Chem. Soc.* **75**, 1490 (1953).

34. V. Nair in A. Hassner (Ed.), *Small Ring Heterocycles*, Part I, Wiley-Interscience, New York, 1983, pp. 225; A. Padwa and F. R. Kinder in H. Suschitzky and E. F. V. Scriven (Eds.), *Progress in Heterocyclic Chemistry* Vol. **2**, Pergamon Press, Oxford, 1990, pp. 32; Vol. **3**, 1991, pp. 54.

35. F. W. Fowler, *Adv. Heterocycl. Chem.* **13**, 45 (1977).

36. F. W. Fowler and A. Hassner, *J. Am. Chem. Soc.* **90**, 2875 (1968).

37. V. Nair, *Org. Magn. Res.* **6**, 483 (1974).

38. G. Smolinsky, *J. Org. Chem.* **27**, 3557 (1962).

39. A. Padwa and Hao Ku, *J. Org. Chem.* **44**, 255 (1979).

40. G. Smolinsky and C. A. Pryde, *J. Org. Chem.* **33**, 2411 (1968); P. Wipf and H. Heimgartner, *Helv. Chim. Acta* **7**, 140 (1988).

41. P. W. Neber and A. Burgard, *Ber.* **193**, 281 (1932).

42. R. F. Parcell, *Chem. & Ind.* 1396 (1963).

43. T. Nishiwaki, T. Kitamura and A. Nakano, *Tetrahedron,* **26**, 453 (1970).

44. E. F. Ullman and B. Singh, *J. Am. Chem. Soc.* **88**, 1844 (1966).

45. A. Hassner, *Heterocycles* **14**, 1517 (1980).

46. A. Laurent, P. Mison, A. Nafti and N. Pellissier, *Tetrahedron Lett.* 3955 (1979).

47. A. G. Hortmann and D. A. Robertson, *J. Am. Chem. Soc.* **94**, 2758 (1972).

48. A. Hassner and D. J. Anderson, *J. Org. Chem.* **39**, 3070 (1974).

49. A. Padwa, and J. K. Rasmussen, *J. Am. Chem. Soc.* **97**, 5913 (1975).

50. M. Bartok and K. L. Lang in S. Patai (Ed.), *The Chemistry of Ethers, Crown Ethers, Hydroxyl Groups and Their Sulfur Analogues.* Wiley- Interscience, New York, 1980, pp. 609; A. Padwa and R. L. Chinn in H. Suschitzky and E. F. V. Scriven (Eds.), *Progress in Heterocyclic Chemistry* Vol. **1**, Pergamon Press, Oxford, 1989, pp. 82; A. Padwa and F. R. Kinder, Vol. **2**, 1990, pp. 22; Vol. **3**, 1991, pp. 42; A. Padwa and S. S. Murphree, Vol. **4**, 1992, pp. 34; Vol. **5**, 1993, pp. 54; Vol **6**, 1994, pp. 56.

51. A. S. Roa, S. K. Parniker and J. G. Kirtane, *Tetrahedron* **39**, 2323 (1983); E. G. Lewars in A. R. Katritzky and C. W. Rees (Eds.), *Comprehensive Heterocyclic Chemistry* Vol. **7**, Pergamon Press, Oxford, 1984, pp. 95; M. Bartok and K. L. Lang in A. Hassner (Ed.), *Small Ring Heterocycles*, Part III, Wiley-Interscience, New York, 1985, pp. 1.

52. S. Greenberg and J. F. Liebman, *Strained Organic Molecules*, Academic Press, New York, 1978, pp. 28.

53. K. Pihlaja and E. Taskinen in A. R. Katritzky (Ed.). *Physical Methods in Heterocyclic Chemistry* Vol. **6**, Academic Press, New York, 1974, pp. 199.

54. H. O. House, *Modern Synthetic Reactions,* 2nd Edn., Benjamin, New York, 1972.

55. N. N. Schwartz and J. H. Blumbergs, *J. Org. Chem.* **29**, 1976 (1964).

56. B. M. Lynch and K. H. Pausacker, *J. Chem. Soc.* 1525 (1955).

57. W. Huckel and V. Worfeffel, *Chem. Ber.* **88**, 338 (1955).

58. H. O. House and R. S. Roa, *J. Am. Chem. Soc.* **80**, 2428 (1958).

59. C. A. Bunton and G. J. Minkoff, *J. Chem. Soc.* 655 (1949).

60. Y. Ogata and Y. Sawaki, *Tetrahedron* **20**, 2065 (1964).

61. E. J. Corey and M. Chaykovsky, *J. Am. Chem. Soc.* **87**, 1345 (1965).

62. C. E. Cook, R. C. Corley and M. E. Wall, *Tetrahedron Lett.* 891 (1965).

63. V. Franzen and H. E. Driessen, *Tetrahedron Lett.* 661 (1962).

64. C. D. Gutsche, *Org. React.* **8**, 364 (1954).

65. M. Ballester, *Chem. Rev.* **55**, 283 (1955).

66. M. Ballester and P. D. Bartlett, *J. Am. Chem. Soc.* **75**, 2042 (1953).

67. R. H. Hunt, L. J. Chinn and W. S. Johnson, *Org. Synth.* **4**, 459 (1963).

68. M. S. Kharasch and O. Reinmuth, *Grignard Reactions of Nonmetallic Compounds*, Prentice-Hall, New York, 1954, pp. 181.

69. V. Mark, *J. Am. Chem. Soc.* **85**, 1884 (1963).

70. F. H. Dickey, W. Fickett and H. J. Lucas, *J. Am. Chem. Soc.* **74**, 944 (1952).

71. E. L. Eliel in M. S. Newman (Ed.), *Steric Effects in Organic Chemistry*, Wiley-Interscience, New York, 1956, pp. 106.

72. R. Fuchs and C. A. Vander Werf, *J. Am. Chem. Soc.* **76**, 1631 (1954).

73. M. N. Rerick and E. L. Eliel, *J. Am. Chem. Soc.* **84**, 2356 (1962).

74. E. L. Eliel and M. N. Rerick, *J. Am. Chem. Soc.* **82**, 1362 (1960).

75. N. G. Gaylord and E. I. Becker, *Chem. Rev.* **49**, 413 (1951).

76. R. J. Anderson, *J. Am. Chem. Soc.* **92**, 4978 (1970).

77. W. S. Wadsworth, Jr., and W. D. Emmons, *J. Am. Chem. Soc.* **83**, 1733 (1961).

78. D. B. Denney, J. J. Vill and M. J. Boskin, *J. Am. Chem. Soc.* **84**, 3944 (1962).

79. L. Horner, H. Hoffmann and V. G. Toscano, *Chem. Ber.* **95**, 536 (1962).

80. L. Tomoskozi, *Chem. & Ind.* 689 (1965).

81. A. Orekhoff amd M. Tiffeneau, *Bull. Chem. Soc. Fr.* **146**, 697 (1908).

82. F. Arndt, J. Amende and W. Ender, *Monatsh. Chem.* **58**, 202 (1932).

83. E. L. Gustus and P. G. Stevens, *J. Am. Chem. Soc.* **55**, 378 (1933).

84. M. T. Bogert and R. O. Roblin, Jr., *J. Am. Chem. Soc.* **55**, 3741 (1933).

85. B. Rickborn and R. O. Thummel, *J. Org. Chem.* **34**, 3583 (1969).

86. J. K. Crandall and M. Apparu, *Org. React.* **29**, 345 (1983).

87. M. J. Borkin and D. B. Denny, *Chem. & Ind.* 330 (1959).

88. D. E. Bissing and A. J. Spezial, *J. Am. Chem. Soc.* **87**, 2683 (1965); H. N. C. Wong, M. Y. Honn, C. W. Tse, Y. C. Yip, J. Tanko and T. Hudlicky, *Heterocycles* **26**, 1345 (1987).

89. V. N. Yandovskii and B. A. Ershov, *Russ. Chem. Rev.* **41**, 403 (1972).

90. R. E. Parker and N. S. Isaacs, *Chem. Rev.* **59**, 737 (1959).

91. H. Alper, D. Des Roches, T. Durst and R. Legault, *J. Org. Chem.* **41**, 3611 (1976).

92. D. Milstein, O. Buchman and J. Blum, *J. Org. Chem.* **42**, 2299 (1977).

93. M. Suzuki, A. Watanabe and R. Noyori, *J. Am. Chem. Soc.* **102**, 2095 (1980).

94. E. G. Lewars, *Chem. Rev.* **83**, 519 (1983).

95. O. P. Strausz, R. K. Gosavi, A. S. Denes and I. G. Csizmadia, *J. Am. Chem. Soc.* **98**, 4784 (1976).

96. E. Lewars and G. Morrison, *Tetrahedron Lett.* 501 (1977).

97. K. P. Zeller, *Tetrahedron Lett.* 707 (1977).

98. U. Zoller in A. Hassner (Ed.), *Small Ring Heterocycles*, Part I, Wiley-Interscience, New York, 1983, pp. 333; D. C. Dittmer in A. R. Katritzky and C. W. Rees (Eds.), *Comprehensive Heterocyclic Chemistry* Vol 7, Pergamon press, Oxford, 1984, pp. 131; A. V. Fokin, M. A. Allakhverdiev and A. F. Kolomiets, *Usp. Chim.* **59**, 705 (1990); A. Padwa and R. L. Chinn in H. Suschitzky and E. F. V. Scriven (Eds.), *Progress in Heterocyclic Chemistry* Vol. **1**, Pergamon Press, Oxford, 1989, pp. 93; A. Padwa and F. R. Kinder, Vol. **2**, 1990, pp. 34; Vol. **3**, 1991, pp. 55.

99. C. C. Culvenor, W. Davies and W. E. Savige, *J. Chem. Soc.* 4480 (1952) and references cited therein.

100. G. L. Braz, *J. Gen. Chem.* (USSR) **21**, 757 (1951).

101. A. V. Fokin and A. F. Kolomiets, *Russ. Chem. Rev.* **44**, 138 (1957).

102. T. Fujisawa and T. Kobori, *Chem. Lett.* 935 (1972).

103. F. Capozzi, G. Capozzi and S. Menichetti, *Tetrahedron Lett.* 4177 (1988).

104. E. J. Corey and M. Chaykovsky, *J. Am. Chem. Soc.* **87**, 1353 (1965) and references cited therein.

105. H. Staudinger and J. Siegwart, *Helv. Chim. Acta* **3**, 833 (1970).

106. J. M. Beiner, D. Lecadet, D. Paquer and A. Thuillier, *Bull. Soc. Chim. Fr.* 1983 (1973).

107. R. Huisgen and E. Langhals, *Tetrahedron Lett.* 5369 (1989).

108. S. Searles, Jr., H. R. Hays and E. F. Lutz, *J. Org. Chem.* **27**, 2832 (1962).

109. E. E. Van Tamelen, *J. Am. Chem. Soc.* **73**, 3444 (1951).

110. W. Davies and E. W. Savige, *J. Chem. Soc.* 317 (1950).

111. N. P. Neureites and F. G. Bordwell, *J. Am. Chem. Soc.* **81**, 578 (1959).

112. P. E. Sonnet, *Tetrahedron,* **36**, 557 (1980).

113. L. Goodman and E. J. Reist in M. S. Kharasch and C. Y. Meyers (Eds.), *The Chemistry of Organic Sulfur Compounds*, Pergamon Press, Oxford, 1966, pp. 93.

114. G. K. Helmkamp and D. J. Pettitt, *J. Org. Chem.* **25**, 1754 (1960).

115. C. C. J. Culvenor, W. Davies and N. S. Heath, *J. Chem. Soc.* 282 (1949).

116. M. Kamata and T. Miyashi, *J. Chem. Soc. Chem. Commun.* 557 (1989).

117. A. Krantz and J. Laureni, *J. Am. Chem. Soc.* **103**, 486 (1981) and references cited therein.

118. G. Capozzi, V. Luccnini, G. Modena and P. Scrimin, *Tetrahedron Lett.* 911 (1977).

119. D. Clark, *Int. J. Sulfur Chem.* (C) **7**, 11 (1972).

120. G. R. Newkome and W. W. Paudler, *Contemporary Heterocyclic Chemistry*, Wiley-Interscience, New York, 1982.

121. J. C. Philips, J. V. Swisher, D. Haidukewych and I. Morales, *Chem. Commun.* 22 (1971).

122. E. Schmitz, *Adv. Heterocycl. Chem.* **2**, 83 (1963); E. Schmitz in A. R. Katritzky and C. W. Rees (Eds.), *Comprehensive Heterocyclic Chemistry* Vol. 7, Pergamon Press, Oxford, 1984, pp. 195.

123. H. W. Hine in A. Hassner (Ed.), *Small Ring Heterocycles* Part II, Wiley-Interscience, New York, 1983, pp. 547.

124. E. Schmitz, *Angew. Chem.* **71**, 127 (1959).

125. H. J. Abendroth and G. Henrich, *Angew. Chem.* **71**, 283 (1959).

126. E. Schmitz, R. Ohme and D. R. Schmidt, *Chem. Ber.* **95**, 2714 (1962).

127. M. Schulz and G. West, *J. Prakt. Chem.* **312**, 161 (1970).

128. T. Akiyama, T. Kitamura, T. Isida and M. Kaunisi, *Chem. Lett.* 185 (1974).

129. E. Schmitz and D. Habisch, *Chem. Ber.* **95**, 680 (1962).

130. W. H. Graham, *J. Org. Chem.* **30**, 2108 (1965).

131. W. H. Graham, *J. Am. Chem. Soc.* **84**, 1063 (1962).

132. E. Schmitz and R. Ohme, *Chem. Ber.* **94**, 2166 (1961).

133. S. K. Richardson and R. J. Ife, *J. Chem. Soc. Perkin Trans. 1*, 1172 (1989).

134. E. Schmitz, *Adv. Heterocycl. Chem.* **24**, 63 (1979); M. J. Haddadin and J. P. Freeman in A. Hassner (Ed.), *Small Ring Heterocycles*, Part III, Wiley-Interscience, New York, 1985, pp. 283; F. A. Davis and A. C. Sheppard, *Tetrahedron* **45**, 5703 (1989); F. A. Davis, A. Kumar and B. -C. Chen, *J. Org. Chem.* **56**, 1143 (1991); E. Schmitz, *Synthesis*, 327 (1991); A. Padwa and R. L. Chin in H. Suschitzky and E. F. V. Scriven (Eds.), Progress in *Heterocyclic Chemistry* Vol. **1**, Pergamon Press, Oxford, 1989, pp. 93; A. Padwa and F. R. Kinder, Vol. **2**, 1990, pp. 33; A. Padwa and S. S. Murphree, Vol. **4**, 1992, pp. 43; Vol. **5**, 1993, pp. 64.

135. D. R. Boyd, *Tetrahedron Lett.* 4561 (1968).

136. W. H. Pirkle and P. L. Rinaldi, *J. Org. Chem.* **42**, 2080 (1977).

137. E. Schmitz, R. Ohme and D. Murawski, *Chem. Ber.* **98**, 2516 (1965).

138. E. Schmitz, R. Ohme and S. Schramm, *Angew. Chem.* **76**, 2521 (1964).

139. E. Schmitz, R. Ohme and D. Murawski, *Angew. Chem.* **73**, 708 (1961).

140. J. Spitter and M. Calvin, *J. Org. Chem.* **23**, 651 (1958).

141. J. F. Garvey and J. A. Hashmall, *J. Org. Chem.* **43**, 2380 (1978).

142. W. D. Emmons. *J. Am. Chem. Soc.* **78**, 6208 (1956).

143. L. Horner and E. Jurgens, *Chem. Ber.* **90**, 2184 (1957).

144. A. R. Butler and B. C. Challis, *J. Chem. Soc.* (B) 778 (1971).

145. J. Bjrgo, D. R. Boyd, R. M. Campbell. and D. C. Neill, *Chem. Commun.* 162 (1976).

146. K. Suda, F. Hino and C. Yijima, *J. Org. Chem.* **51**, 4232 (1986).

147. L. Henn, D. M. B. Hickey, C. J. Moody and C. W. Rees, *J. Chem. Soc. Perkin Trans.* 1, 2189 (1984).

148. W. H. Rastetter, W. R. Wanger and M. A. Findeis, *J. Org. Chem.* **47**, 419 (1982).

149. H. Suginome and F. Yagihashi, *J. Chem. Soc. Perkin Trans.* 1, 2488 (1977).

150. M. E. Oliveros–Desherces, M. M. Riviere, J. Parello and A. Lattes, *Tetrahedron Lett.* 851 (1975).

151. F. A. Davis, N. F. Abdul–Malik, S. B. Awad and M. E. Harakal, *Tetrahedron Lett.* 917 (1981).

152. F. A. Davis, M. E. Harakal and S. B. Awad, *J. Am. Chem. Soc.* **105**, 3123 (1983).

153. R. D. Bach and G. J. Wolber, *J. Am. Chem. Soc.* **106**, 1410 (1984).

FOUR-MEMBERED HETEROCYCLES

CONTENTS

FOUR-MEMBERED HETEROCYCLES

Four-membered heterocycles are the heterocyclic analogs of cyclobutane and are considered to be derived by replacing a $-CH_2$ (methylene group) by a heteroatom (NH, O or S). The four-membered saturated heterocycles containing nitrogen, oxygen and sulfur are known as azetidines **1**, oxetanes **2** and thietanes **3**, respectively. Four-membered heterocyclic rings are less strained, and hence more

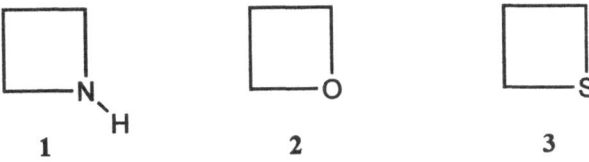

    1   **2**   **3**

stable than the three-membered rings and, therefore, the ring cleavage is less likely. Moreover, four-membered heterocycles are more difficult to synthesize by direct intramolecular cyclization than the three-membered heterocycles because ring forming ability falls off with the chain length. The molecules of oxetane **2** and thietane **3** are planar, but not square because of relatively larger size of the oxygen and sulfur atoms than the carbon atom. The planarity of these heterocycles **2** and **3**, as compared to cyclobutane which is puckered, has been attributed to the reduction in the number of non-bonded interactions between methylene groups.

1 FOUR-MEMBERED AZAHETEROCYCLES

1.1 Azetidines

1.1.1 General

The azetidine ring is much less strained than that of aziridine and, therefore, azetidine and its N-substituted derivatives[1,2] behave in many respects as secondary and tertiary amines. Azetidine is considerably stronger base (pK_a= 11.29) than aziridine (pK_a = 8.04). In IR-spectra, the asymmetric C–H strectching frequency decreases (3047 cm^{-1} to 2966 cm^{-1}) with increasing the ring size. The effect of angle strain on the barrier to nitrogen inversion in azetidines is much smaller than in aziridines. However, the effects of substituents on the nitrogen inversion are the same as in aziridines. N-Alkyl-, N-acyl- and A-arylazetidines undergo nitrogen inversion faster as compared in the azetidines containing electron-withdrawing substituents with an adjacent lone pair (N-halo, N-amino, N-alkoxy and N-nitroso). The halo substituents on the nitrogen atom drastically slow the inversion rate so that N-chloro-2-methylazetidine can be resolved into two diastereomers.

1.1.2 Synthesis

1.1.2.1 Intramolecular Cyclization

Azetidines are frequently synthesized by intramolecular cyclization of γ-haloalkyl-amines **4** in the presence of a base (scheme-1). The rate of cyclization is very low in the case of unsubstituted γ-haloalkylamines **4**, but the chain substitution facilitates cyclization (scheme-2)[3].

The intramolecular cyclization proceeds by nucleophilic displacement of a suitable leaving group in the γ-position of a three carbon chain by an amino group.

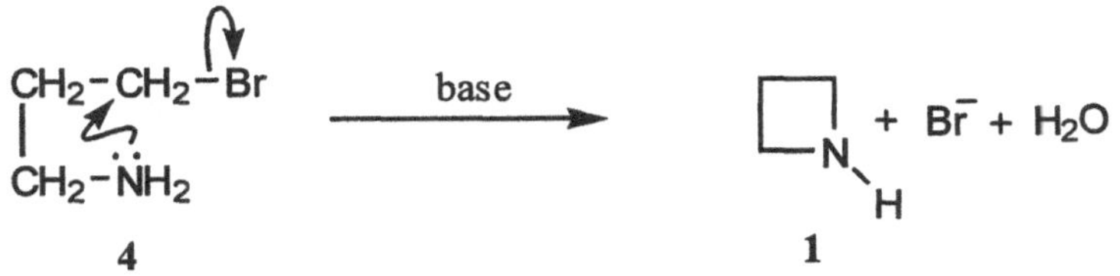

CH₂–CH₂–Br
|
CH₂–NH₂

4

base →

+ Br⁻ + H₂O

1

Scheme-1

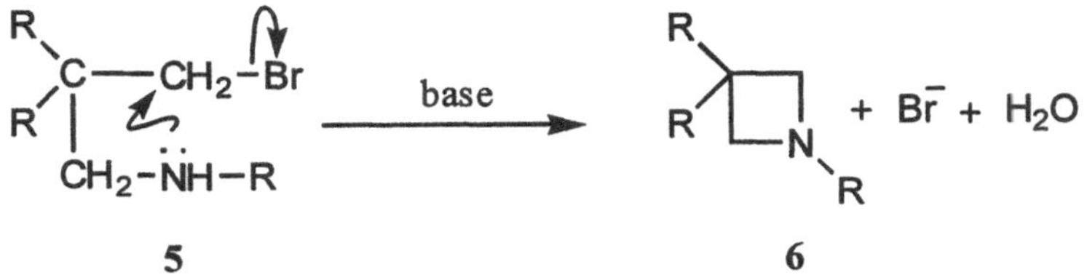

R
\
C––CH₂–Br
/
R |
CH₂–NH–R

5

base →

+ Br⁻ + H₂O

6

Scheme-2

1.1.2.2 Cycloaddition

Azetidine **1** is best prepared by the cycloaddition of trimethylene chlorobromide **7** with *p*-toluenesulfonamide, followed by the reduction with sodium and n-pentanol (scheme-3)[4,5].

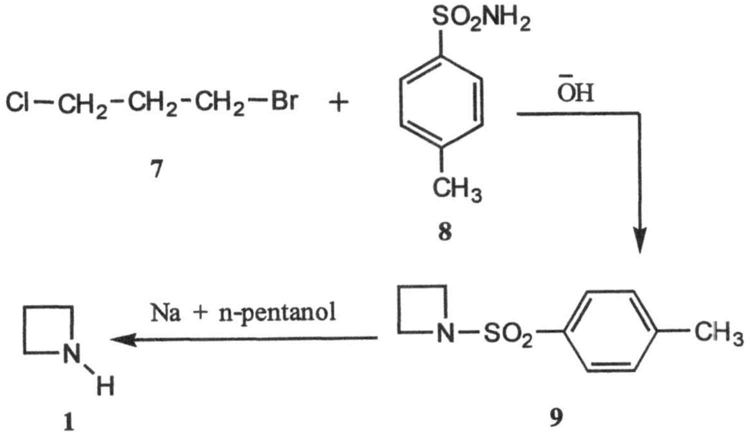

Cl–CH₂-CH₂-CH₂–Br +

SO₂NH₂

CH₃

7

8

⁻OH →

N–SO₂–⬡–CH₃

9

Na + n-pentanol ←

N–H

1

Scheme-3

1.1.2.3 From Isoxazolines

Isoxazolines **10** undergo ring opening reaction with the cleavage of N–O bond on treatment with sodium in *n*-pentanol and subsequently with tosyl chloride and pyridine providing tosylates **12** which on cyclization and reduction yield azetidines (scheme-4)[6].

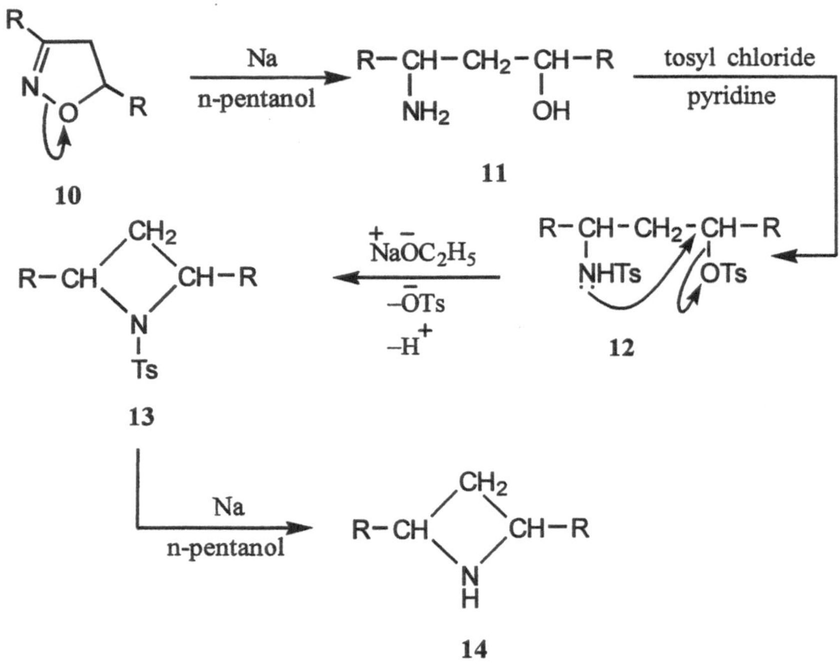

Scheme-4

1.1.2.4 From Aziridines

The reaction of aziridines **15** with sulfur ylides **16** proceeds with the transfer of a methylene group (> CHR or > CR$_2$) from the sulfur ylide and results in the formation of azetidines (scheme-5)[7].

1.1.2.5 From γ-Lactones

Following sequence of the reactions involving ring opening and ring closure in γ-lactones provides azetidines **24** (scheme-6)[8,9].

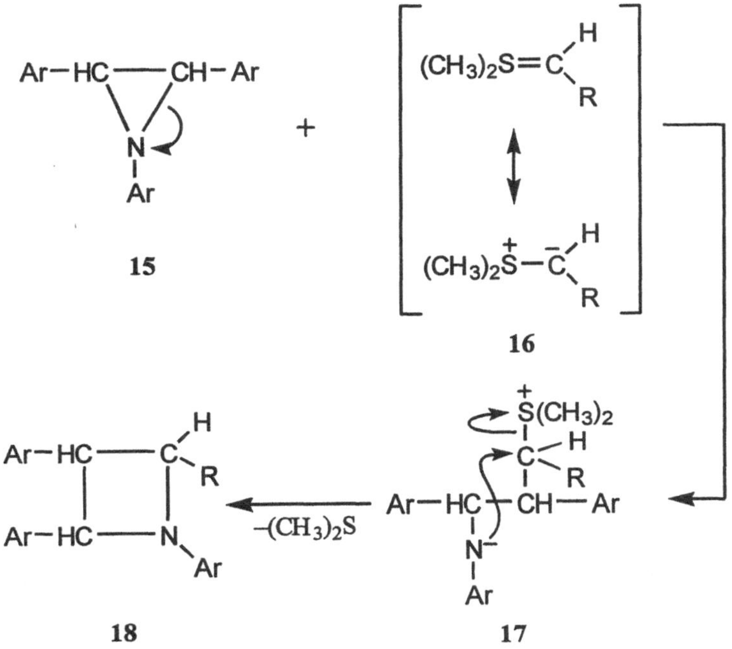

Scheme-5

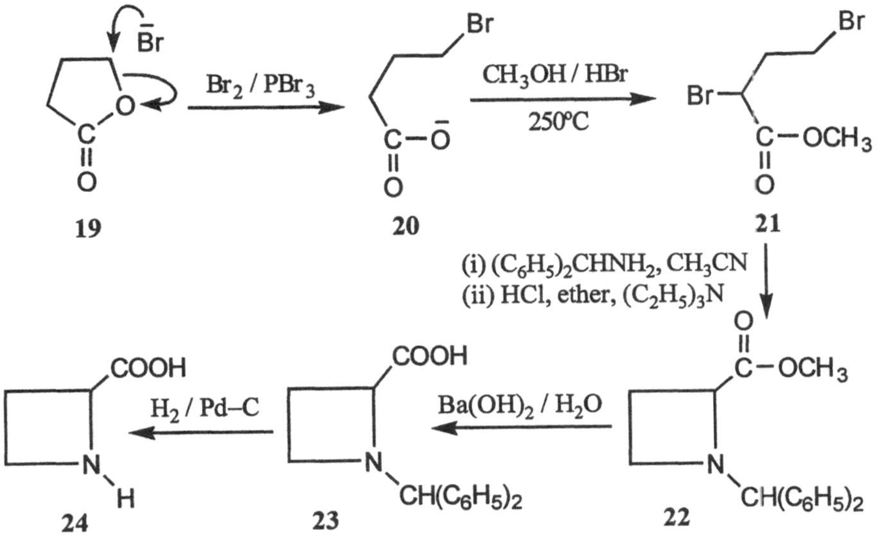

Scheme-6

1.1.3 Reactions

1.1.3.1 Ring Opening Reactions

Azetidines, being less strained than the three-membered aziridines, show lesser degree of reactivity and the cleavage of azetidine ring is relatively sluggish. Nucleophiles react with azetidines at very reduced rate as compared to aziridines. Azetidines are quite resistant to the action of bases and nucleophiles. However, hydrogen peroxide cleaves the ring with the formation of acrolein and ammonia (scheme-7).

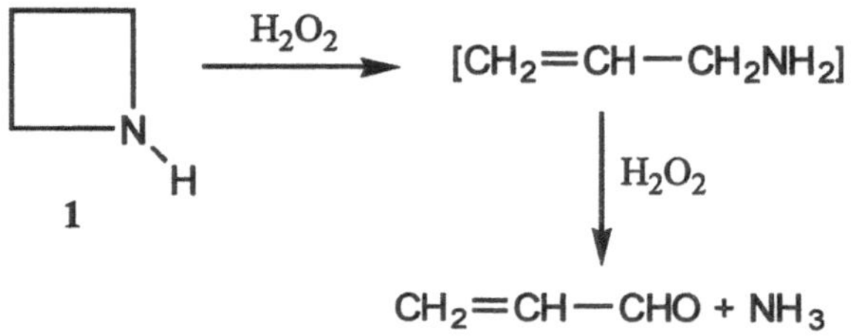

$$\text{azetidine} \; \mathbf{1} \xrightarrow{\text{H}_2\text{O}_2} [\text{CH}_2{=}\text{CH}{-}\text{CH}_2\text{NH}_2]$$

$$\downarrow \text{H}_2\text{O}_2$$

$$\text{CH}_2{=}\text{CH}{-}\text{CHO} + \text{NH}_3$$

Scheme-7

Azetidines are susceptible to attack by electrophiles and the ring cleavage is considerably accelerated in the acid catalyzed reactions. The reaction with hydrochloric acid provides 3-chloropropylamine hydrochloride[10] (scheme-8).

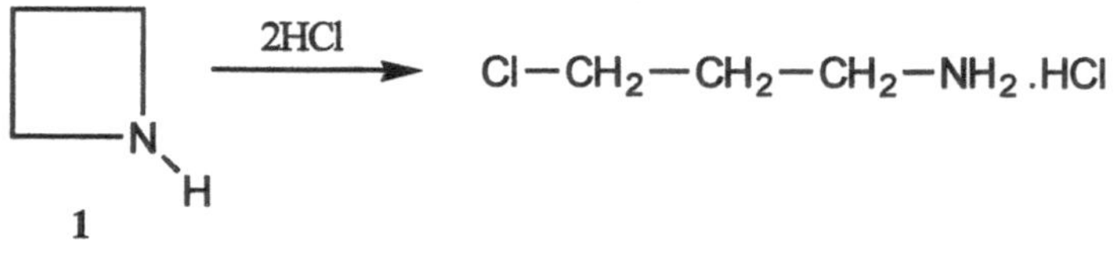

$$\text{azetidine} \; \mathbf{1} \xrightarrow{2\text{HCl}} \text{Cl}{-}\text{CH}_2{-}\text{CH}_2{-}\text{CH}_2{-}\text{NH}_2 \cdot \text{HCl}$$

Scheme-8

1.1.3.2 Functionalization at Nitrogen

Azetidine behaves like secondary aliphatic amines and undergoes similar reactions. The reaction with carbon disulfide forms a salt and with nitrous acid gives N-nitrosoazetidine. The reaction with formaldehyde provides N-hydroxymethyl-azetidine (scheme-9).

N-Alkylazetidine reacts similarly as tertiary amine and forms quaternary salt (scheme-10).

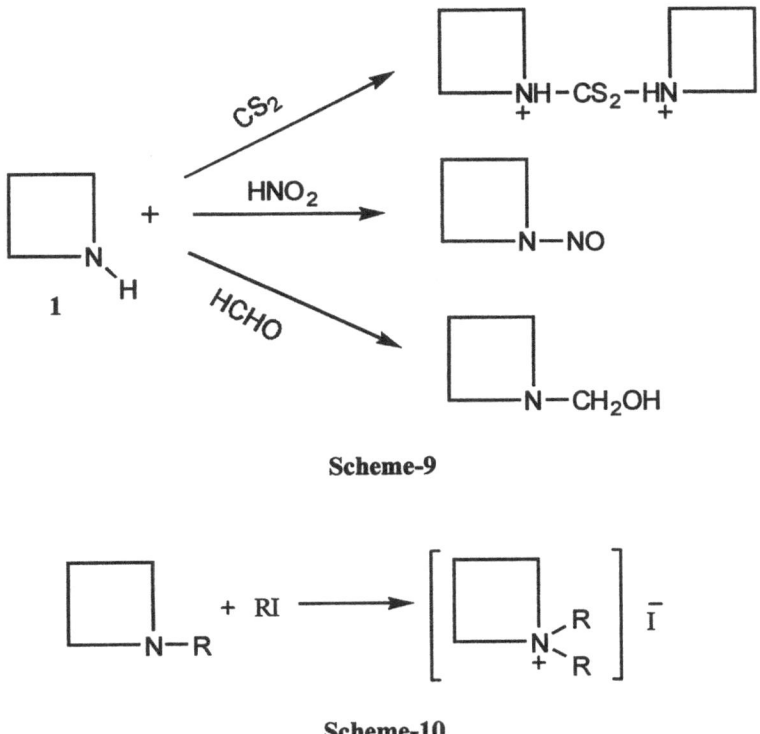

Scheme-9

Scheme-10

1.1.3.3 Rearrangements

Azetidines undergo rearrangement reactions similar to three-membered aziridines. Since less relief of strain is involved in the ring opening reaction of azetidine, more severe reaction conditions are required (scheme-11).

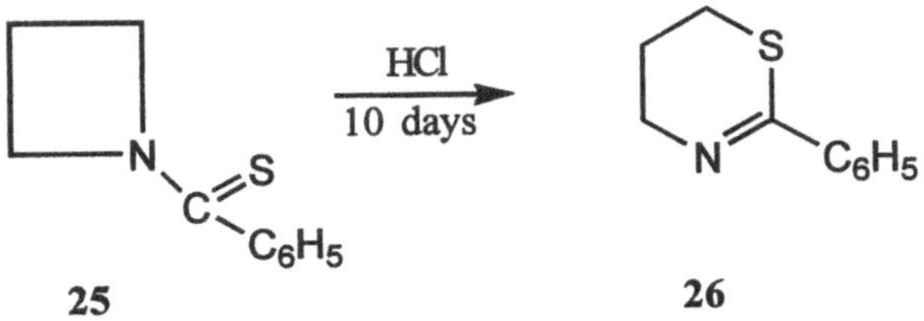

Scheme-11

Azetidine ketones **27** undergo photochemical rearrangement resulting in ring expansion to pyrroles **31** and **34** (scheme-12)[11]. The reaction proceeds to involve intramolecular hydrogen shift with the generation of 1,3-biradical intermediate **29** which undergoes ring closure to form bicyclic unstable compounds **30** and **33** which, in turn, finally lead to the formation of pyrroles **31** and **34**, respectively with the loss of water molecules.

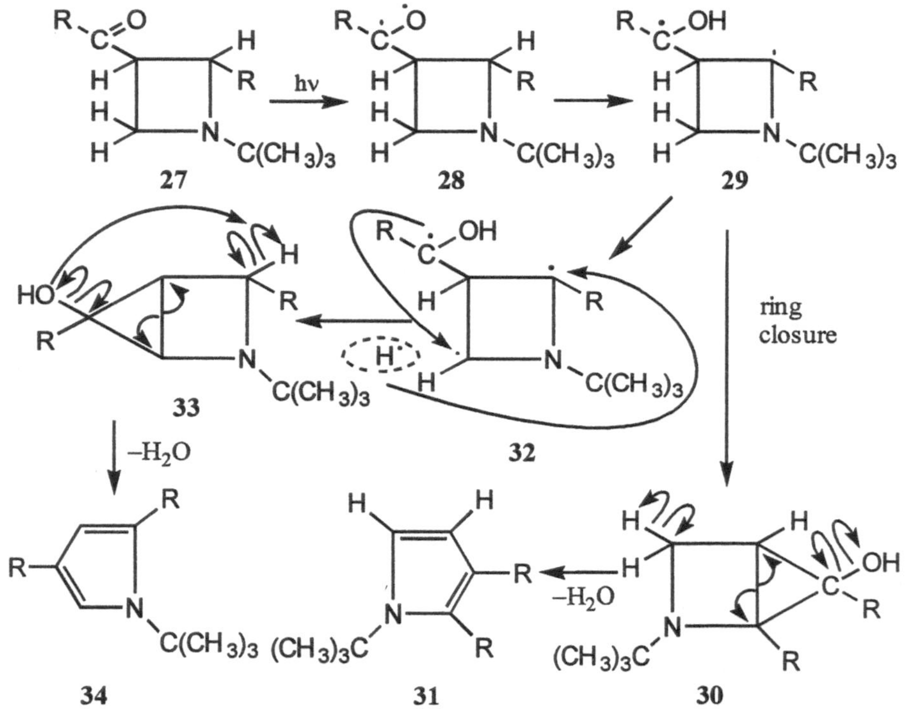

Scheme-12

1.2 Azetidinones (β-Lactams)

1.2.1 General

Azetidinones[13,14] are the carbonyl derivatives of azetidines containing carbonyl group at the position-2. These are also known as 2-azetidinones or more commonly β-lactams **35**. Azetidinone or β-lactam chemistry is of great importance because of the use of β-lactam derivatives as antibacterial agents[15,16].

35

The discovery of penicillin structure that contains β-lactam system led extensive investigations to obtain β-lactam antibiotics with a wider spectrum of activities and greater resistance to enzymatic cleavage. β-Lactam antibiotics contain two basic structural units; penam **36** and cepham **37** (both contain β-lactam unit), and include two powerful antibiotics; penicillins **38** and cephalosporins **39**.

36

37

38

39

Infrared absorption spectrum of monocyclic β-lactams (absorption at 1735–1765 cm⁻¹ as compared to that of unstrained amides at 1660 cm⁻¹) reflects the behaviour of carbonyl group as an ester linkage which accounts for its higher reactivity in the ring (making carbonyl carbon more electrophilic than in acyclic amides). In penicillin **38** and cephalosporin **39**, the fusion of β-lactam ring with heterocyclic ring has the effect of shifting the amide carbonyl absorption to 1770–1780 cm⁻¹ suggesting increased electrophilicity and, in turn, more reactivity of carbonyl group. The increased reactivity of carbonyl group which has been considered to be associated with the antibiotic activity in penicillin **38** and cephalosporin **39** is attributed to the fact that the ring fusion does not allow the amide nitrogen (bridgehead nitrogen) to achieve the planarity, since sp^2-hybridized nitrogen imposes greatly increased angle strain on the system.

1.2.2 Synthesis

1.2.2.1 Cyclization Reactions

1.2.2.1.1 Intramolecular Cyclization of β-Amino Acids

Intramolecular cyclization of β-amino acids in the presence of certain reagents including acyl chloride, phosphorus trichloride and thionyl chloride provides β-lactams (scheme-13)[17,20-25]. However, β-aminopropionic acids are not cyclized to

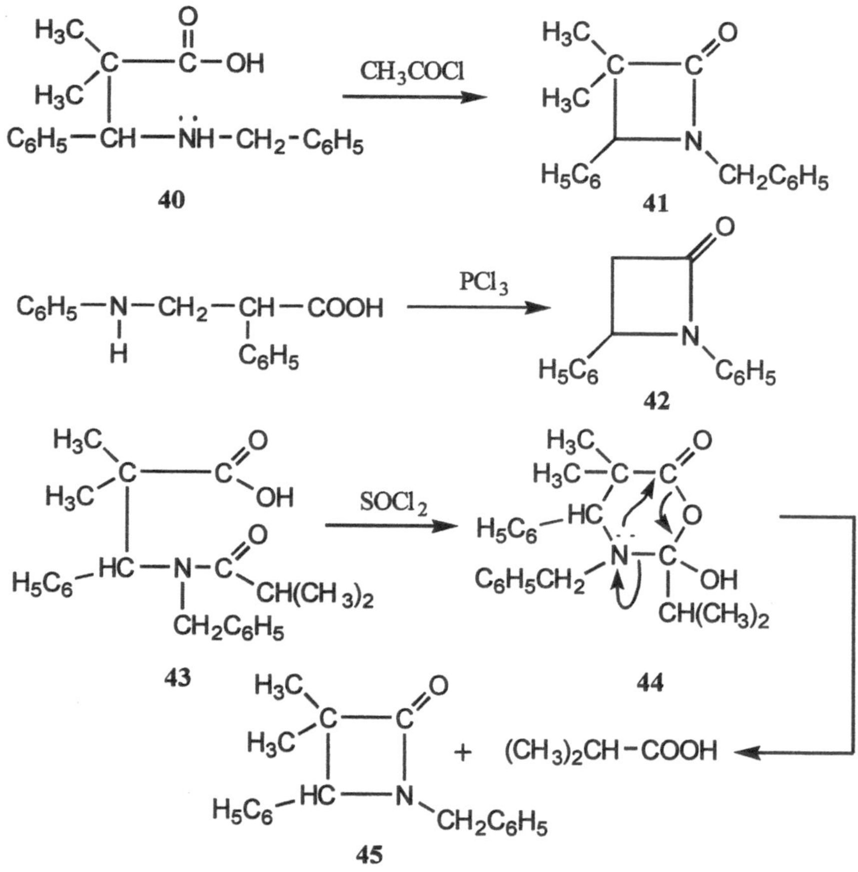

Scheme-13

β-lactams on heating, but undergo elimination reaction providing amines and acids (scheme-14).

$$C_6H_5-NH-CH_2-CH_2-COOH \xrightarrow{\Delta} C_6H_5-NH_2 + CH_2=CH-COOH$$

Scheme-14

1.2.2.1.2 Cyclization of Amino Esters

The reaction of β-amino esters with Grignard reagents **46** leads to the formation of azetidinones (β-lactams) **35** via the formation of N-anion **47** (scheme-15)[21-23]. Further reaction of mesityl magnesium bromide **46** with β-lactam at carbonyl site is prevented due to its steric nature.

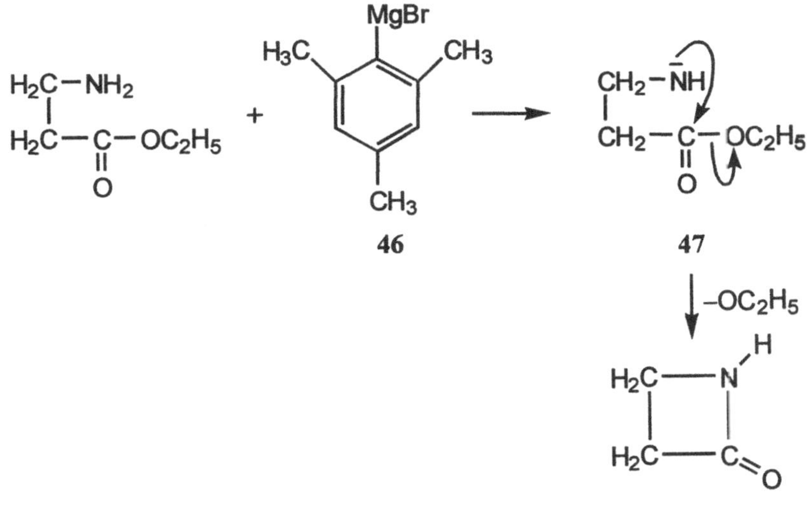

Scheme-15

N-Substituted diethylmalonates **48** undergo base catalyzed ring closure to yield β-lactams (scheme-16)[21-24].

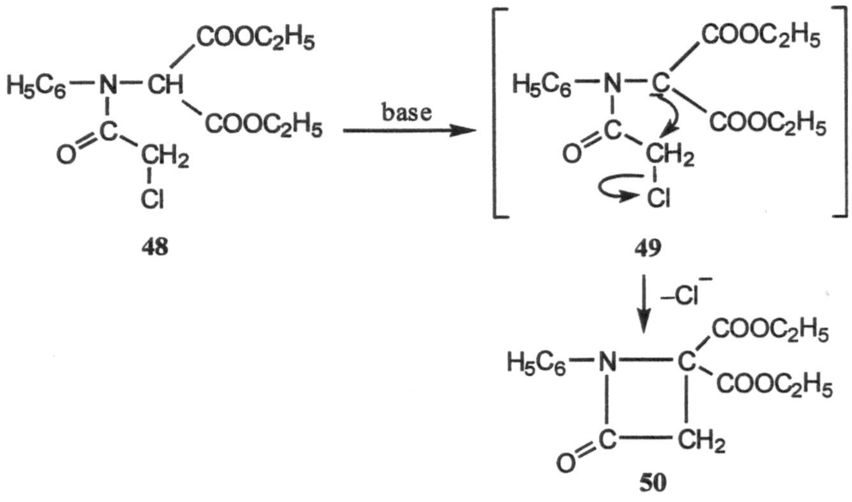

Scheme-16

1.2.2.1.3 Cyclization of β-Halo Amides

N-Substituted β-halo amides **51** are cyclized in the presence of a base to β-lactams **53** via an intermediate **52** (scheme-17)[26-28].

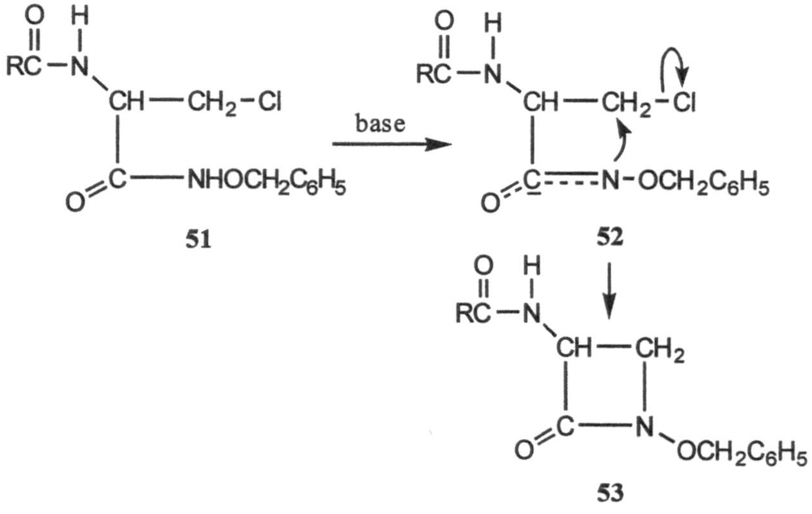

Scheme-17

1.2.2.1.4 Cyclization of β,γ-Unsaturated Hydroxamates

The bromine induced cyclization of O-acyl-β,γ-hydroxamates **54** provides β-lactams **56** via the formation of bromonium ion intermediate **55** (shceme-18). The presence of a phenyl group at the γ-position fails to provide β-lactams because the regioselectivity of opening of the bromonium ion intermediate **55** is reversed due to the formation of stabilized benzylic carbonium ion[29].

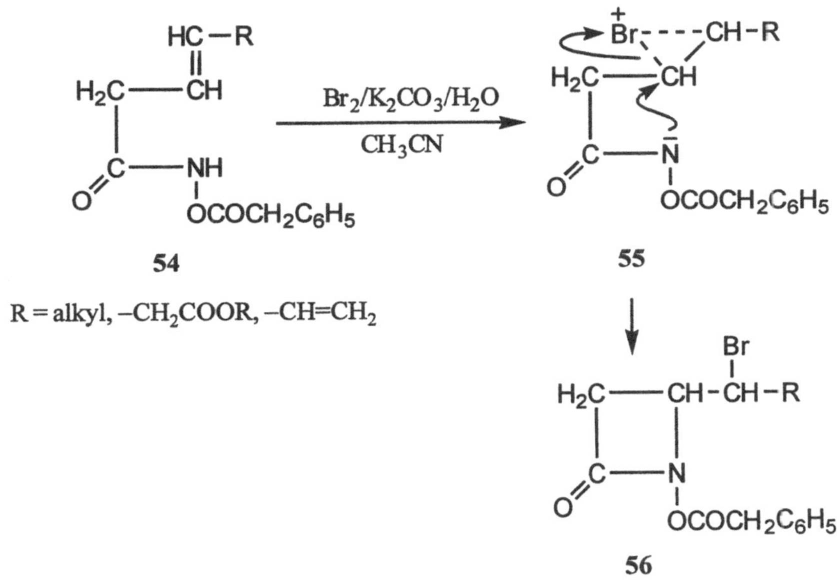

Scheme-18

1.2.2.2 Cycloaddition Reactions

1.2.2.2.1 Cycloaddition of Olefins to Isocyanates

The reaction of nucleophilic olefins with isocyanates provides β-lactams involving [2 + 2] cycloadditions. Chlorosulfonyl isocyanate reacts with olefins to form the corresponding chlorosulfonyl β-lactams **58**. The chlorosulfonyl group can be easily removed from the cycloadducts by hydrolysis to provide N-unsubstituted β-lactams **59** (scheme-19)[30]. The reaction proceeds with the formation of dipolar intermediate **57** involving electrophilic attack at the olefinic site by an isocyanate group. The resulting intermediate **57** readily collapses to yield β-lactam **59**.

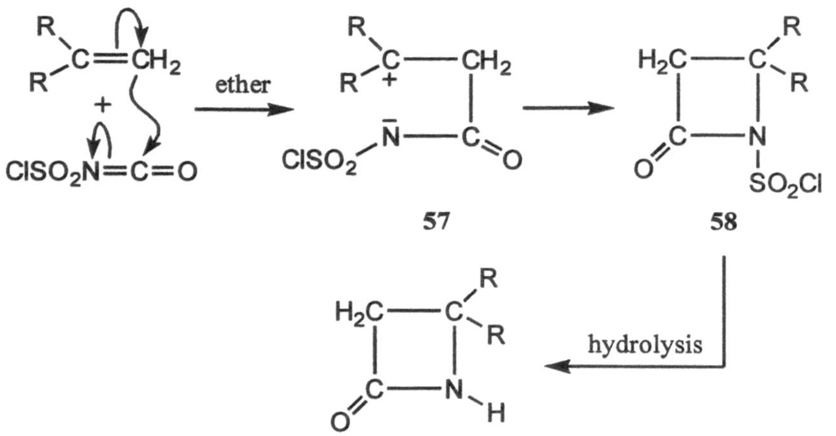

Scheme-19

1.2.2.2.2 Cycloaddition of Imines with Ketenes

The cycloaddition of ketoketenes **60** with imines **61** also results in the formation of β-lactams **62** (scheme-20)[31-35].

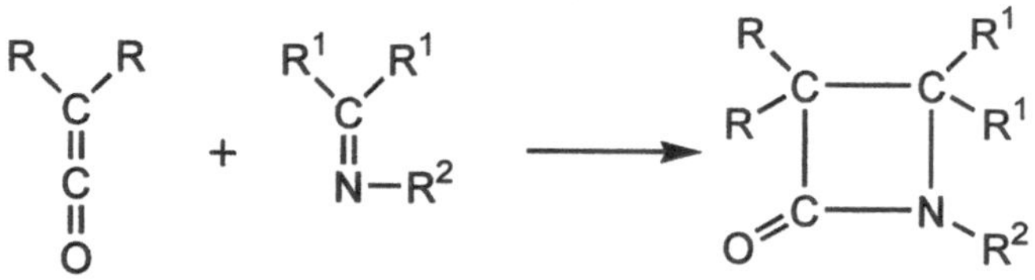

Scheme-20

1.2.2.2.3 Cycloaddition of Imines with Acid Chlorides

The reaction of acid chlorides with imines in the presence of a base provides β-lactams (scheme-21)[34-40].

The reaction mechanism involves direct acylation of imine with acid chloride giving N-acylium chloride **65** which is in equilibrium with chloramide **66**. The reaction of **65** or **66** with a base gives β-lactams **67**. However, an alternative mechanism has also been proposed involving prior formation of a ketene **68** by the reaction of acid chloride **63** with a base and subsequent cycloaddition with imine via zwitterionic intermediate **69** (scheme-22)[40].

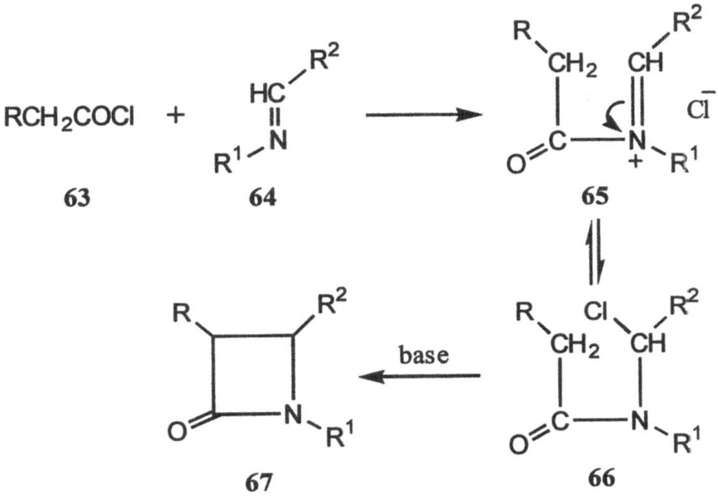

Scheme-21

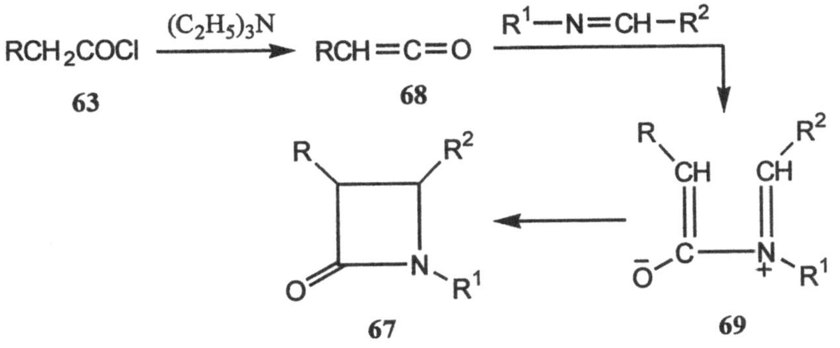

Scheme-22

2-Aza-1,3-dienes **70**, obtained by the reaction of aldehydes with allylamine, also undergo [2 + 2] cycloaddition with acid chlorides yielding N-1-propenyl-β-lactams **71**. N-Propenyl group can be removed by ozonolysis followed by the oxidation of resulting N-formyl lactam **72** with potassium permanganate. It provides a convenient route for the preparation of N-unsubstituted β-lactams **73** (scheme-23)[41].

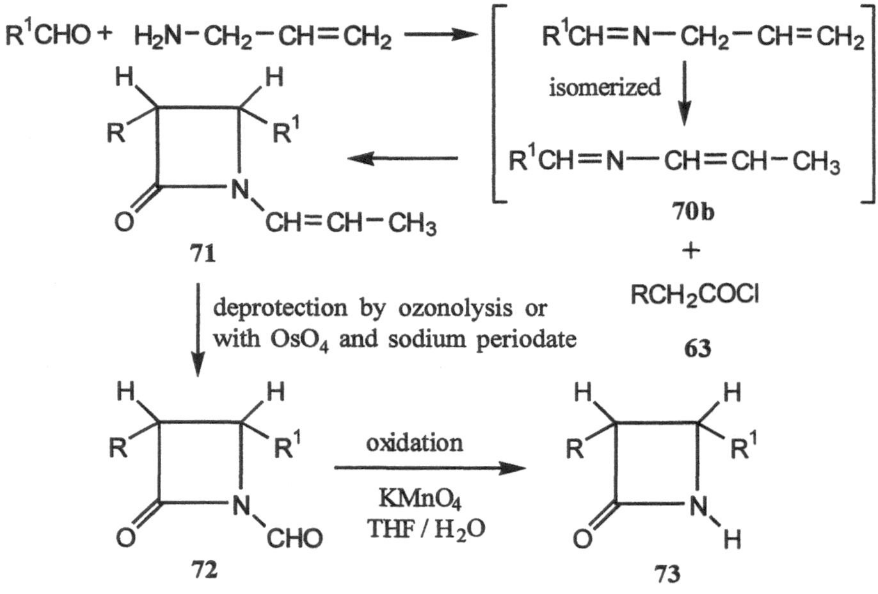

Scheme-23

1.2.2.2.4 Cycloaddition of Imines with α-Amino Esters

The reaction of α-amino esters **74** with imines in the presence of zinc chloride affords stereoselective β-lactams **76** involving the formation of zinc enolates **75** (scheme-24)[42,43]. The nature of the solvent and the substituents present on an imine affect the reaction stereochemically.

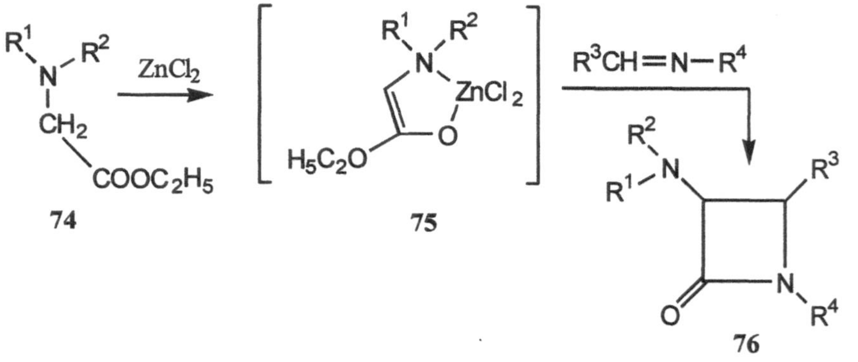

Scheme-24

1.2.2.3 Ring Expansion Reactions

Rhodium (I) catalyzed carbonylation of aziridines results in the ring expansion to β-lactams with the insertion of CO into the more substituted C–N bond (scheme-25)[44-47]. The process is stereospecific and enantiospecific and proceeds with the retention of configuration. However, the ring expansion using nickel carbonyl occurs with the insertion of CO into the less substituted C–N bond[48].

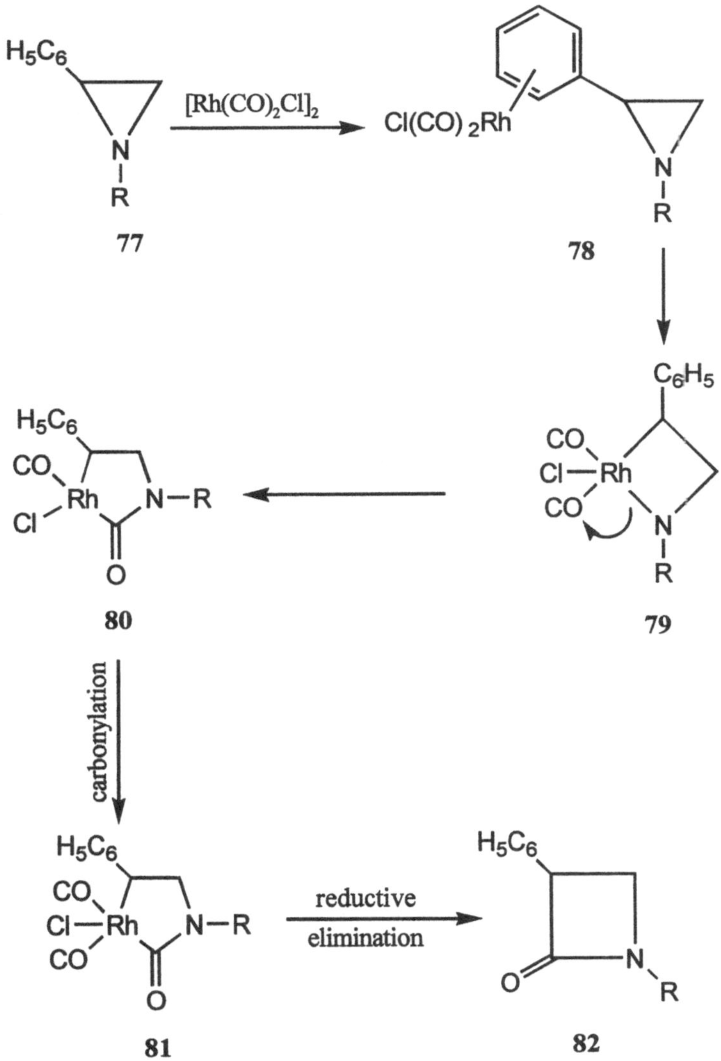

Scheme-25

Cyclopropanes also undergo ring expansion reaction with the formation of β-lactams **86** (scheme-26).

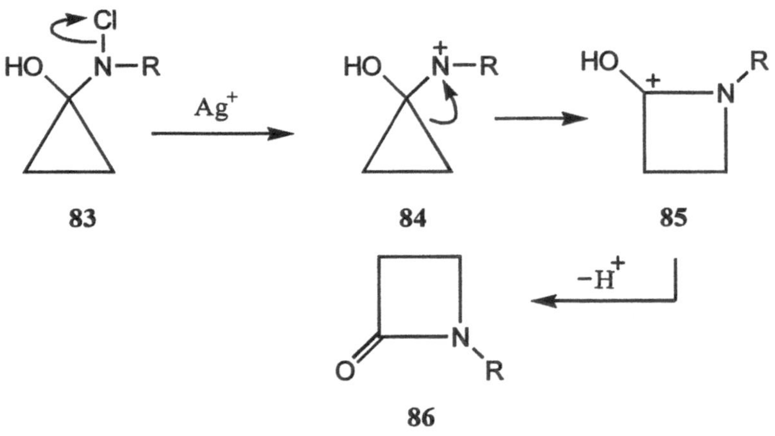

Scheme-26

1.2.2.4 Methylene Insertion Reactions

The photolysis of N,N-disubstituted diazoacetamides **87** proceeds with the insertion of methylene group in the carbon–carbon bond affording β-lactams **89** (scheme-27)[49,50].

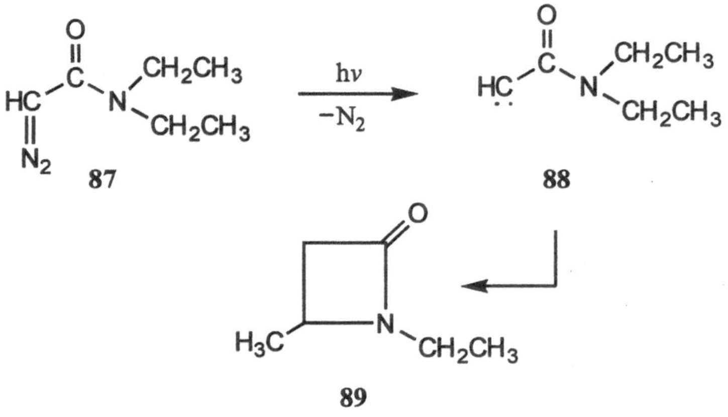

Scheme-27

Rhodium (II) acetate catalyzed decomposition of diazoacetoactamides **90** with bulky N-substituents also involves methylene insertion in the carbon–carbon bond providing β-lactams (scheme-28)[51].

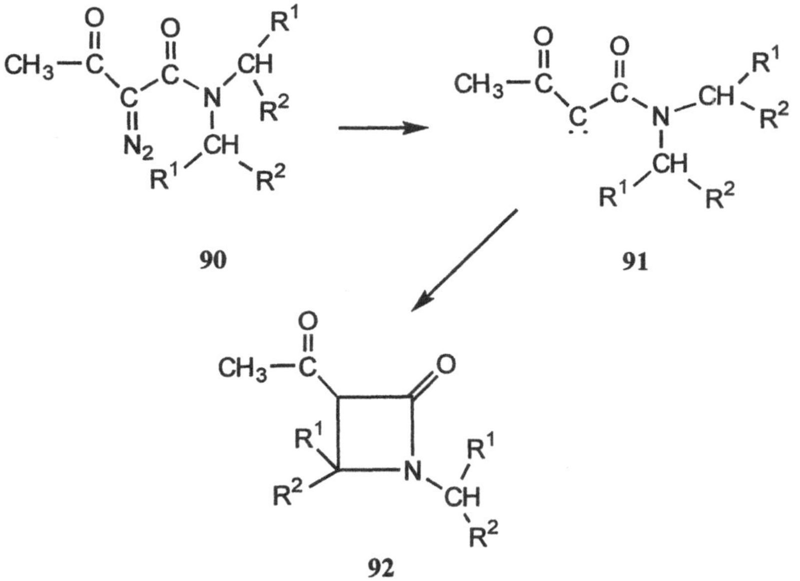

Scheme-28

β-Lactams **94** are also obtained by the photolysis of chromiumcarbene complexes **93** in the presence of imines involving methylene group insertion (scheme-29)[52].

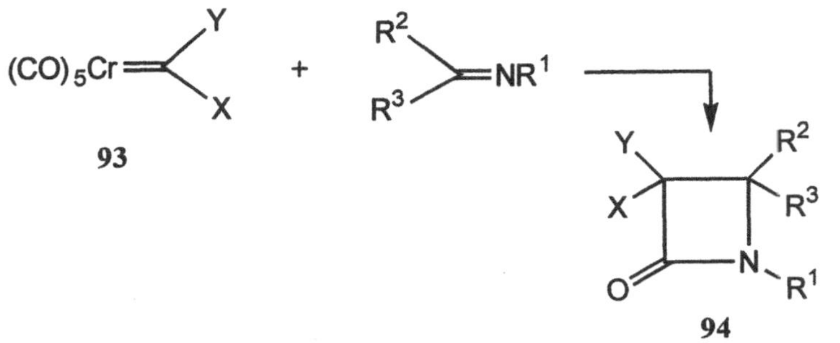

Scheme-29

1.2.2.5 From Substituted Azetidines

The reaction of N-substituted azetidine-2-carboxylic acids **95** with lithium diisopropylamide results in the formation of β-lactams involving oxidative decarboxylation of the resulting dicarbanion intermediate **96** (scheme-30)[53].

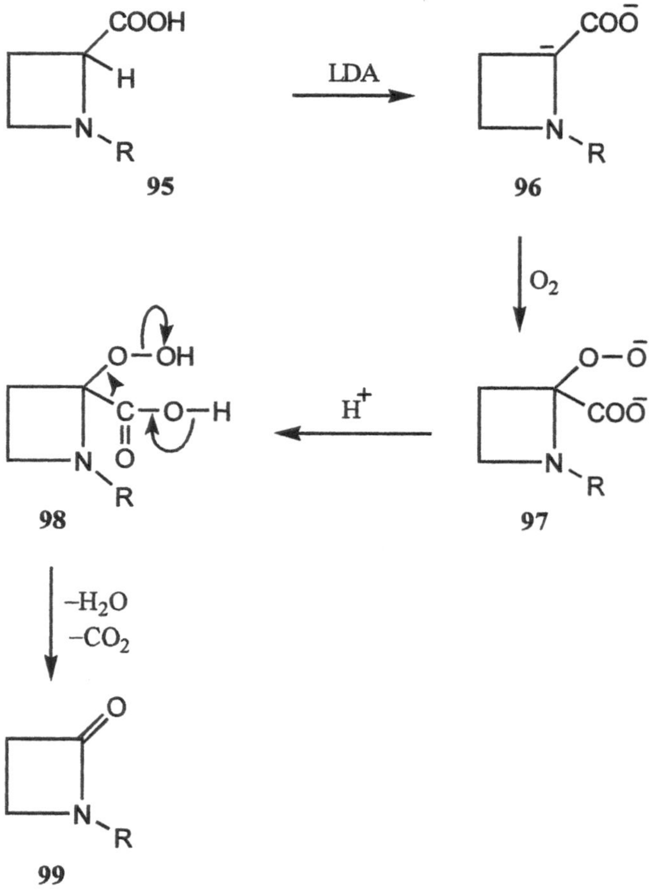

Scheme-30

Alternatively, decarboxylation of azetidine-2-carboxylic acids with oxalyl chloride followed by peroxidation with *m*-chloroperbenzoic acid provides β-lactams **99** (scheme-31).

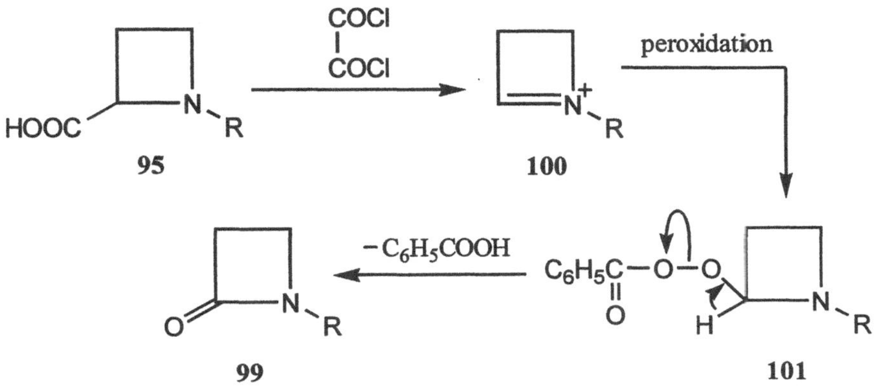

Scheme-31

1.2.3 Reactions

1.2.3.1 Ring Opening Reactions

β-Lactams are considerably more reactive than the acyclic amides and the larger ring lactams because of the ring strain. The additional ring strain in the resonating form **102a** compared to in the form **102b** makes the carbonyl group of β-lactam more like an ester carbonyl group. The resonating form **102a** makes the β-lactam carbonyl more electrophilic than an acyclic amide and thus carbonyl group in β-

lactams is more susceptible towards the nucleophilic attack. The ring strain after the addition of nucleophile facilitates the cleavage of ring (acyl–nitrogen bond fission). β-Lactams undergo ring opening reaction in the presence of a base providing β-amino acids (scheme-32). Similarly, the reaction of β-lactams with amines involves the nucleophilic attack at the carbonyl carbon and proceeds with the cleavage of ring providing β-amino amides **103** (scheme-33).

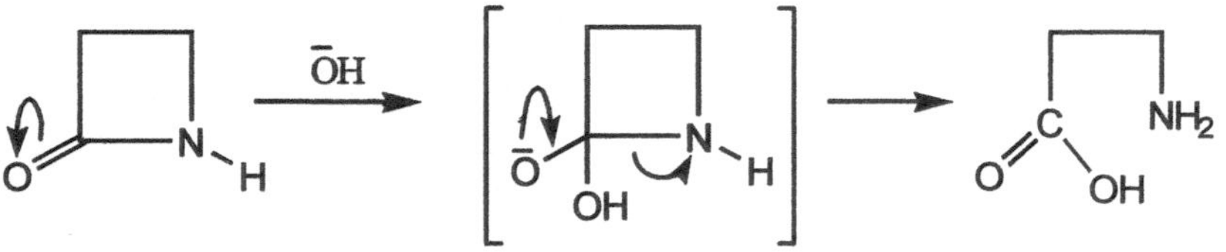

Scheme-32

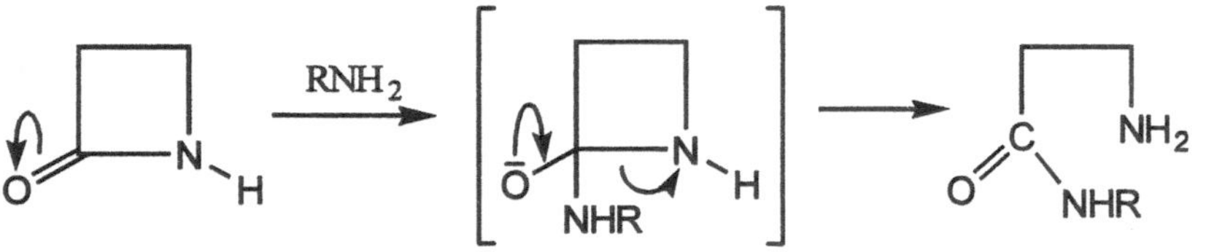

103

Scheme-33

The acylating property of β-lactams is believed to be related to the antibacterial function of penicillins. The reaction of alcoholic hydrochloric acid with β-lactams also causes ring cleavage providing β-amino esters **104** (scheme-34).

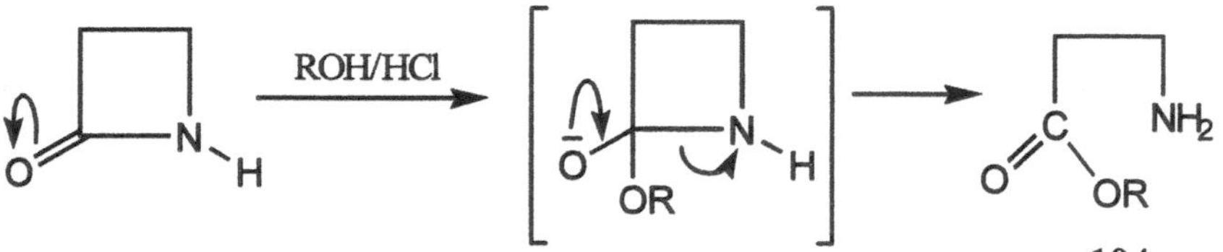

104

Scheme-34

The reduction of β-lactams with lithium aluminium hydride results in the cleavage of ring yielding γ-amino alcohols **105** (scheme-35).

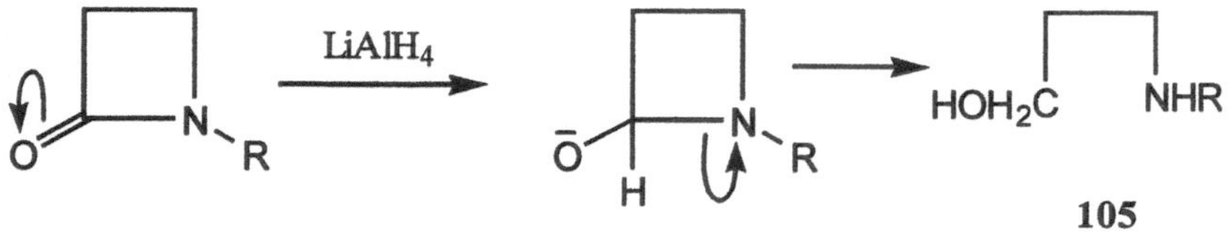

105

Scheme-35

1.2.3.2 Functionalization at Nitrogen

The nitrogen atom in N-unsubstituted β-lactams can be substituted by electrophiles in the absence of nucleophiles which can cause ring cleavage (scheme-36).

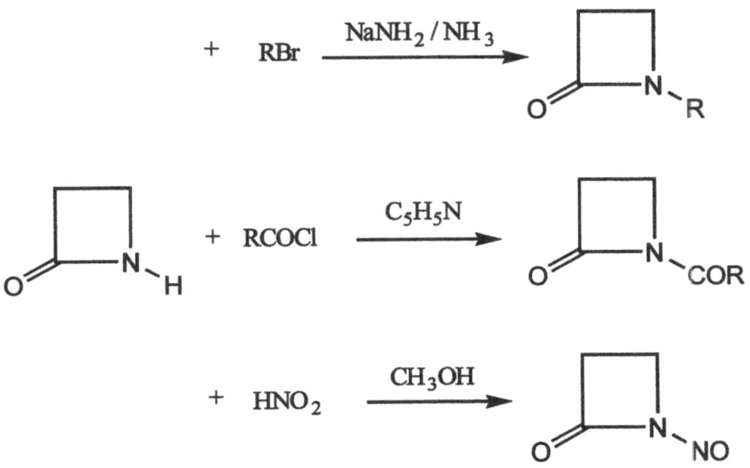

Scheme-36

1.2.3.3 Reaction with Phosphorus Pentasulfide

The reaction of β-lactams with phosphorus pentasulfide results in the formation of thiolactams **107** (scheme-37)[54].

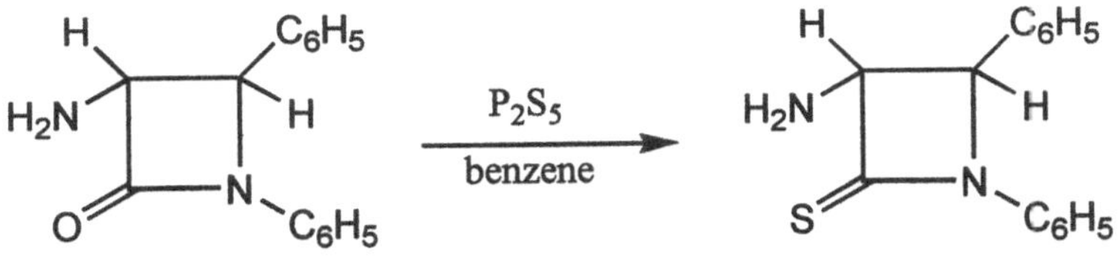

Scheme-37

1.2.3.4 Photochemical Reactions

Photolysis of N-phenyl-β-lactam **108** proceeds with the cleavage of C–N bond of
the β-lactam ring via free radical mechanism (scheme-38)[55].

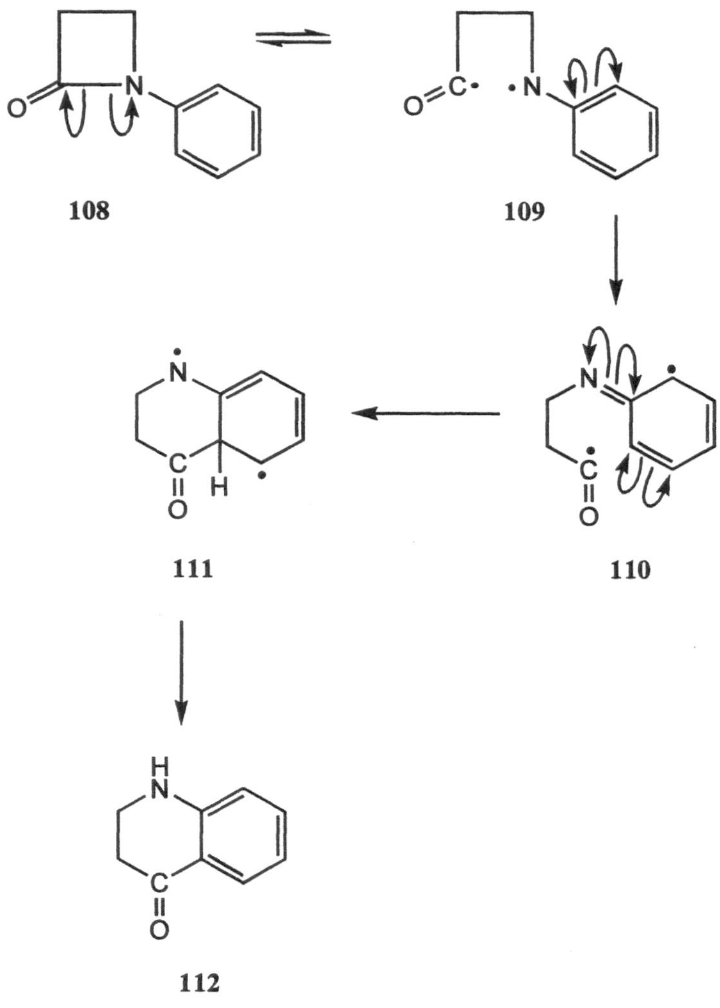

Scheme-38

2 FOUR-MEMBERED OXAHETEROCYCLES

2.1 Oxetanes

2.1.1 General

Four-membered saturated heterocycles containing oxygen as heteroatom are known as oxetanes **2**[56]. These are also named as oxacyclobutanes assigning position-1 to the oxygen atom.

Oxetane ring is planar, contrary to the puckered cyclobutane ring, because the replacement of the methylene group by a divalent oxygen atom causes the reduction in the number of non-bonding interactions between the neighbouring hydrogen atoms.

2

The C–C bond length (1.54Å) is greater than the C–O bond length (1.46Å), thus indicating the molecule not to be perfect square. The higher calculated value of the dipole moment of oxetane (2.01D) than that of ether (1.22D) or dimethyl ether (1.31D) indicates more electron density at the oxygen atom in oxetane than in acyclic aliphatic ethers. Oxygen atom in oxetane can donate electrons much more easily than in three-membered oxirane.

2.1.2 Synthesis

2.1.2.1 Intramolecular Cyclization

Oxetanes are prepared by the intramolecular cyclization of 1,3-halohydrins **113** in the presence of a base (scheme-39)[57,58].

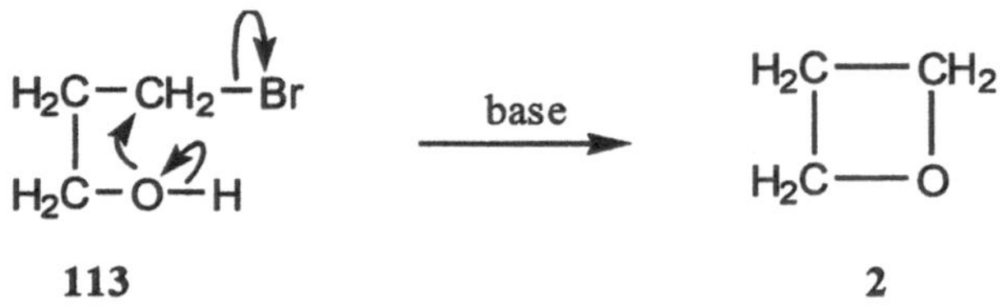

$$\text{H}_2\text{C}-\text{CH}_2-\text{Br} \quad \xrightarrow{\text{base}} \quad \begin{array}{c} \text{H}_2\text{C}-\text{CH}_2 \\ | \qquad | \\ \text{H}_2\text{C}-\text{O} \end{array}$$
$$\text{H}_2\text{C}-\text{O}-\text{H}$$

113   2

Scheme-39

The rate of cyclization is very low, but prior acetylation of 1,3-halohydrin accelerates intramolecular cyclization. The acetyl group is supposed to limit the concentration of alkoxide ions and favours cyclization rather than intramolecular polymerization (scheme-40).

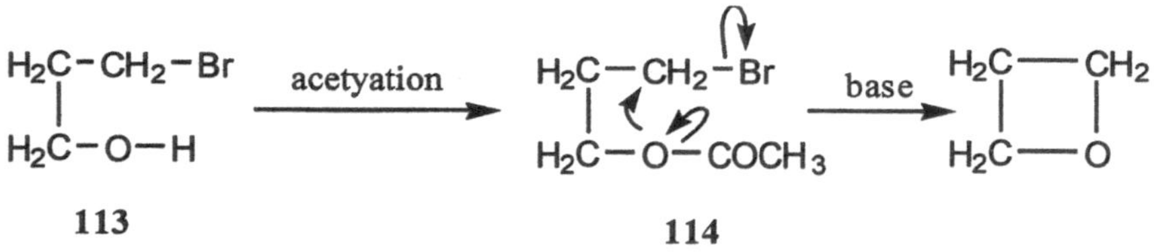

$$\begin{array}{c} \text{H}_2\text{C}-\text{CH}_2-\text{Br} \\ | \\ \text{H}_2\text{C}-\text{O}-\text{H} \end{array} \xrightarrow{\text{acetyation}} \begin{array}{c} \text{H}_2\text{C}-\text{CH}_2-\text{Br} \\ | \\ \text{H}_2\text{C}-\text{O}-\text{COCH}_3 \end{array} \xrightarrow{\text{base}} \begin{array}{c} \text{H}_2\text{C}-\text{CH}_2 \\ | \qquad | \\ \text{H}_2\text{C}-\text{O} \end{array}$$

113   114

Scheme-40

HO   CH$_2$–CH$_2$Cl   CH$_2$
                        $\xrightarrow[\text{ether}]{\text{KOH}}$ O   CH$_2$

115   116

Cl   CH$_2$–CH$_2$–OH   CH$_2$
                        $\xrightarrow[\text{H}_2\text{O}]{\text{KOH}}$ X   O   CH$_2$

117   116
                                        (not possible)

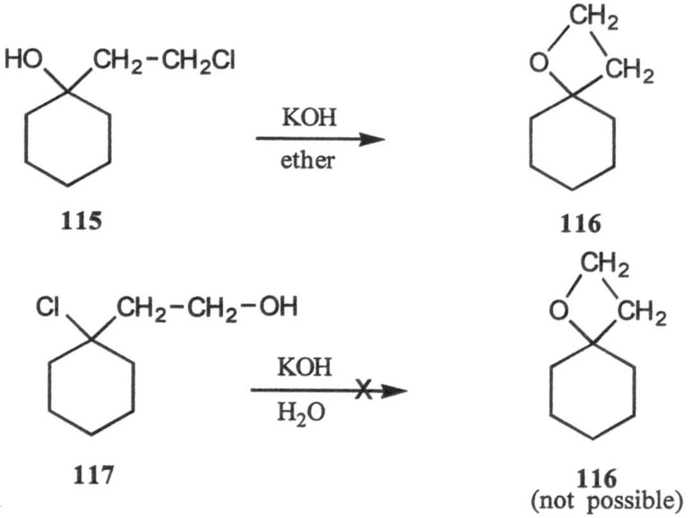

Scheme-41

The rate of cyclization is markedly influenced by the presence of substituents and their positions of the attachment. The presence of an alkyl group at the carbon atom bearing –OH group enhances the rate of cyclization, while an alkyl substituent at carbon atom bearing halogen atom retards the rate of cyclization and causes difficulty in the formation of oxetane (scheme-41) as the elimination predominates over cyclization[59,60]. Trimethylamino- or tosylate group can also be used as leaving group in place of halogen in the intramolecular cyclization reactions (scheme-42).

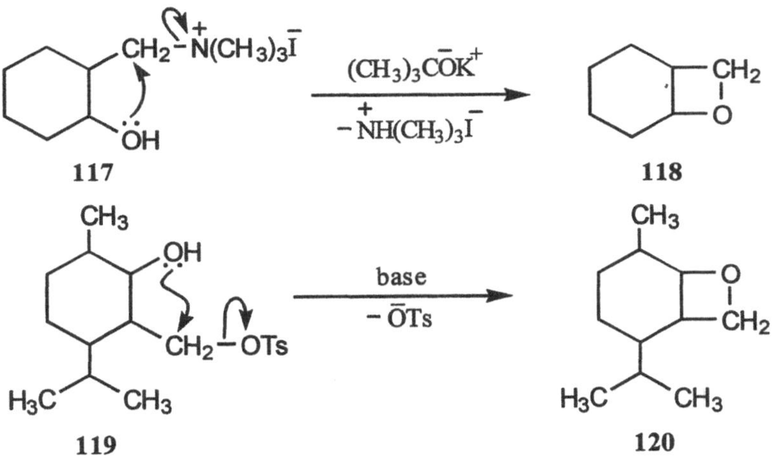

Scheme-42

2.1.2.2 Methylene Insertion Reaction

The reaction of oxiranes **121** with sulfur ylides **122** provides oxetanes **123** involving methylene insertion from the less hindered side (scheme-43)[61].

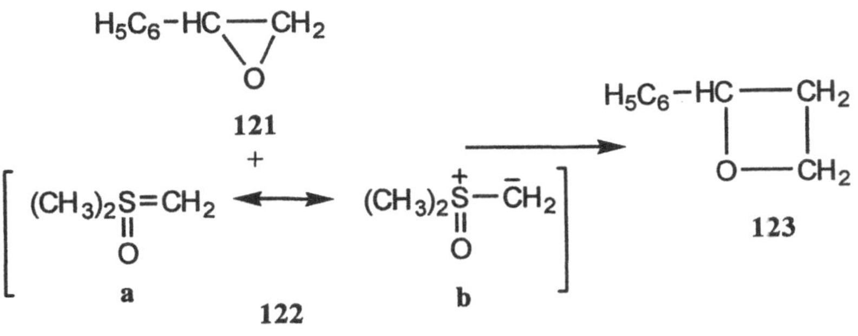

Scheme-43

2.1.2.3 Photochemical Cycloaddition

Photochemical [2 + 2] cycloaddition of carbonyl compounds to olefins gives oxetanes. This reaction is known as Paterno–Büchi reaction (scheme-44)[62-66].

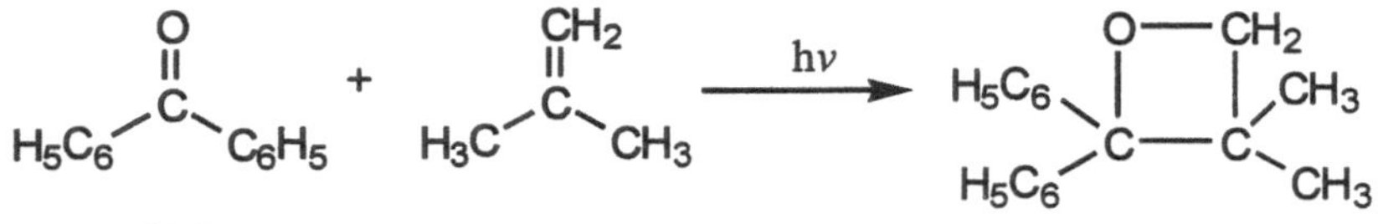

Scheme-44

The mechanism of this reaction involves the addition of an excited state of the carbonyl compound (the singlet or triplet state formed by $n{\rightarrow}\pi^*$ transition) to olefins[67].

The stereochemistry of this photocycloaddition has been proposed to involve the formation of a diradical intermediate by the interaction of electron-deficient carbonyl lone pair orbital with the π-orbitals of electron-rich olefin. The product **126** formed by subsequent cyclization of more stable biradical predominates (scheme-45).

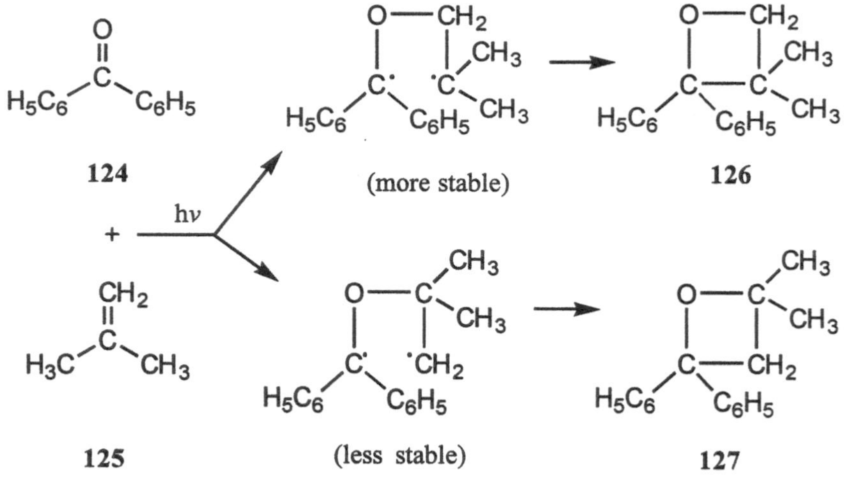

Scheme-45

2.1.3 Reactions

The oxetane ring is less strained than the three-membered oxirane ring. However, its chemical properties are similar to those of oxirane in many respects.

2.1.3.1 Ring Opening Reactions

2.1.3.1.1 Electrophilic Ring Opening Reactions

Oxetanes are susceptible to the attack by electrophiles. Although the structure and hybridizations of the oxetane and oxirane rings are fundamentally different, the reactivity of two ring systems is identical under acidic conditions. The lesser degree of strain in oxetane as compared in oxirane can be compensated for the greater electron donor capability of the oxygen atom in oxetane.

Oxetane readily undergoes ring opening reaction under acidic conditions (scheme-46). But in unsymmetrically substituted oxetane, the direction of ring cleavage depends on the 'push-pull' mechanism and generally two products are formed (scheme-47).

The formation of two products can be explained by the formation of oxonium ion as an intermediate which is attacked both at C-2 and C-4, preferably at the least hindered α-carbon atom (scheme-48)[68-71].

The oxonium ion intermediate does not undergo ring opening to form a stable carbonium ion but directs the attack preferably at the least hindered carbon atom. The presence of carbonium ion stabilizing groups can override this effect and the reaction proceeds to form a single product involving pull-mechanism (scheme-49).

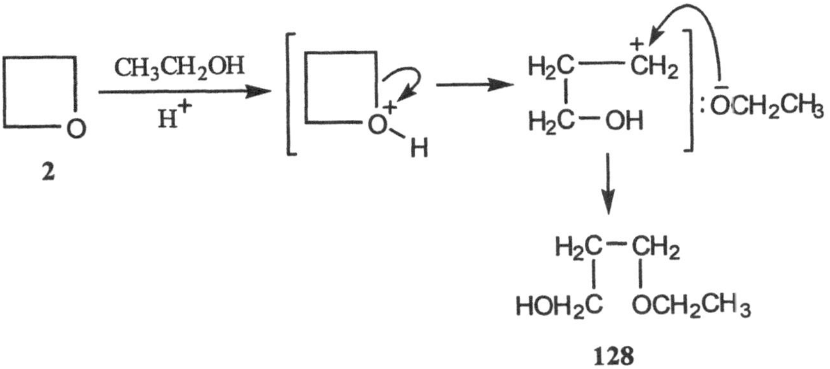

Scheme-46

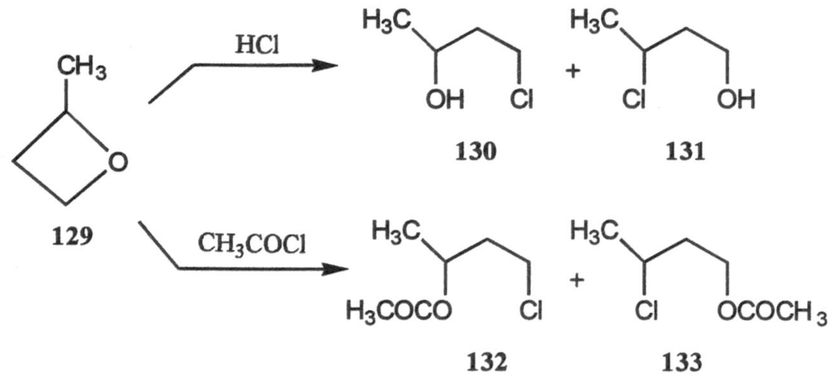

Scheme-47

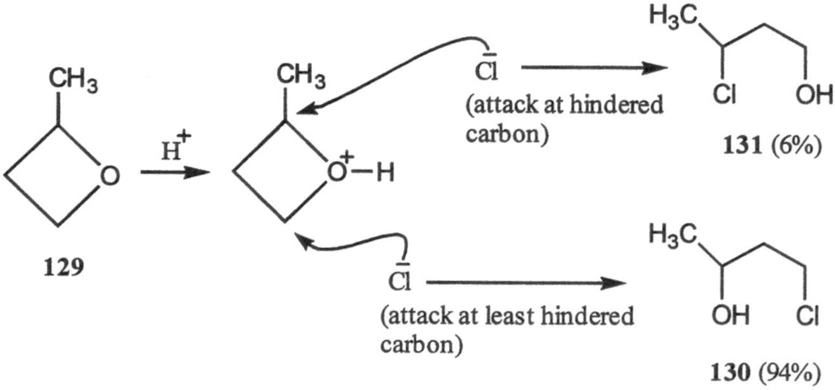

Scheme-48

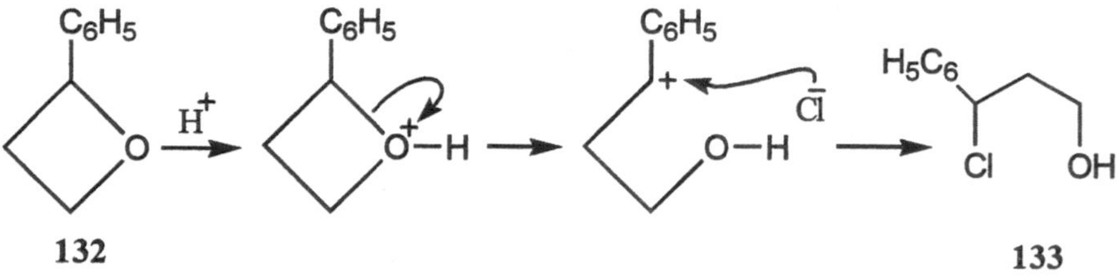

Scheme-49

The formation of single products in Friedel–Crafts reaction of 2-methyloxetane and 2-phenyloxetane is due to the generation of stabilized carbonium ions (scheme-50).

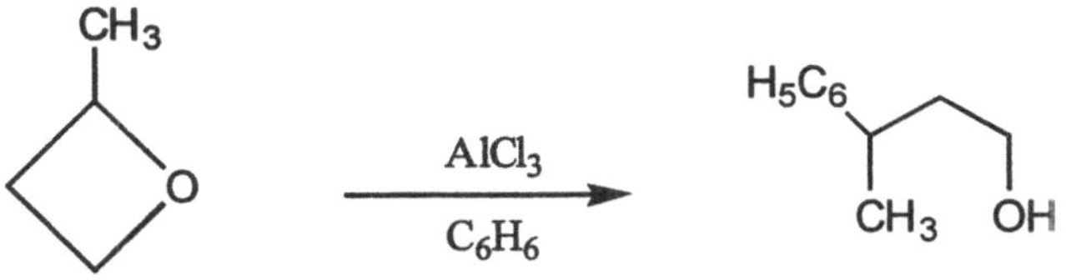

129 **134**

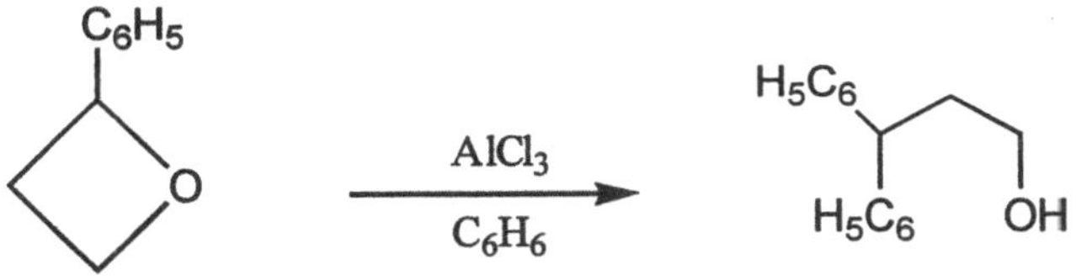

132 **135**

Scheme-50

Oxetane also reacts with carbon dioxide in the presence of tetraphenylantimony iodide providing trimethylene carbonate (scheme-51).

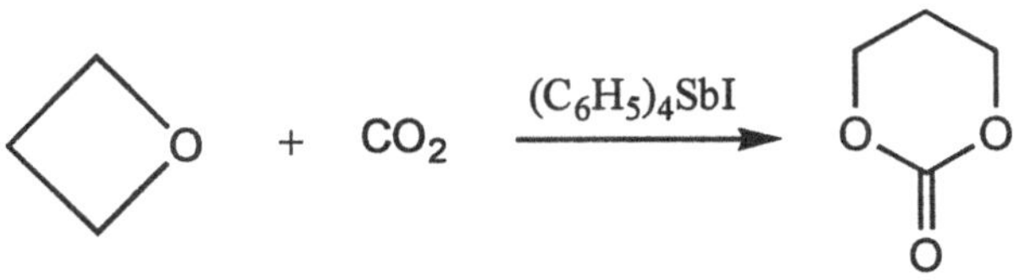

2 **136**

Scheme-51

2.1.3.1.2 Nucleophilic Ring Opening Reactions

Four-membered heterocycles react with nucleophiles at considerably reduced rates as compared to three-membered analogs and, therefore, the cleavage of oxetane ring requires generally more vigorous conditions than those for the ring opening of oxirane. There is less contribution of the leaving group in the ring cleavage reactions which require strong nucleophilic reagents and the reaction occurs by SN^2 or push mechanism (scheme-52).

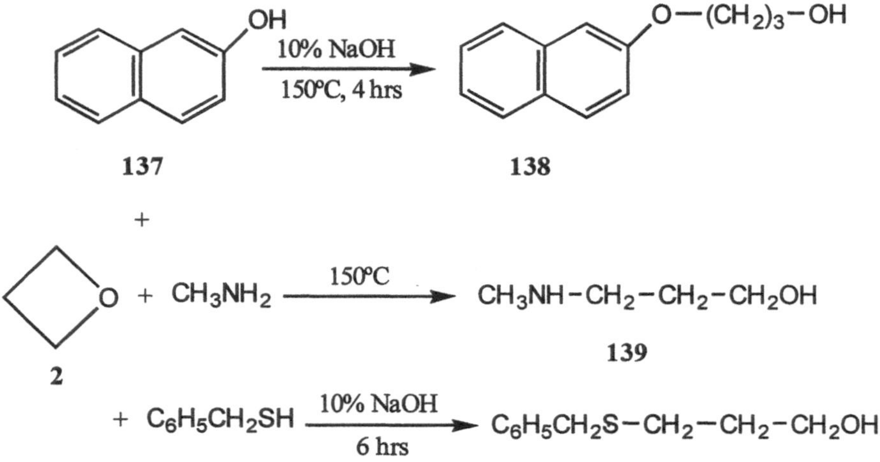
OH
10% NaOH
150°C, 4 hrs
O—(CH2)3—OH
+
O + CH3NH2
150°C
CH3NH-CH2-CH2-CH2OH
+ C6H5CH2SH
10% NaOH
6 hrs
C6H5CH2S-CH2-CH2-CH2OH

137 138 2 139 140

Scheme-52

Oxetane reacts with Grignard reagents at higher temperature with the lengthening of carbon chain (scheme-53). If unsymmetrical oxetane is involved, the less

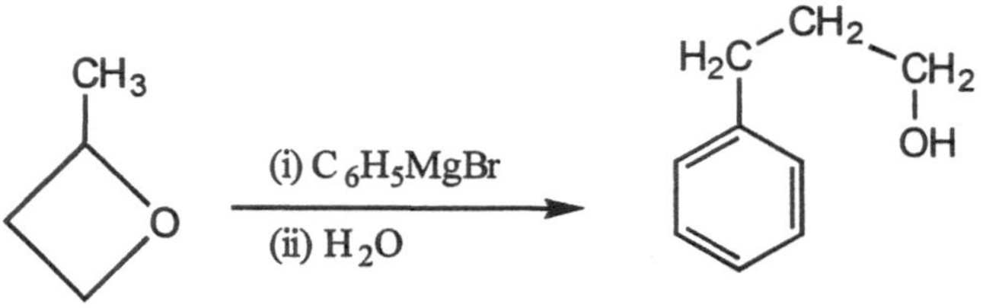
CH3
O
(i) C6H5MgBr
(ii) H2O
H2C
CH2
CH2
OH

2 141

Scheme-53

substituted carbon atom attached to the oxygen atom is attacked by nucleophile and the formation of single product predominates (scheme-54).

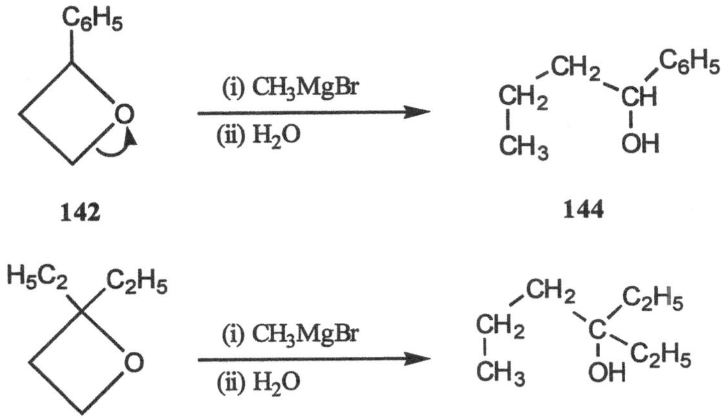

Scheme-54

2.1.3.2 Photochemical Reactions

Oxetanes **146** substituted with chromophoric substituents undergo photochemical transformations providing rearranged products **149** (scheme-55)[64,65].

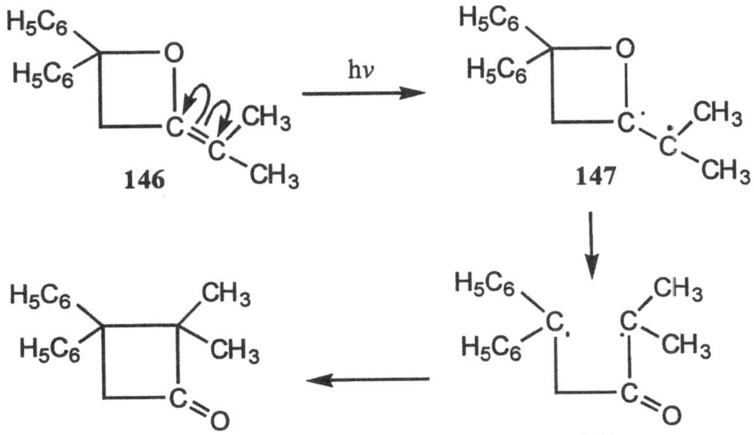

Scheme-55

2.2 Oxetanones (β-Lactones)

2.2.1 General

Oxetanones are the carbonyl derivatives of oxetane and designated as oxetan-2-ones (2-oxetanones) or more commonly β-lactones **150**. These are strained internal esters. The simplest β-lactone, β-propiolactone or 2-oxetanone, first obtained in 1916, possesses carcinogenic properties.

2-Oxetanone is a planar molecule with the following structural parameters :

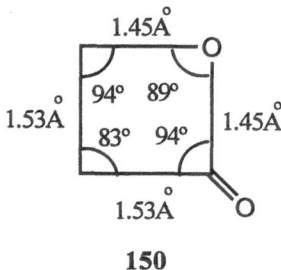

150

The carbonyl stretching frequency in the infrared spectra of β-lactones is higher (1840 cm⁻¹) than in acyclic esters (1736 cm⁻¹) and γ-lactones (1770 cm⁻¹). This has been attributed to the deformation of bond angles or the presence of ring strain.

2.2.2 Synthesis

2.2.2.1 Intramolecular Cyclization Reactions

Intramolecular cyclization of β-hydroxy acids to form β-lactones is generally not successful because β-hydroxy acids undergo β-elimination reaction providing α,β-unsaturated acids. But β-hydroxy acids undergo intramolecular cyclodehydration with the formation of 2-oxetanones when treated with benzenesulfonyl chloride in the presence of pyridine (scheme-56)[72,73]. However, the reaction of β-halo acids with a base under the controlled conditions also causes intramolecular cyclization to yield β-lactones (scheme-57)[74].

$$CH_3-CH-CH-COOH \xrightarrow[\text{pyridine}]{C_6H_5SO_2Cl}$$

Scheme-56

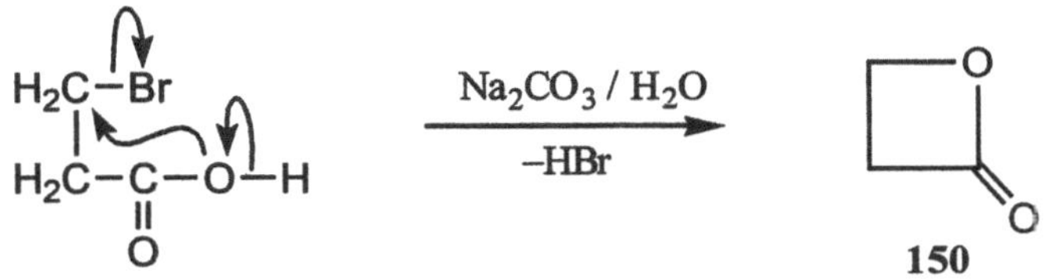

151 **150**

Scheme-57

α,β-Dihalo acids **152** do not favour intramolecular cyclization, but undergo internal decarboxylative elimination resulting in the formation of alkenes (scheme-58)[75].

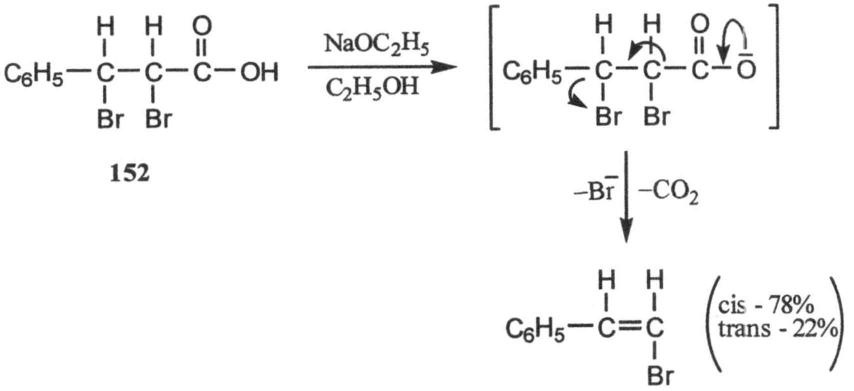

152 **153**

Scheme-58

The cyclizations leading to β-lactones (2-oxetanones) are stereospecific and proceed with the inversion of configuration at the carbon bearing halogen atom.

2.2.2.2 Cycloaddition Reactions

2.2.2.2.1 Reaction of Ketenes with Carbonyl Compounds

The condensation of ketenes with carbonyl compounds in the presence of a lewis acid provides β-lactones (scheme-59)[76].

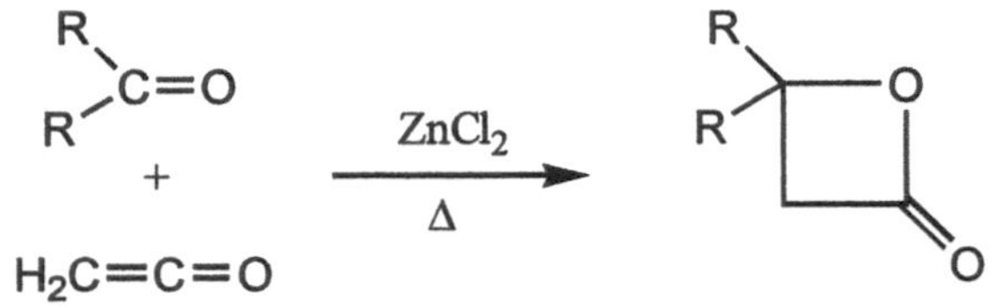

Scheme-59

2.2.3 Reactions

2.2.3.1 Ring Opening Reactions

β-Lactones are considerably more reactive than the larger ring lactones and undergo ring opening reactions due to the ring strain. These are normally stable at room temperature, but dissociate on heating (scheme-60).

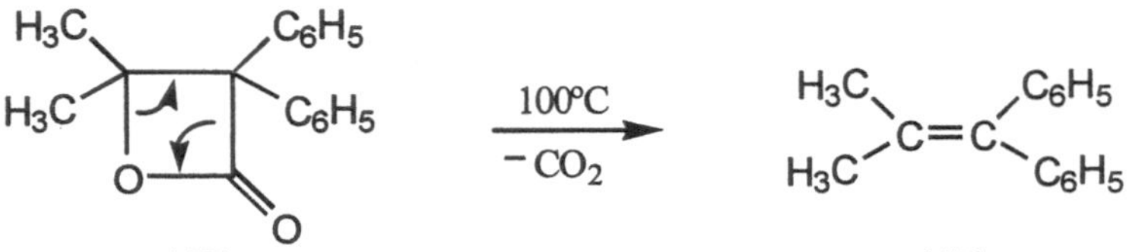

Scheme-60

The reaction is highly stereospecific and proceeds with the retention of configuration.

2.2.3.1.1 Hydrolysis

β-Lactones being cyclic esters can be hydrolyzed to the corresponding open chain derivatives. However, the bond fission depends on the reaction conditions :

(i) In neutral or slightly acidic medium, bimolecular alkyl–oxygen bond fission occurs, and

(ii) In strongly acidic medium, the reaction proceeds via acyl–oxygen bond fission.

The mechanism of bond fission is depicted in (scheme-61)[78-81] :

150

157

158

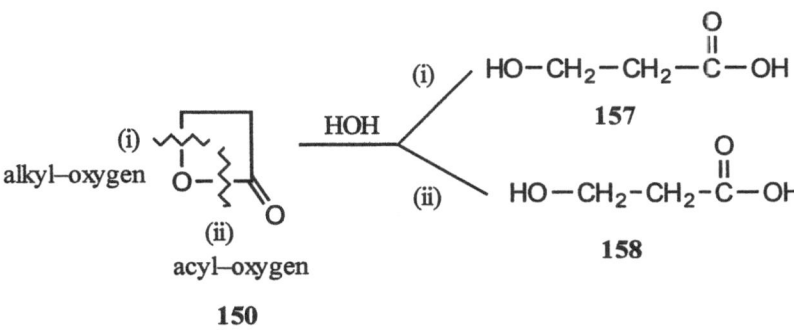

Scheme-61

2.2.3.1.2 Nucleophilic Ring Opening Reactions

The reaction of β-lactones with nucleophiles proceeds to involve bimolecular alkyl–oxygen bond fission mechanism (scheme-62)[82-84].

159 160

150 161

150

162

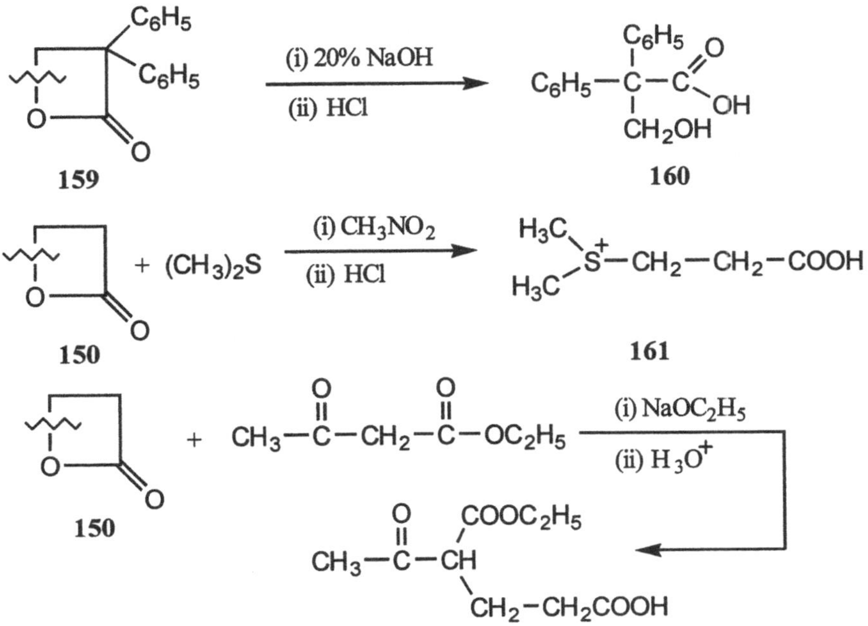

Scheme-62

The reaction of β-lactones with alcohols depends on the reaction conditions (scheme-63)[80,82].

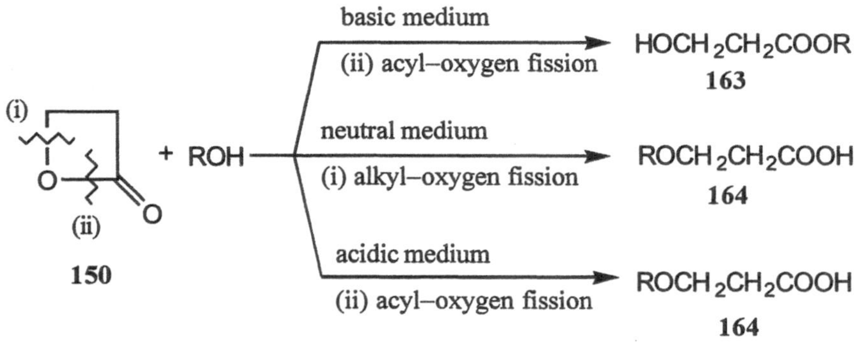

Scheme-63

The reaction of b-lactones with ammonia and amines depends on the nature of the solvent and proceeds with the cleavage of bonds in both the directions resulting in two products (scheme-64). The relative proportions of the products depend upon the nature of amine, reaction medium and the order of addition of the reactant[85].

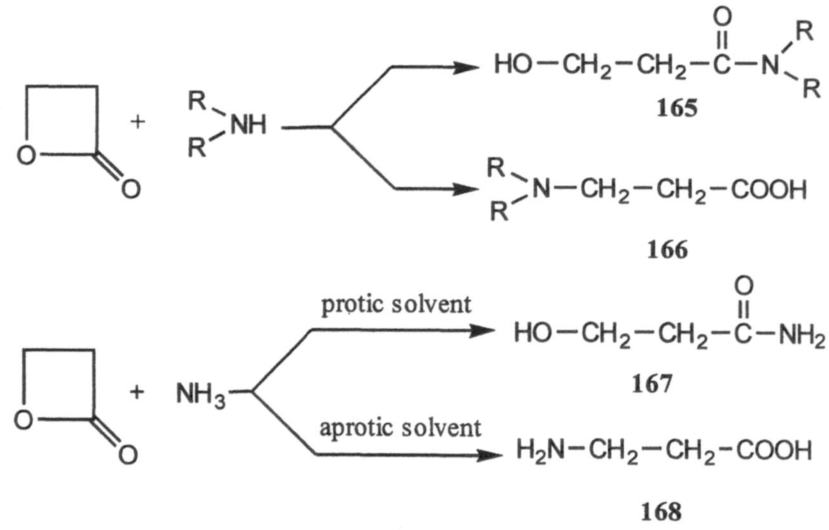

Scheme-64

2.2.3.1.3 Electrophilic Ring Opening Reactions

The ring opening reactions of β-lactones with phenol can not be generalized. Phenol reacts slowly with β-lactone at room temperature to give β-phenoxypropionic acid involving cleavage of alkyl–oxygen bond. But in the presence of catalytic amount of acid, the cleavage of acyl–oxygen bond occurs with the formation of an ester (scheme-65)[86].

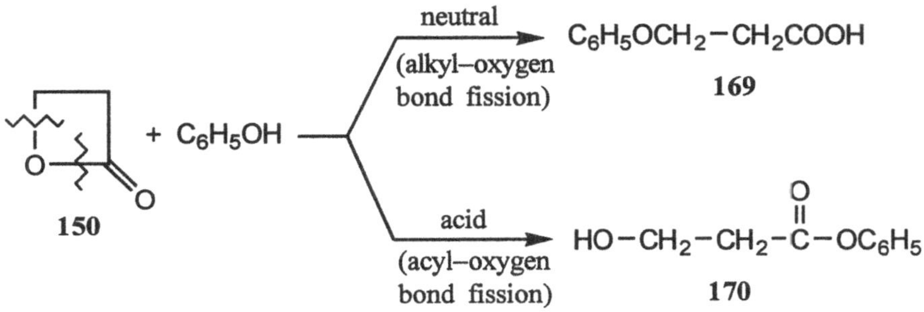

Scheme-65

2.2.3.2 Photochemical Reactions

The irradiation of α-pyrone **171** at −20°C results in the formation of β-lactone **172** which on continued irradiation undergoes decarboxylation providing cyclobutadiene **173** (scheme-66)[87].

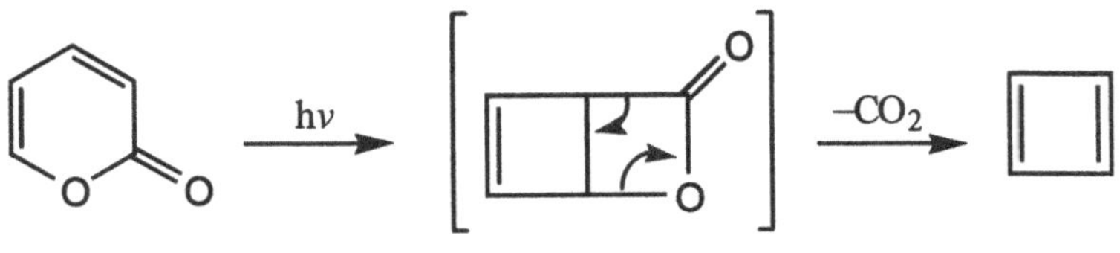

Scheme-66

3 FOUR-MEMBERED THIAHETEROCYCLES

3.1 Thietanes

3.1.1 General

Four-membered saturated heterocycles containing sulfur as heteroatom are known as thietanes **3**[88]. The corresponding unsaturated compounds are called thietenes or thietes **174**.

3   **174**

In thietane, however the puckering angle is 26°, the energy barrier to puckering is lower (3.14 kJ/mol) than the ground vibrational energy level. Thus, thietane molecule is planar with the following structure parameters. The planarity has been attributed to the reduction in the number of non-bonding interactions and the lower torsional energy barrier about the C–S bond.

$$C–S = 1.847\text{Å}$$
$$C–S = 1.549\text{Å}$$

$\theta_1 = 76.8°; \quad \theta_3 = 95.6°$
$\theta_2 = 90.6°; \quad \theta_4 = 90.6°$

The ultraviolet spectrum of thietane shows a sharp maximum at 235 nm in addition to a maximum at 218 nm, typical of cyclic sulfides (215–220 nm). The weak absorption maximum at higher wavelength (275 nm) in comparison to in cyclic sulfides of different ring sizes is attributed to the higher electron density at the sulfur atom in four-membered ring system. The order of electron density at the sulfur atom in different ring systems decreases in the order[89] :

four > five > seven > three

3.1.2 Synthesis

3.1.2.1 Intramolecular Cyclization Reactions

3.1.2.1.1 From 1,3-Dihaloalkanes

The reaction of 1,3-dihaloalkanes **175** with anhydrous sodium sulfide in ethanol provides thietanes **3** in low yield (20–30%) (scheme-67)[88, 90-92], but the reaction with an aqueous solution of sodium sulfide containing hexadecyltriethylammonium chloride as a phase transfer catalyst improves the yield (70%)[93-95].

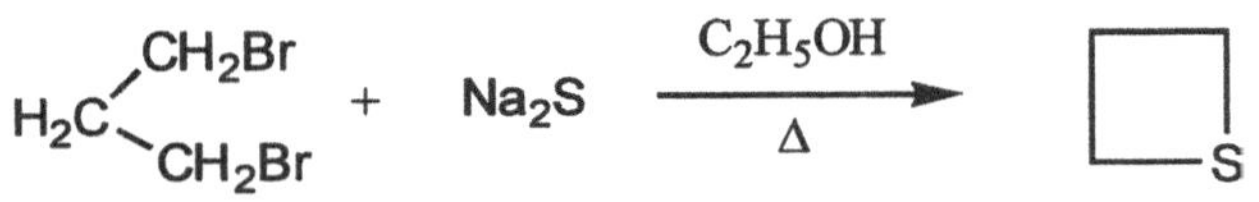

$$H_2C \overset{CH_2Br}{\underset{CH_2Br}{\big<}} \quad + \quad Na_2S \quad \xrightarrow[\Delta]{C_2H_5OH} \quad \square{-}S$$

175   **3**

Scheme-67

Thietane can also be prepared by alkaline decomposition of monothiouronium salt **177** obtained by the reaction of 1-bromo-3-chloropropane **176** with thiourea (scheme-68)[96].

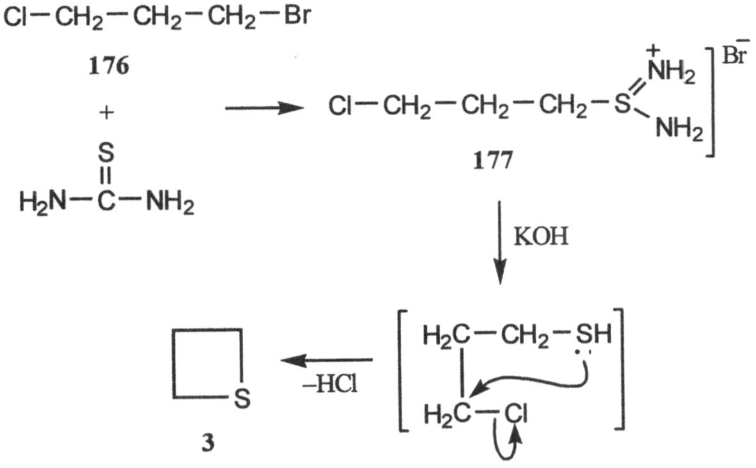

$$Cl-CH_2-CH_2-CH_2-Br$$

176

+

$$\underset{H_2N-C-NH_2}{\overset{S}{\underset{\|}{}}}$$

$$\longrightarrow \quad \left[Cl-CH_2-CH_2-CH_2-S \overset{\overset{+}{N}H_2}{\underset{NH_2}{\big<}} \right] Br^-$$

177

$$\Big\downarrow KOH$$

$$\square{-}S \quad \overset{}{\underset{-HCl}{\longleftarrow}} \quad \left[\begin{array}{l} H_2C-CH_2-\overset{..}{S}H \\ \big| \\ H_2C{-}Cl \end{array} \right]$$

3

Scheme-68

3.1.2.1.2 From Chloromethyloxirane

The cyclization of 2-hydroxy-3-chloropropanethiol **179** (formed by the condensation of chloromethyloxirane **178** with barium hydrosulfide) leads to the formation of 3-hydroxythietane **180** (scheme-69)[97].

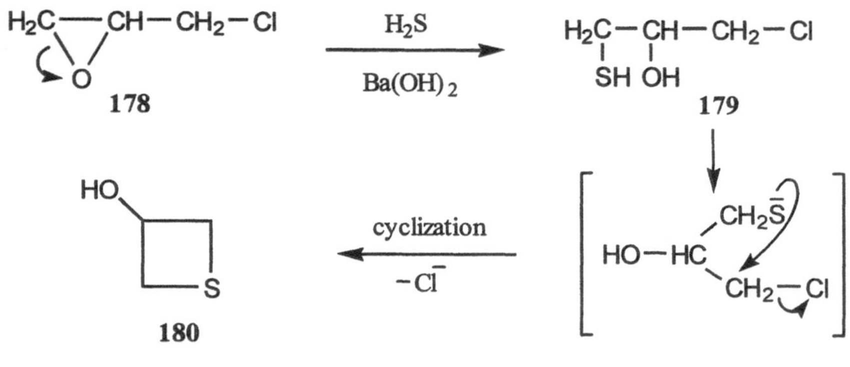

Scheme-69

3.1.2.1.3 From Chloromethylthiirane

2-Hydroxy-3-chloropropanethiol **179** (formed by the ring opening of chloromethylthiirane **181** under alkaline conditions) undergoes cyclization to yield 3-hydroxythietane **180** (scheme-70)[98].

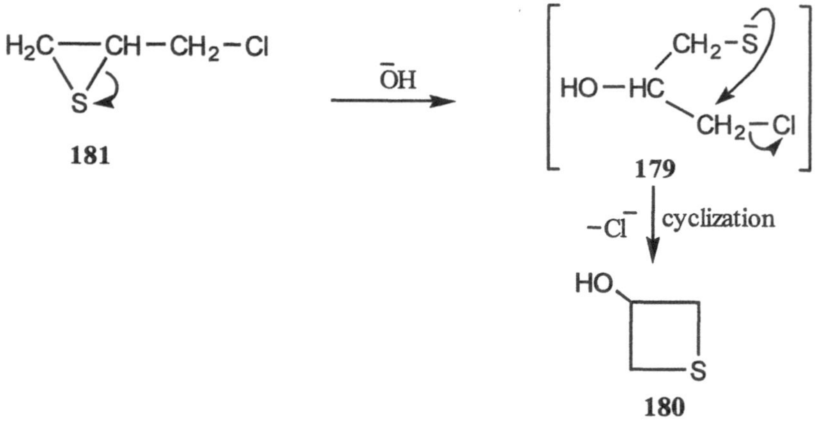

Scheme-70

3.1.2.1.4 From Cyclic Carbonates

The decomposition of cyclic carbonates (1,3-dioxan-2-ones) **183** in the presence of thiocyanate ions at 170–210°C yïelds thietanes **3** (scheme-71)[99].

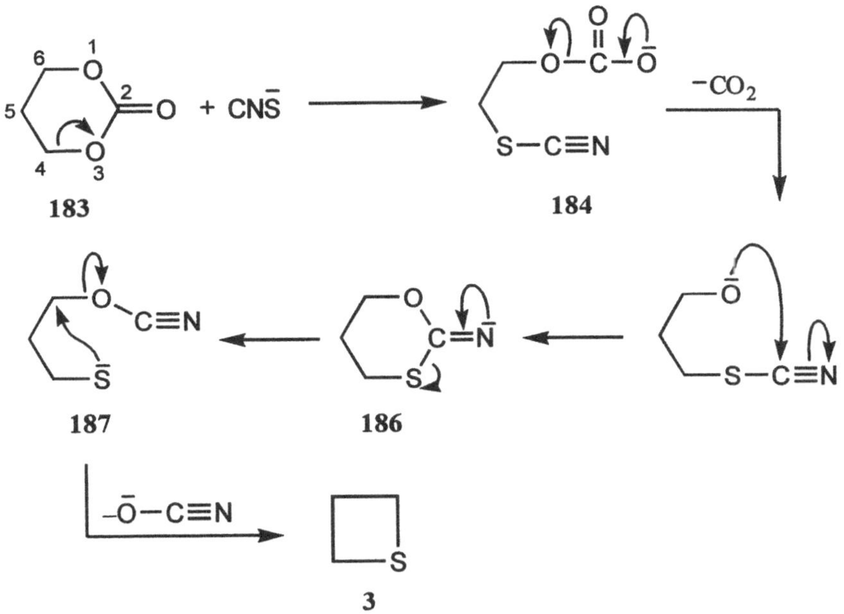

Scheme-71

The reaction involves nucleophilic attack of thiocyanate ion at C–4 with the ring opening in the first step. Subsequent step involves elimination of carbon dioxide followed by the ring closure and ring openinig processes and finally cyclization to provide thietanes.

3.1.3 Reactions

3.1.3.1 Ring Openinig Reactions

Thietane ring system is less reactive than thiirane ring system because of decrease in the ring strain and thus thietanes are less susceptible to ring cleavage. However, thietanes also undergo ring opening reactions, but very slowly and require drastic conditions. Thietane is cleaved on heating in gas phase to give ethylene and thioformaldehyde (scheme-72).

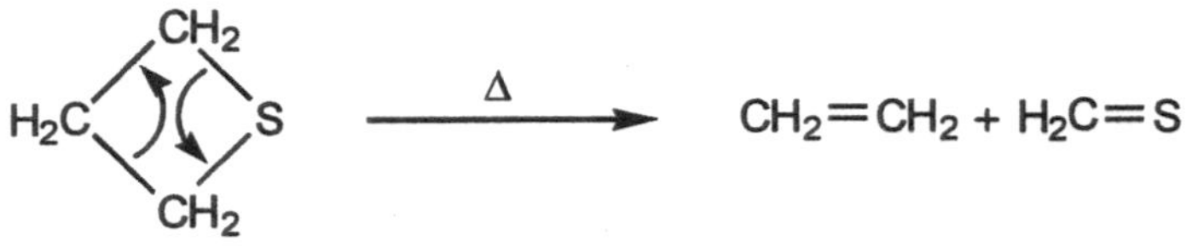

Scheme-72

3.1.3.2 Reaction with Ammonia

Thietane **3** undergoes ring opening reaction on treatment with ammonia at 200°C providing 3-aminopropanethiol (scheme-73).

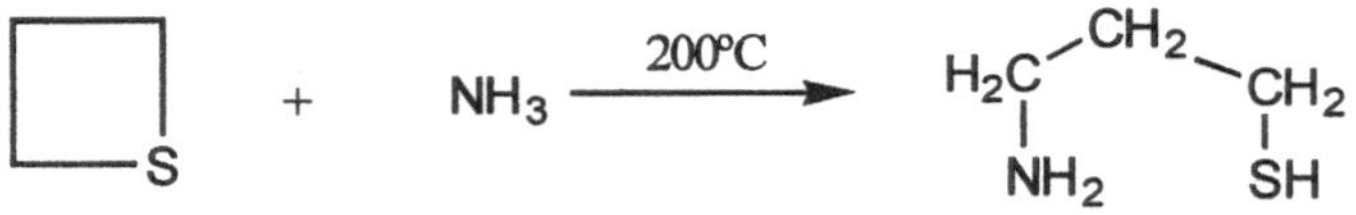

Scheme-73

3.1.3.3 Reaction with Acetyl Chloride

Thietane reacts with acetyl chloride in the presence of stannic chloride and benzene with the cleavage of thietane ring to yield 3-chloropropyl thioacetate (scheme-74).

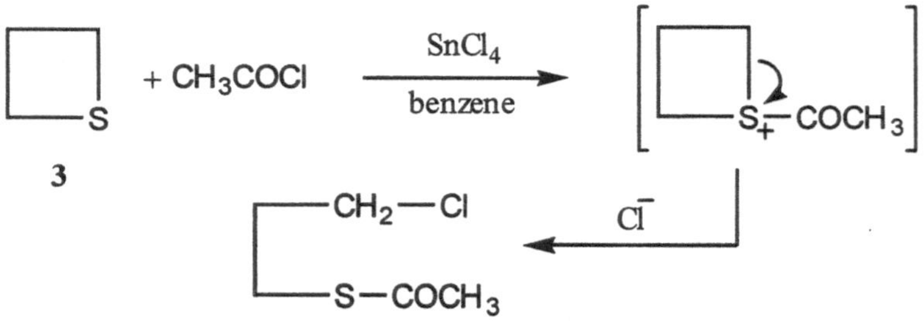

188

Scheme-74

In unsymmetrical thietane (2-methylthietane **189**), 1,2-bond is preferantially attacked by acetyl chloride (scheme-75).

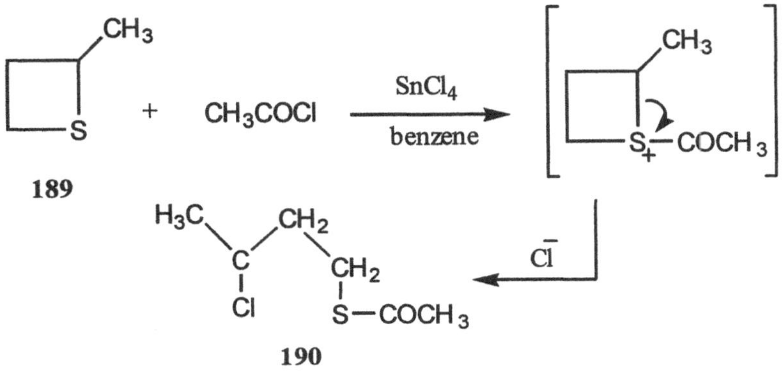

Scheme-75

3.1.3.4 Reaction with Halogens

Thietane reacts with halogens, similar to thiirane, with the cleavage of the ring to provide 3-chloropropyl disulfide **192** (scheme-76)[100].

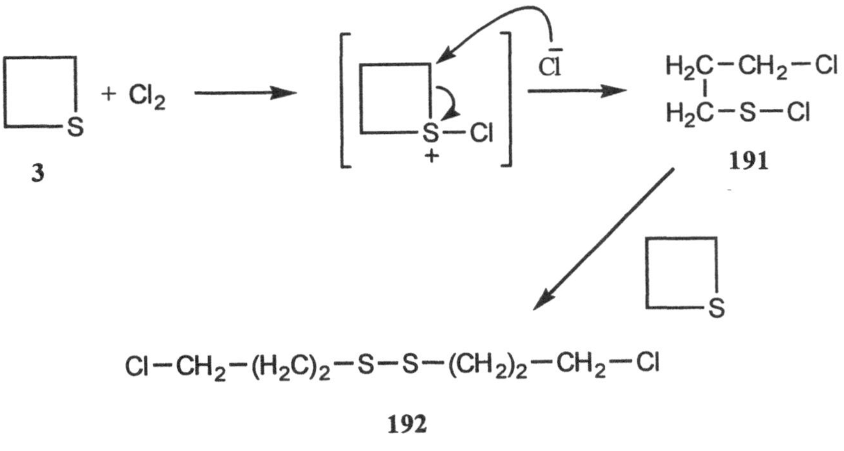

Scheme-76

3.1.3.5 Reaction with Allyl Bromide

The reaction of thietane with allyl bromide appears to involve electrophilic attack at sulfur with the cleavage of thietane ring (scheme-77)[101].

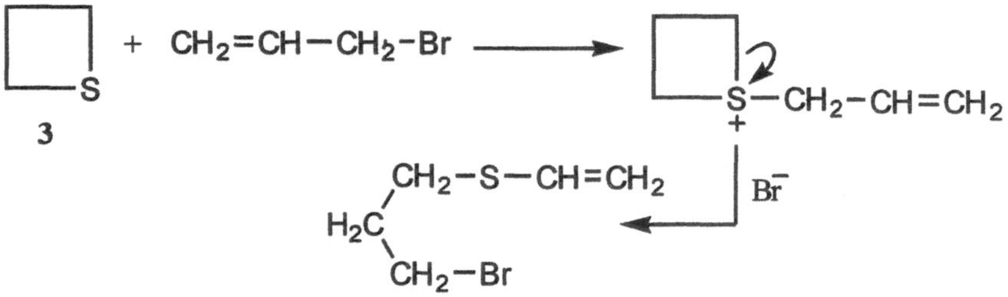

Scheme-77

3.1.3.6 Reaction with Alkyl Iodide (Salt Formation)

Thietane forms quaternary thietanium salt with methyl iodide which immediately undergoes ring opening reaction (scheme-78)[102].

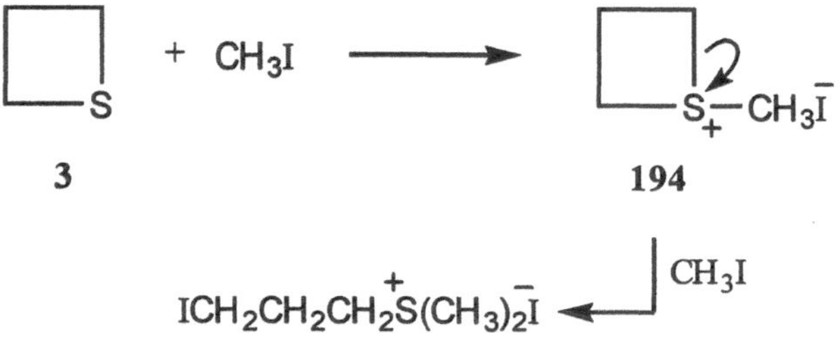

The quaternary thietanium salt is decomposed to cyclopropane in the presence of *n*-butyllithium (scheme-79)[103].

3.1.4 Oxidation

Thietanes can be easily oxidized by hydrogen peroxide to sulfoxide **199** and finally to sulfones **200** (scheme-80).

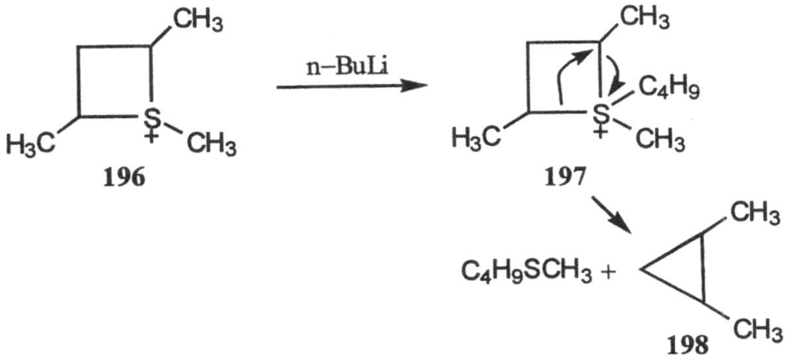

Scheme-79

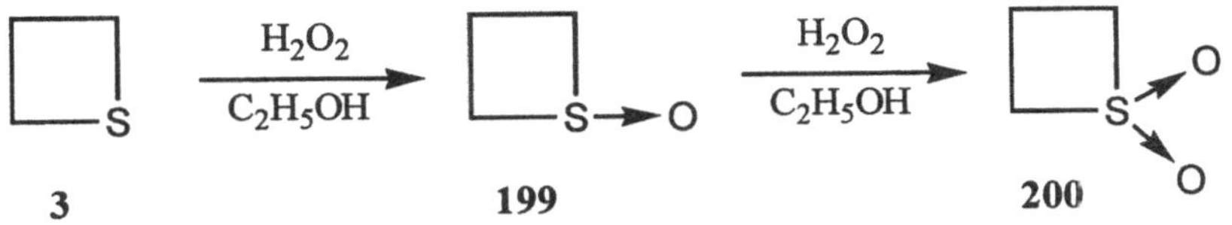

Scheme-80

3.1.5 Photochemical Reactions

Photochemical reactions proceed via the formation of biradical intermediates with the retention of configuration. Photolysis of *cis-* and *trans-*3-ethyl-2-phenyl-thietane **201** provides the corresponding *cis-* and *trans-*alkenes (scheme-81)[104].

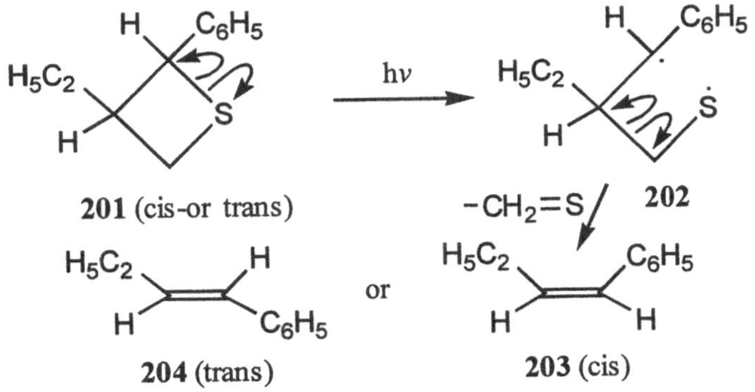

Scheme-81

REFERENCES

1. N. H. Cromwell and B. Phillips, *Chem. Rev.* **79**, 331 (1979); D. E. Davies and R. C. Storr in A. R. Katritzky and C. W. Rees (Eds.), *Comprehensive Heterocyclic Chemistry* Vol. **7**, Pergamon Press, Oxford, 1984, pp. 238; R. storr in H. Suschitzky and E. F. V. Scriven (Eds.), *Progress in Heterocyclic Chemistry* Vol. **1**, Pergamon Press, Oxford, 1989, pp. 98; J. Parrick, Vol. **2**, 1990, pp. 37; Vol. **3**, 1991, pp. 58; J. Parrick and L. K. Mehta, Vol. **4**, 1992, pp. 49; Vol. **5**, 1993, pp. 69; Vol. **6**, 1994, pp. 74.

2. J. A. Moore and R. S. Ayers in A. Hassner (Ed.), *Small Ring Heterocycles*, Part II, Wiley-Interscience, New York, 1983, pp. 1.

3. C. Mannich and G. Baumgarten, *Ber.* **70**, 210 (1937).

4. S. Searles, M. Tamres, F. Block and L. A. Quarterman, *J. Am. Chem. Soc.* **78**, 4917 (1956).

5. F. C. Schaefer, *J. Am. Chem. Soc.* **77**, 5928 (1955).

6. J. P. Freeman, D. G. Pucci and G. Binsch, *J. Org. Chem.* **37**, 1894 (1972).

7. M. Vaultier, R. Danion–Bougot, D. Danion, J. Hamelin and R. Carrie, *J. Org. Chem.* **40**, 2990 (1975).

8. N. H. Cromwell and R. M. Rodebaugh, *J. Heterocycl. Chem.* **6**, 435 (1969).

9. H. H. Wasserman, W. T. Han, J. M. Schaus and J. W. Faller, *Tetrahedron Lett.* 3111 (1984).

10. S. Searles and C. F. Butler, *J. Am. Chem. Soc.* **76**, 56 (1954).

11. A. Padwa, R. Gruber and L. Hamilton, *J. Am. Chem. Soc.* **89**, 3077 (1967).

12. A. Padwa and R. Gruber, *J. Am. Chem. Soc.* **92**, 100 (1970).

13. Y. Etienne and N. Fischer in A. Weissberger (Ed.), *The Chemistry of Heterocyclic Compounds* **XIX**, Part II, Wiley-Interscience, New York, 1964, pp. 729–880; D. E. Davies and R. C. Storr in A. R. Katritzky and C. W. Rees (Eds.), *Comprehensive Heterocyclic Chemistry* Vol. **7**, Pergamon Press, Oxford, 1984, pp. 247.

14. G. A. Koppeal in A. Hassner (Ed.), *Small Ring Heterocycles*, Part II, Wiley-Interscience, New York, 1983, pp. 219; R. storr in H. Suschitzky and E. F. V. Scriven (Eds.), *Progress in Heterocyclic Chemistry* Vol. **1**, Pergamon Press, Oxford, 1989, pp. 103; J. Parrick, Vol. **2**, 1990, pp. 42; Vol. **3**, 1991, pp. 63; J. Parrick and L. K. Mehta, Vol. **4**, 1992, pp. 55; Vol. **5**, 1993, pp. 74; Vol. **6**, 1994, pp. 81.

15. R. Bucourt in G. I. Gregory (Ed.), *Recent Advances in the Chemistry of β- Lactam Antibiotics*, Royal Society of Chemistry, London, 1981, pp. 1.

16. E. H. Flyn (Ed.), *Cephalosporins and Penicillins–Chemistry and Biology*, Academic Press, New York, 1972; D. N. McGregor in A. R. Katritzky and C. W. Rees (Eds.), *Comprehensive Heterocyclic Chemistry* Vol. 7, Pergamon Press, Oxford, 1984, pp. 299.

17. S. Searles and R. E. Wann, *Chem. & Ind.* 2097 (1967).

18. S. Kim, P. HO Lee and T. Au Lee, *Synth. Commun.* **18**, 247 (1988).

19. S. Kim, P. HO Lee and T. Au Lee, *J. Chem. Soc. Chem. Commun.* 1242 (1988).

20. T. Kunieda, T. Nagamatsu, T. Higuchi and M. Hirobe, *Tetrahedron Lett.* 2203 (1988).

21. J. C. Sheehan and A. K. Bose, *J. Am. Chem. Soc.* **73**, 1761 (1951).

22. B. G. Chatterjee, V. V. Rao and N. G. Mazumdar, *J. Org. Chem.* **30**, 4101 (1965).

23. D. H. Hua and A. Verma, *Tetrahedron Lett.* 547 (1985).

24. A. K. Bose, M. S. Manhas and R. B. Romer, *Tetrahedron* **21**, 449 (1965).

25. N. Miyachi, F. Kanda and M. Shibasaki, *J. Org. Chem.* **54**, 3511 (1989).

26. A. R. Katritzky, *Handbook of Heterocyclic Chemistry*, Pergamon Press, Oxford, 1985, pp. 393.

27. M. J. Miller, P. G. Mattingly, M. A. Morrison and J. F. Kerwin, Jr., *J. Am. Chem. Soc.* **102**, 7026 (1980).

28. H. H. Wasserman, D. J. Hlasta, A. W. Tremper and J. S. Wu, *Tetrahedron Lett.* 549 (1979).

29. G. Rajendra and M. J. Miller, *Tetrahedron Lett.* 6257 (1987).

30. J. K. Rasmussen and A. Hassner, *Chem. Rev.* **76**, 389 (1976).

31. H. Staudinger, *Ann. Chem.* **365**, 51 (1907).

32. D. R. Wagle, C. Garai, J. Chiang, M. G. Monteleone, B. E. Kurys, T. W. Strohmeyer, V. R. Hegde, M. S. Manhas and A. K. Bose, *J. Org. Chem.* **53**, 4227 (1988).

33. D. A. Evans and J. M. Williams, *Tetrahedron Lett.* 5065 (1988).

34. T. Kawabata, Y. Kirmura, Y. Ito, S. Terashima, A. Sasaki and M. Sunagawa, *Tetrahedron*, **44**, 2149 (1988).

35. D. R. Wagle, C. Garai, M. C. Monteleone and A. K. Bose, *Tetrahedron Lett.* 1649 (1988).

36. J. S. Sandhu and B. Sain, *Heterocycles*, **27**, 777 (1987).

37. T. R. Govindachari, P. Chinnasamy, R. Rajeswari, S. Chandrashekharan, M. S. Pramila, S. Nataraja, K. Nagarajan, and B. R. Pai, *Heterocycles*, **22**, 585 (1984).

38. A. K. Mukerjee and A. K. Singh, *Tetrahedron* **34**, 1731 (1978).

39. N. S. Isaacs, *Chem. Soc. Rev.* **4**, 181 (1976).

40. J. E. Lynch, S. M. Riseman, W. L. Laswell, R. P. Valante, G. B. Smith, I. Shinkai and D. M. Tschaen, *J. Org. Chem.* **54**, 3792 (1989).

41. G. I. Geerg, J. Kant, P. He, A. M. Iy and L. Limpe, *Tetrahedron Lett.* 2409 (1988).

42. F. H. Van der Steen, H. Kleijn, J. T. B. H. Jastrazebski and G. Van Koten, *Tetrahedron Lett.* 765 (1989).

43. F. H. Van der Steen, J. T. B. H. Jastrazebski and G. Van Koten, *Tetrahedron Lett.* 2467 (1988).

44. H. Alper, *Isr. J. Chem.* **21**, 203 (1981).

45. J. L. Davidson and P. N. Prestion, *Adv. Heterocycl. Chem.* **30**, 319 (1982).

46. S. Calet, F. Usro and H. Alper, *J. Am. Chem. Soc.* **111**, 931 (1989).

47. H. Alper, F. Usro and D. J. H. Smith, *J. Am. Chem. Soc.* **105**, 6737 (1983).

48. W. Chamchaang and A. R. Pinhas, *J. Chem. Soc. Chem. Commun.* 710 (1988).

49. R. R. Rando, *J. Am. Chem. Soc.* **92**, 6707 (1970).

50. R. R. Rando, *J. Am. Chem. Soc.* **94**, 1629 (1972).

51. M. P. Doyle, M. S. Shanklin, S. M. Oon, H. Q. Pho, F. R. Van der Heide and W. R. Veal, *J. Org. Chem.* **53**, 3384 (1988).

52. L. S. Hegadus and S. D. Andrea, *J. Org. Chem.* **53**, 3113 (1988).

53. E. J. Moriconi and P. H. Mazzocchi, *J. Org. Chem.* **31**, 1372 (1966).

54. K. R. Henery–Logan, H. P. Konepfel and J. Rodricks, *J. Heterocycl. Chem.* **5**, 433 (1965).

55. M. Fischer and A. Mattheus, *Chem. Rev.* **69**, 342 (1969).

56. G. Dittus in E. Mueller (Ed.), *Methoden der Organischen Chemie*, Houben–Weyl, 1965, pp. 493; S. Searles in A. R. Katritzky and C. W. Rees (Eds.), *Comprehensive Heterocyclic Chemistry* Vol. 7, Pergamon Press, Oxford, 1984, pp. 263

57. S. Searles, Jr., K. A. Pollart and F. Block, *J. Am. Chem. Soc.* **79**, 952 (1957).

58. D. C. Dittmer, W. R. Hertler and H. Winicov, *J. Am. Chem. Soc.* **79**, 4431 (1957).

59. S. Searles, and M. J. Gortatowski, *J. Am. Chem. Soc.* **75**, 3030 (1953).

60. S. Searles, Jr., R. G. Nickerson and W. K. Witsiepe, *J. Org. Chem.* **24**, 1839 (1959).

61. K. Okuma, Y. Tanaka, S. Kaji and H. Ohta, *J. Org. Chem.* **48**, 5133 (1983).

62. E. Pasterno and G. Chieffi, *Gazz. Chim. Ital.* **39**, 342 (1909).

63. D. R. Arnold, *Adv. Photochem.* **6**, 301 (1968).

64. H. Gotthardt, R. Steinmetz and G. S. Hammond, *J. Org. Chem.* **33**, 2774 (1968).

65. D. R. Arnold and A. H. Glick, *Chem. Commun.* 813 (1966).

66. J. Matty and K. Buchkremer, *Heterocycles* **27**, 2153 (1988); M. Braun, *Nachr. Chem. Techn. Lab.* **33**, 213 (1985)

67. D. R. Arnold, R. L. Hinman and A. H. Glick, *Tetrahedron Lett.* 1425 (1964).

68. S. Searles, Jr. in A. Weissberger (Ed.), *The Chemistry of Heterocyclic Compounds* **XIX**, Part II, Wiley-Interscience, New York, 1964, pp. 983.

69. M. Yamaguchi, Y. Nobayashi and I. Hirao, *Tetrahedron* **40**, 4261 (1984).

70. T. Suzuki, H. Saimoto, H. Tomioka, K. Oshima and H. Nozaki, *Tetrahedron Lett.* 3597 (1982).

71. P. F. Hudrlik and C. N. Wan, *J. Org. Chem.* **40**, 2963 (1975).

72. H. E. Zaugg, *Org. React.* **8**, 305 (1954).

73. A. Pomier and J. -M. Pons, *Synthesis*, 441, (1993).

74. H. E. Zaugg, *J. Am. Chem. Soc.* **72**, 2998 (1950).

75. E. Grovenstein, Jr., and D. E. Lee, *J. Am. Chem. Soc.* **75**, 2639 (1953).

76. H. J. Hagemeyer, Jr., *Ind. Eng. Chem.* **41**, 765 (1949).

77. D. S. Noyce and E. H. Banitt, *J. Org. Chem.* **31**, 4043 (1966).

78. F. A. Long and M. Purchase, *J. Am. Chem. Soc.* **72**, 3267 (1950).

79. P. D. Bartlett and G. Small, *J. Am. Chem. Soc.* **72**, 4867 (1950).

80. A. R. Olson and P. V. Youle, *J. Am. Chem. Soc.* **73**, 2468 (1951).

81. J. O. Edwards, *J. Am. Chem. Soc.* **76**, 1540 (1954).

82. H. E. Zaugg. *J. Am. Chem. Soc.* **72**, 3001 (1950)

83. N. F. Blau and C. G. Stuckwisch, *J. Am. Chem. Soc.* **73**, 2355 (1951).

84. T. L. Gresham, J. E. Jansen, F. W. Shaver and J. T. Gregory, *J. Am. Chem. Soc.* **70**, 999 (1948).

85. T. L. Gresham, J. E. Jansen, F. W. Shaver, R. A. Bankert and F. T. Fiedorek, *J. Am. Chem. Soc.* **73**, 3168 (1951).

86. T. L. Gresham, J. E. Jansen, R. A. Bankert, W. L. Beears and M. G. Prendergast, *J. Am. Chem. Soc.* **71**, 661 (1949).

87. E. J. Corey and J. Streith, *J. Am. Chem. Soc.* **86**, 950 (1964).

88. M. Sander, *Chem. Rev.* **66**, 341 (1966); E. Block in A. R. Katritzky and C. W. Rees (Eds.), *Comprehensive Heterocyclic Chemistry* Vol. 7, Pergamon Press, Oxford, 1984, pp. 403; D. C. Dittmer and T. C. Segerman in A. Hassner (Ed.), *Small Ring Heterocycles*, Part III, Wiley-interscience, 1985, pp. 431.

89. R. E. Davis, *J. Org. Chem.* **33**, 1380 (1968).

90. H. J. Baker and K. J. Keunig, *Rec. Trav. Chem.* **53**, 808 (1934).

91. Y. Etienne, R. Soulas and H. Lumbroso in A. Weissberger (Ed.), *The Chemistry of Heterocyclic Compounds* **XIX**, Part II, Wiley-interscience, New York, 1964, pp. 647.

92. R. W. Bost and M. W. Conn., *Oil Gas J.* **32**, 17 (1933).

93. M. Lancaster and D. J. H. Smith, *Synthesis*, 582 (1982).

94. E. V. Dehmlow, *Angew. Chem. Int. Edn. Engl.* **13**, 170 (1974).

95. J. Docks, *Synthesis*, 441 (1973).

96. F. G. Bordwell and B. M. Pitt. *J. Am. Chem. Soc.* **77**, 572 (1955).

97. D. C. Dittmer and M. E. Christy, *J. Org. Chem.* **26**, 1324 (1961).

98. E. P. Adams, K. N. Ayad, F. P. Doyle, D. O. Holland, W. H. Hunter, J. H. C. Nayler and A. Queen, *J. Chem. Soc.* 2665 (1960).

99. S. Searles Jr., H. R. Hays and E. F. Lutz, *J. Org. Chem.* **27**, 2828 (1962).

100. J. M. Stewart and C. H. Burnside, *J. Am. Chem. Soc.* **75**, 243 (1953).

101. E. Vedejs and J. P. Hagen, *J. Am. Chem. Soc.* **97**, 6878 (1975).

102. D. C. Palmer and E. C. Taylor, *J. Org. Chem.* **51**, 846 (1986).

103. B. M. Trost, W. L. Schinski and I. B. Mantz, *J. Am. Chem. Soc.* **91**, 4320 (1969).

104. D. R. Rice and R. D. Stier, *Chem. Commun.* 166 (1973).

SUBJECT INDEX

Springer
and the
environment

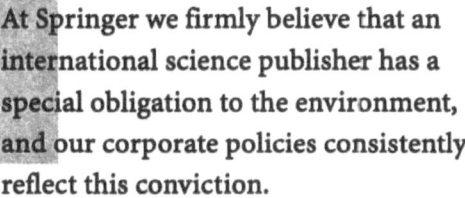

Springer

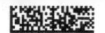